Modelling Geographical Systems

The GeoJournal Library

Volume 70

The titles published in this series are listed at the end of this volume.

Modelling Geographical Systems

Statistical and Computational Applications

edited by

BARRY BOOTS

Wilfrid Laurier University, Waterloo, Ontario, Canada

ATSUYUKI OKABE

Center for Spatial Information Science,
University of Tokyo, Japan

and

RICHARD THOMAS

University of Manchester, United Kingdom

KLUWER ACADEMIC PUBLISHERS
DORDRECHT / BOSTON / LONDON

A C.I.P. Catalogue record for this book is available from the Library of Congress

ISBN 1-4020-0821-X

Published by Kluwer Academic Publishers,
P.O. Box 17, 3300 AA Dordrecht, The Netherlands.

Sold and distributed in North, Central and South America
by Kluwer Academic Publishers,
101 Philip Drive, Norwell, MA 02061, U.S.A.

In all other countries, sold and distributed
by Kluwer Academic Publishers,
P.O. Box 322, 3300 AH Dordrecht, The Netherlands.

Printed on acid-free paper

Printed in the Netherlands.

Table of Contents

Part II: Computational Methods

SECTION A: SIMULATION MODELS

SECTION B: GIS MODELS

SECTION C: THE INTERNET

Preface

Within the realm of quantitative geography, systems modelling is specifically concerned with understanding those relationships that influence the attributes of phenomena located in space and time. The intention is to replicate the main processes influencing a system's behaviour and, thereby, assist its management through a capability to estimate future change. Over the last few decades, one of the major institutional initiatives for promoting such research has been provided by specialised Study Groups and Commissions established by the International Geographical Union (IGU). These scholarly networks have aimed to co-ordinate international research agendas for geographical systems modelling and their activities have been recorded in both edited volumes (Fischer and Getis, 1997) and special issues of learned journals (Wilkinson and Boots, 2000; Leung and Okabe, 2001). Presently, this facilitative task is the charge of the Commission on Modelling Geographical Systems (CMGS) appointed at the IGU Hague Congress in 1996 and chaired by Barry Boots (1996–2000) and Richard Thomas (2000–present).

Set against this background, this book provides a perspective on the work of the CMGS from 1996 until the IGU Seoul Congress in August 2000 through a collection of papers first presented to our sessions at this event. Moreover, a number of Japanese delegates were attracted to this Asian venue and their contributions provide many new ideas concerning the implementation of systems analysis. The selection of the complete set of materials, however, was guided by our intention of bringing together some of the main innovative ideas currently shaping both the development and testing of geographical systems models. To reflect this state of flux, the book is organised in two parts. The first is focussed on the more long-standing mathematical and statistical approaches to the analysis of geographical systems, while the second concentrates on the more recent opportunities offered by advances in geocomputation and the development of relatively new information technologies like the Internet.

In addition to this methodological stance, we were also conscious of the need to demonstrate the applicability of systems modelling to the wider geographical community. Reflecting current concerns for relevance, each paper has an applied component relating to one or more contemporary issues. Accordingly, the topics addressed by the contributors span both human and physical geography and many tackle key environmental concerns. Themes underscoring this variety of application include the design of warning systems for anticipating physical hazards (earthquakes, floods and new agents of infectious disease) and a contrasting emphasis on the analysis of urban systems (the management of sprawl and energy consumption, the timing of shopping behaviour, and the local impact of socio-economic policy initiatives). More subtly, the geography of information technology is itself becoming increasingly scrutinised by systems analysts. As with many other phenomena of interest, computing facilities are also distributed unevenly in space and time which is creating fresh issues related to accessibility that are explored in the second part of the book.

Like most edited volumes, this book provides an overview of recent research that is not necessarily directed at beginning students. Instead, the intended readership is researchers, postgraduates, final-year undergraduates and professionals in the areas of quantitative geography, spatial analysis and modelling, and geographical information sciences. Given this audience, the collection of papers will provide a useful supplementary text for courses on quantitative geography and geographical systems modelling in both human and physical geography, and GIS and geocomputation. This role is particularly relevant for the many master's degree courses in these subject areas that have proliferated around the world during the past decade.

Finally, we would like to acknowledge the efforts of the authors contributing to this volume both for keeping to our stringent editorial requirements and for their advice in the form of referee's reports. In this respect, we also thank John Beebe, Marie-Josee Fortin, Houston Saunderson and Eric Sheppard who acted as external reviewers on some of the papers. Invaluable assistance with the production of the electronic version of the manuscript was given by Tim Jenkins (Manchester) and Tomoko Kuroiwa (Tokyo). Further information relating to this book, including coloured versions of many of the figures, is available on www.ua.t.u-tokyo.ac.jp/okabelab/atsu/mgs.html.

January, 2002

Barry Boots
Atsuyuki Okabe
Richard Thomas

REFERENCES

Fischer, M.M. and A. Getis (1997), *Recent Developments in Spatial Analysis: Spatial Statistics, Behavioural Modelling and Computational Intelligence*. Springer, Berlin.

Leung, Y. and A. Okabe (2001), 'The International Conference on Modeling Geographical and Environmental Systems with Geographical Information Systems.' *Geographical and Environmental Modelling*, **5**, 109–110.

Wilkinson, G.G. and B. Boots (2000), 'Guest Editorial: Advances in the Modelling and Interpretation of Remotely Sensed Data.' *Geographical and Environmental Modelling*, **4**, 3–5.

Barry Boots, Atsuyuki Okabe and Richard Thomas

Introduction

1. Context

Definitions of what constitutes a *system* are legion, but arguably the most enduring is that given by Hall and Fagen (1956). Their system is a set of objects together with the relationships between these elements and their attributes. In geography, these objects are usually physical entities like towns, regions or stream links that form larger units like urban, national or drainage basin systems. The attributes of such objects are often measures of their capacity (population size, link length etc.) which serve to constrain the flows of people, information or energy between these constituent parts. At any point in time, these flows define the behaviour of the system and the purpose of their investigation is to develop an analytical capability to either reconstruct past behaviour or to anticipate future change. What makes geographical systems particularly distinctive is that their analysis usually requires the specification of relationships measuring *locational* characteristics of the objects like their degree of *spatial separation*. Typically, flows will be inhibited by such separation either through increasing travel costs or the laws of energy conservation.

Accordingly, the magnitude of any system of geographical interest may range from a few metres to thousands of kilometres. In addition, systems at large scales may nest within those at smaller geographic scales. The features considered may be physical (*natural geographical systems*), social and economic (*human geographical systems*), or both (*integrated geographical systems*). Thus, by their very nature, geographical systems are extremely complex and ever changing.

Inevitably, the representation of geographical systems depends on models, which employ various levels and degrees of abstraction, to simplify and analyze geographic reality. The particular form of abstraction chosen depends on the available data, our knowledge of the processes involved and how they operate, and the spatial extent of the area to be studied. They also depend on the objectives of the analyst so that the models may be *exploratory*, *explanatory*, or *predictive*. Further, they may consider a system at one point in time (*static model*) or explore changes in its behaviour over discrete or continuous time (*dynamic model*). Given these considerations, no single method or approach dominates geographical systems modelling. The response adopted by many modellers is to integrate appropriate concepts, procedures and techniques drawn from multiple sources. Consequently, the array of models for analyzing geographical systems and schemes for classifying them are almost as diverse and extensive as the systems themselves (Wegener, 1999). In this book, we focus

B. Boots et al. (eds.), Modelling Geographical Systems, 1–8.

on just two types, statistical and computational models, which are becoming increasingly important in the analysis of geographical systems.

As their name suggests, *statistical models* primarily use statistical concepts to represent the real world components and their interactions. *Computational models* employ the computer as an integral part of the modelling procedure. As such they are part of a larger body of activity referred to as geocomputation (Longley *et al.*, 1998). Computational models typically involve computationally intensive procedures such as simulation and their outputs are often highly visual and interactive in nature. In practice the two types of model are not mutually exclusive, in part, because both types have exploited recent conceptual and technical advances described below to make more efficient use of data and to stimulate new ways of exploring data. Thus, the two are best thought of as representing convenient labels for two complimentary modelling styles that may be mixed according the nature of the system being analyzed and the goals of the researcher.

One of the major recent conceptual advances has been the move within the spatial sciences towards analysis in a local context. This stems from the recognition that as well as examining regularities in system-wide behaviour (*global analysis*), it is important to search for significant, site specific variations from the global trends within the system (*local analysis*) (Fotheringham, 1997, 1999; Fotheringham and Brunsdon, 1999; Unwin and Unwin, 1998). This is particularly important for geographical systems which are likely to be inhomogeneous in terms of environmental and human conditions at most scales of analysis.

Local analysis has arisen in response to several influences. Fundamental developments in key information collection technologies have meant that large, complex, and frequently updated spatial data sets have become increasingly available. Some of these are the result of larger spatial coverage, finer spatial resolution, or both, associated with developments in automated monitoring devices such as orbiting earth satellites. Others are a product of the increasing penetration of computerization into virtually all facets of society's public and private sector administrative activities (Openshaw, 1998). Regardless of its origin, the larger the size of the data set studied, the more likely it will be heterogeneous (Unwin, 1996).

Another influence on the development of local analysis is changes in computer environments, in particular, increases in computing power and memory capacity, coupled with cheaper hardware costs. Related to this is the on-going development of more sophisticated geographical information systems (GIS) which, amonsgt other things, provide ideal environments for visualizing the multiple results of local analyses in a variety of ways.

Such developments in computing, in general, and in GIS, in particular, have also been influential elsewhere in modelling geographical systems (Fotheringham and Wegener, 2000). Advances in computing technology have made computationally intensive operations both viable and practical. Another ramification is that, conditional on data availability, analysts now have the opportunity to turn theoretical models into operational ones. The papers by Lee and Konagawa (this volume) provide examples of this tendency.

GIS were conceived as sophisticated spatial data processing and display systems and they continue to be important in this regard (Hearnshaw and Unwin, 1994). As such they provide rich environments in which modellers can apply their skills (as in Lee and Zhao and Tamura's papers in this volume). However, most GIS originally had only rudimentary analytical abilities. Much recent effort has been devoted to remedying this shortcoming. This activity includes incorporating the capability to handle effectively tasks such as formal analysis and human reasoning required to implement spatial decision support systems (see Feick and Hall, this volume). The combination of traditional and new roles for GIS has lead Sui and Goodchild (2001) to proclaim the advent of GIS as media. They suggest that GIS has joined with other mass media as a way of communicating information, especially geographical information, to the general public. An illustration of this is the VENTEN System (see Ogawa *et al.*, this volume). This view of GIS as media is also consistent with growing interest in spatial multimedia and virtual reality (Camara and Raper, 1999).

Another increasingly influential external force on the modelling of geographical systems is the rapid growth of the Internet. Not only has this opened up new possibilities in terms of data availability and storage, it has also provided an alternate platform for GIS through network GIS (Coleman, 1999). Together with other communications networks, the World-Wide Web, and other electronic "places", the Internet forms cyberspace (Dodge and Kitchin, 2000, 2001). While we have much to learn about the geography of such digital spaces, cybergeography (see Baker, Shiode, and Takeyama, this volume), it is already clear that the Internet has been responsible for a new generation of "time-space convergence" (Janelle, 1969) whereby the uneven distribution of Internet development has caused shrinkages in the time distance between specific places. There are indications that these new forms of space will have fundamental effects on social and economic behavior ranging from inter-personal communication (Takeyama, this volume) to retailing (Baker, this volume).

The various changes identified above have resulted in greater emphasis being placed on problem and application oriented approaches. Gone are the days of general, theory-led models in search of suitable applications. Today's models are much more likely to be data-driven responses to specific environmental and social problems such as natural hazards (Zhao and Tamura, this volume) and the transmission of HIV in space and time (Smith and Thomas, this volume). They are also more likely to be involved with predictions (Konagawa, this volume) and to have proactive concerns, especially when modelling outputs have obvious implications for future social or environmental policy decisions (Ballas *et al.* and Watanabe *et al.*, this volume).

2. CONTENTS

In this book we present 17 papers which illustrate a range of innovative ideas involved in analyzing geographical systems using statistical and computational models. These papers were selected from those presented at the sessions of the Commission on Modelling Geographical Systems at the 29th Congress of the International Geographical Union held in Seoul, Korea in August 2000. Collectively, the papers span all geographic scales, deal with both individuals and aggregates, and represent natural, human, and integrated spatial systems. Reflecting current concerns for relevance, each paper has an applied component relating to one or more contemporary issues.

The book is organized into two parts corresponding to the two types of models, statistical and computational. In turn, each part is divided into sections.[1] Part I has two sections. Section A focuses on spatial statistics and all papers in this section show how the local/global tension described above may be profitably used to facilitate integrative understanding of spatial processes operating at different scales. They also demonstrate how the emphasis in much of this work has swung away from confirmatory analyses towards exploratory ones. This is a partial reflection of the increasing size and complexity of available data sets as well as the recognition of a reduced role for classical statistical inference (Fotheringham *et al.*, 2000). Further, the techniques used in all three papers are computationally intense and would have not been possible without the recent developments in computer environments outlined above.

In the first of these, Lee shows how a statistical procedure, join-counts, originally developed to analyze static patterns at a global scale, can also be used to explore local changes in spatial processes over time. He illustrates his approach by analyzing residential development at the land parcel level for Geauga County in northeastern Ohio, USA over the past two hundred years. In a similar fashion, the second paper by Boots shows how concepts and procedures involved in identifying distinctive observations in a global data set (local measures of spatial association) can be developed into a means of evaluating polygon boundaries recorded in choropleth maps of thematic properties such as forest characteristics. In the final paper in this section Nakaya develops an existing local modelling procedure, geographically weighted regression (GWR) to produce local versions of a global spatial interaction model. These are then used to gain more detailed insight into migration flows in Japan during the latter half of 1980.

Section B is composed of papers that all combine space and time in the specification of the system but they do so at markedly different geographical scales. They all offer support to the emerging view that time need not be viewed in the traditional linear fashion. The first paper in this section, by Joh *et al.*, adopts an activity-based modelling approach in which travel is viewed as a consequence of participating in activities rather than being pursued for its own sake. Using a simulation approach, they develop a conceptualization and specification of processes by which individuals make adjustments to the repetitive components of their short-term activity schedules in response to unforeseen events. In contrast, in the second paper in this section, Mizuno explores circular and cumulative causation processes over a longer time horizon and at a larger geographic scale to examine urban agglomeration in East Asian countries. The "geographical metropolis model" he creates combines components of both economic and geographic models of urban industrial aggregates. The last paper in this section considers a continental scale. In it Smith and Thomas add a time component to the principles of spatial interaction theory to represent regional contact rates and use these to fit an epidemic model to recorded national AIDS incidence in Europe for two different scenarios of population mixing.

There are three sections in Part II. The first section contains simulation models. Although the three papers in this section span geographic scales ranging from neighbourhoods within cities, through individual cities, to nation states, they share a common concern with predicting the effects of changing inputs on the behaviour of the system under investigation. Using microsimulation modelling applied to UK Census Samples of Anonymised Records (SARs), Ballas *et al.* create a spatially disaggregated data set of attributes of individuals and households for the Leeds (UK) Metropolitan District. In turn, this micro-data is used to examine the impacts of alternative social policies upon the economic well-being of the District's residents. Watanabe *et al.*'s approach is an integrated one which simulates the diurnal climate of variously configured urban regions to examine the effects of different land use patterns on energy consumption. Konagaya also combines physical and economic conditions to create a probabilistic version of the Von Thünen-Ricardo model that permits a consideration of land use mixes. This model is used to predict the impact of increasing urbanization on the distribution of agricultural land use in China.

As noted above, most contemporary GIS packages were designed primarily as spatial data management systems and, as such, lack much of the functionality necessary for analyzing geographical systems. However, the computational capabilities of GIS software offer substantial potential for the development of such activities. The papers in Section B illustrate three novel ways in which this potential can be exploited. In Japan, pre-World War II statistical data is lacking and so Koike and Arai devise a method for estimating past population distributions using old, paper, pre-GIS era maps manipulated within a GIS environment. Zhao and Tamura take standard equations for describing physical systems and place them in a spatial context to create a spatial hydrological model. Then they combine this model with a digital elevation model and soil and vegetation databases within a GIS to estimate landslide, mudflow and flood hazards in the Sendai watershed in Japan. Hall and Fieck extend traditional, non-spatial decision analysis and model this using a GIS to provide an interactive, multi-criteria decision process for exploring the consensus and conflict surrounding tourist land development in an environmentally sensitive area in the Cayman Islands.

The final section is perhaps the most eclectic, reflecting its topic, the Internet. This is one area where scientific developments currently lag behind technological ones. The papers here illustrate some of the ways we can use the Internet's potential. Recognizing that the Internet facilitates increased accessibility and opportunity for various social and economic activities, Shiode considers its geography at an intra-national scale (UK and Japan). He visualizes this information in the context of urban geography to reveal where the Internet is impacting and where information power resides, both in terms of physical location and sectors of activity. By its nature the Internet relaxes many of the spatial and temporal constraints on human activity. With this in mind Baker explores the application of a RASTT-type model (time-dependent differential equations with a second order time operator) to Internet transactions. This model introduces the idea of relative time where flows of data can go either forwards or backwards in time relative to the rotation of the Earth. The Internet also provides a convenient place to store large amounts of information. Ogawa *et al.* have exploited this to create an Internet based GIS, VENTEN, (Vehicle through Electronic Network of disasTer gEographical informatioN, also the name of the Asian goddess of disaster management), which provides a means of sharing geographical data and disaster information for the East Asian region. VENTEN enables spatial analysis to be undertaken, using the Internet, by those who might otherwise be prohibited from doing so because they lack the money and skills to establish their own GIS systems. This is particularly important for potential users in less-developed parts of the region. Finally, as noted above, the Internet can be considered as one component of electronically mediated spaces, cyberspace, which may be integrated seamlessly with physical places to create smart places. As prototypes of such spaces, Takeyama describes experimental projects to design a smart zoo and a smart town. This paper also illustrates how the integration of computing devices into everyday life will require us to re-think the nature of the relationships between mind and body and space and time.

Together, the papers in this book demonstrate some of the innovative ways of thinking about spatial phenomena and spatial processes that have characterised recent advances in geographical systems modelling. The emphasis is placed on explicating these ideas through their application to substantive contemporary issues which, inevitably, will change in the future as new data become available and new problems arise. In this context of flux, we should also note that the interplay between mathematical methods and their computational environments generates new modelling styles that enhance our capabilities to schematise the real world. Moreover, given the dynamic nature of geographical systems we do not anticipate this situation is ever likely to change.

Barry Boots
Department of Geography and Environmental Studies,
Wilfrid Laurier University,
Waterloo, Ontario N2L 3C5,
Canada

Atsuyuki Okabe
Center for Spatial Information Science,
University of Tokyo,
7-3-1 Hongo,Bunkyo-ku, Tokyo 113-8656,
Japan

Richard Thomas
School of Geography,
University of Manchester,
Oxford Road, Manchester M13 9PL,
UK

NOTE

[1] However, because of the complementarity between the two model types, this organization is partly one of convenience. This is reflected in several of the papers that are multifaceted and could have been included in more than one section and, in some cases, in either part. For example, while Section B of Part II Computational Models is entitled GIS Models, GIS is used as an analytical and as an organizational environment in the papers by Lee in Part I and Suzuki *et al.* in Part II, respectively. Also, the papers by Joh *et al.* and Mizuno which appear in Part I Section B both use simulation approaches and so could equally well be included in Section A of Part II. Clearly, geographical systems modelling is not an activity that lends itself easily to classification.

REFERENCES

Camara, A.S. and J. Raper (1999), *Spatial Multimedia and Virtual Reality GIS*. Taylor and Francis, London.

Coleman, D.J. (1999), 'GIS in networked environments.' In P.A. Longley, M.F. Goodchild, D.J. Maguire and D.W. Rhind (eds.), *Geographical Information Systems: Principles, Techniques, Applications, and Management*, 2nd ed., Vol. 1, Wiley, New York, pp. 317–329.

Dodge, M. and R. Kitchin (2000), *Mapping Cyberspace*. Routledge.

Dodge, M. and R. Kitchin (2001), *The Atlas of Cyberspace*. Addison-Wesley.

Fotheringham, A.S. (1997), 'Trends in quantitative methods I: stressing the local.' *Progress in Human Geography*, **21**(1), 88–96.

Fotheringham, A.S. (1999), 'Guest editorial: local modelling.' *Geographical & Environmental Modelling*, **3**(1), 5–7.

Fotheringham, A.S. and C. Brunsdon (1999), 'Local forms of spatial analysis.' *Geographical Analysis*, **31**(4), 340–358.

Fotheringham, A.S., C. Brunsdon and M.E. Charlton (2000), *Quantitative Geography: Perspectives on Spatial Data Analysis*. Sage Publications, London.

Fotheringham, A.S. and M. Wegener (eds.) (2000), *Spatial Models and GIS: New and Potential Models*. Taylor & Francis, Andover, Hants.

Hall, A.D. and R.E. Fagen (1956), 'Definition of system.' *General Systems Yearbook*, **1**, 18–28.

Hearnshaw, H.M. and D.J. Unwin (eds.) (1994), *Visualization in Geographical Information Systems*. John Wiley & Sons, Chichester, UK.

Janelle, D.G. (1969), 'Spatial reorganization: model and concept.' *Annals of the Association of American Geographers*, **59**, 348–364.

Longley, P.A., S.M. Brooks, R. McDonnell and B. Macmillan (eds.) (1998), *Geocomputation: a Primer*. John Wiley & Sons, Chichester, UK.

Openshaw, S. (1998), 'Towards a more computationally minded scientific human geography.' *Environment and Planning A*, **30**(2), 317–332.

Sui, D. and M.F. Goodchild (2001), 'GIS as media?' *International Journal of Geographical Information Science*, **15**(5), 387–390.

Unwin, A. and D. Unwin (1998), 'Exploratory spatial data analysis with local statistics.' *Journal of the Royal Statistical Society: Series D (The Statistician)*, **47**(3), 415–421.

Unwin, D. (1996), 'GIS, spatial analysis and spatial statistics.' *Progress in Human Geography*, **20**, 540–541.

Wegener, M. (1999), 'Spatial models and GIS.' In A.S. Fotheringham and M. Wegener (eds.), *Spatial Models and GIS. New and Potential Models*. Taylor & Francis, Andover, Hants.

Part I

Statistical Models of Spatial Systems

Section A

Spatial Statistics

JAY LEE

GEOGRAPHICAL PATTERNS OF URBAN RESIDENTIAL DEVELOPMENT

1. INTRODUCTION

In an urbanized environment, the geographical pattern of residential development is a complex phenomenon to model quantitatively. This is because it is a comprehensive and dynamic phenomenon that involves a wide spectrum of social, economic, cultural, and geographical variables. Conventional approaches to studying this phenomenon have been focusing on using one or just a few variables, while holding others constant, to obtain a sketchy impression of how the development of urban residential land use has changed over space and in time. Some examples can be found in Morcombe (1984), Donovan and Neiman (1993), Hitt (1994), Fulford (1996), Fader (2000), and Levia and Page (2000). As such, results from the research and literature on this topic offer only partial understandings of how urban residential lands develop geographically and temporally. For practical application, we would need to develop integrated models with definable quantitative measures.

Urbanization expands the size of a city or a population settlement until the city becomes too crowded for further development at its core. At that time, sub-urbanization occurs. Both population and economic activities move from the city's core to its peripheral areas. With advancement in transportation technology, sub-urbanization actually expands urbanized areas and connects the surrounding scattered developments into metropolitans. During such process, the development of residential lands takes a variety of geographic patterns at different stages of urbanization.

In the early stage of urbanization, residential lands develop in a more compact form, as new developments tend to occur adjacent to existing ones within and surrounding the urban core. During sub-urbanization, the population settlement continues to expand, new residential developments may occur in a more haphazard manner and at faster rate – a phenomenon often referred to as urban sprawl. The sprawling of urbanized area surrounding its core typically occurs by first spilling new developments over peripheries and then followed by in-filling those vacant lands in between. As peripheries become saturated, the next stage of expansion begins to ignite another cycle.

We illustrate here a set of models that can be used to measure the geographical patterns of how residential lands develop spatially and temporally. The actual pattern can be measured to fit into one of the defined models to see if it is closely related to a compact expansion, a sprawling growth, or an in-fill process.

B. Boots et al. (eds.), Modelling Geographical Systems, 13–31.

Specifically, we use Join-Count Statistics, a simple spatial statistic, to measure the geographical patterns of the development of residential lands. Coupled with temporal trends, these models can be used to quantitatively study residential developments.

The parcel-level data from Geauga County, Ohio, USA, were used to show how a simple statistical method could be applied to quantitatively model the geographical pattern of residential development over time. With similar data at parcel level, modeling of urbanization process can be carried out in other regions in the same way.

2. Join-Count Statistics as a Measure of Geographical Patterns

We suggest that the Join-Count Statistics can be used to measure quantitatively the clusterness and dispersion of the geographical patterns of residential lands. Join-Count Statistic is the simplest form of spatial autocorrelation. While not being the most powerful statistic for measuring spatial patterns, it is appropriate for this study because of its ability to handle polygonal binary nominal data.

In statistical concepts, autocorrelation is the relationship between successive values of residuals along a regression line. In most cases, a strong autocorrelation indicates successive values are strongly related, which implies that data values being regressed may have a systematic trend among them. Spatial autocorrelation is a simple extension of the autocorrelation concept into two dimensions:

- A strong, positive spatial autocorrelation means that the characteristics of geographic objects are very similar to those of nearby objects. This is normally referred to as a *clustered* pattern.
- Alternatively, a strong negative (or inverse) spatial autocorrelation suggests that geographic objects may have very distinctive properties between adjacent objects. This is often known as a *dispersed* or *uniform* pattern.
- When there is no measurable spatial autocorrelation, the geographic objects are said to be in a *random* pattern.

The three patterns, clustered, random, and dispersed, serve as three mileposts on a spectrum along which many other possible patterns may exist. Because of this, a value of spatial autocorrelation coefficient by itself is not useful unless it is tested for its statistical significance of how different it is from a coefficient value indicating a particular pattern. For example, a spatial autocorrelation coefficient measured from an observed pattern will need to be tested to see if it is statistically significantly different from the coefficient value of a random pattern – giving the same spatial structure of the geographic objects.

Join-Count statistics provides a simple and efficient way of quantitatively measuring the degree of clusterness or dispersion among a set of spatially adjacent polygons. It bases its measurements on how different the geographical pattern being observed is different from a theoretically constructed random pattern, given the same number of geographical areas and the same spatial structure. This statistic allows users to measure and test if a geographic pattern is statistically significantly different from a random pattern, and if so, whether the geographic pattern is more clustered than a random pattern or it is more dispersed than a random pattern.

Join-Count Statistics only work with binary nominal data associated with polygons. Given a geographic phenomenon with a yes/no, presence/absence, black/white, or other forms of dichotomy characteristics, each geographic area, or termed as a *polygon*, is associated with either one of the two possible characteristics. In our case, a land parcel can be classified as built or vacant. A *built* parcel is one where it has been built with a construction unit while a *vacant* parcel is without one and is available for future development.

Once the polygons of land parcels are defined as either built or vacant, the Join-Count Statistic counts the numbers of various types of Joints between adjacent polygons. These numbers are then compared with those of a random pattern to determine if the pattern being observed is significantly different from a random pattern or if it is more clustered or dispersed than a random pattern.

A *joint* is a segment of shared boundary between two adjacent polygons. It can be a *BV* joint if the joint connects a built polygon and a vacant polygon. A joint can be a *BB* joint if it is shared by two adjacent polygons that both have been built for residential use. Similarly, a *VV* joint is one that is shared by two vacant parcel polygons.

Following Lee and Wong (2001, pp. 147–156; also in Cliff and Ord, 1981; Upton and Fingleton, 1985; Goodchild, 1986; Griffith, 1987), let O_{BV} be the number of observed *BV* joints, E_{BV} be the number of expected *BV* joints from a random pattern, and σ_{BV} be the estimated standard deviation of E_{BV}, a Z score can be calculated as:

$$Z = \frac{O_{BV} - E_{BV}}{\sigma_{BV}} \tag{1}$$

If the observed number of *BV* joints is greater than the number of expected *BV* joints, it means that the observed pattern has more *BV* joints than that of a random pattern with same polygonal structure. In this case, the observed pattern is likely more dispersed than a random pattern because it has more occurrences of built polygons adjacent to vacant polygons. Alternatively, a pattern whose number of *BV* joints is less than that of a random pattern would imply that it is a more clustered pattern because the built polygons tend to be located next to built polygons. A word of caution should be given here regarding the appropriate use of Z score in testing the statistical significance of Join-Count Statistics. There should be at least 30 (or more) parcels in each data set and that the ratio between built and vacant parcels should not be too close.

A graphic example for a clustered pattern (Figure 1a), a random pattern (Figure 1b), and a disperse pattern (Figure 1c) is shown in Figure 1. In this

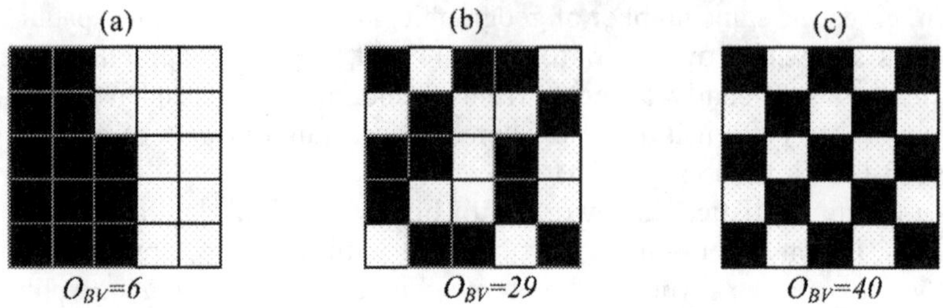

Figure 1. Hypothetic polygonal structure of 13 built parcels and 12 vacant parcels.

hypothetic polygonal structure, black squares denote built parcels while white squares denote vacant parcels.

Over time, a geographical pattern of residential land use may change from one form to another by the added new development. For example, Figure 2 shows two examples for changing geographical patterns by adding a new development. In the first example (Figure 2a), the newly added residential parcel reduces the number of *BV* joints thereby making the pattern more clustered. In the second example (Figure 2b), the newly added residential parcel increases the number of *BV* joints thereby making the pattern more dispersed.

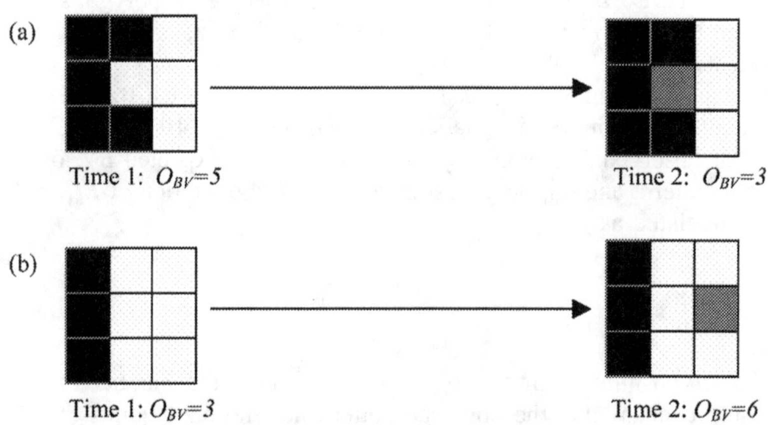

Figure 2. Changes of geographical patterns with new residential development.

The changes of O_{BV} will be reflected in the *Z* scores. When O_{BV} increases, *Z* score tends to increase. Similarly, *Z* score decreases when O_{BV} decreases. Given this relationship between the numbers of *BV* joints and geographic patterns, it is possible, then, to examine how geographic patterns change over time.

The spatial statistic, Join-Count Statistic, has been implemented in ESRI's Avenue scripting language as a part of the accompanying package in Lee and Wong (2001). With this computer script, a total of two steps are needed to calculate the *Z* scores of the Join-Count Statistics for any data set. The first step is to calculate a distance matrix to record the adjacency between polygons. In this step each parcel was examined in turn to determine its adjacent neighbor polygons. These neighbor polygons are then recorded by their identifiers in the distance matrix in a binary form: 0 for non-neighboring polygons and 1 for neighboring polygons.

The second step calculates the Join-Count Statistics based on B (built parcels) and V (vacant parcels). The testing of the statistical significance can be based on a free sampling or a non-free sampling hypothesis. The *free-sampling* hypothesis assumes that the probability of a parcel being built or remain vacant is known without reference to the study region. In other words, the built-vacant division is not based on local condition but rather on the trend from a much larger region that contains the study area. The *non-free sampling* hypothesis assumes that the numbers of built parcels and vacant parcels remain the same but their arrangement can be different in the study area. This hypothesis is based on local conditions. The statistical testing based on this hypothesis allows analyst to assume that the observed pattern is only one of many possible arrangements using the same numbers of built parcels and vacant parcels, without making any reference to outside factors.

3. QUANTITATIVE MODELS WITH TEMPORAL TRENDS

There are three general categories of geographical patterns conventionally used as benchmarks to describe how polygons (as areas) structure spatially. The first category is cluster patterns when polygons of similar properties cluster together. Next, the disperse pattern is when polygons of similar properties are apart from each other. In the extreme case of disperse pattern, a uniform pattern is a pattern in which every polygon is surrounded by polygons of different property. Finally, the third category is the random pattern in which there does not seem to be any structural pattern.

Using the three categories as benchmarks and together with Z scores of Join-Count Statistics calculated from a geographical pattern of polygons, it is possible to construct a number of patterns that characterize different types of urbanization process. Specifically, by calculating Z scores of Join-Count Statistics of a polygonal structure over time, we would plot the changes of Z scores to see how they change so to gain detailed understanding of its temporal trends.

First, a *compact expansion* of residential development is a growth pattern in which new residential lands are developed next to existing built parcels. The growth, in a compact expansion pattern, tends to be slow and gradual whereby the overall geographical pattern shows a compact form of increasing territory. When a population settlement starts as a small village and gradually grows into a city, it often takes the form of the compact expansion. Quantitatively, this pattern is characterized by negative Z scores of Join-Count Statistics because of its apparent clusterness, i.e., $O_{BV} < E_{BV}$. Over time, a compact expansion tends to have decreasing Z scores, but only in a slow rate.

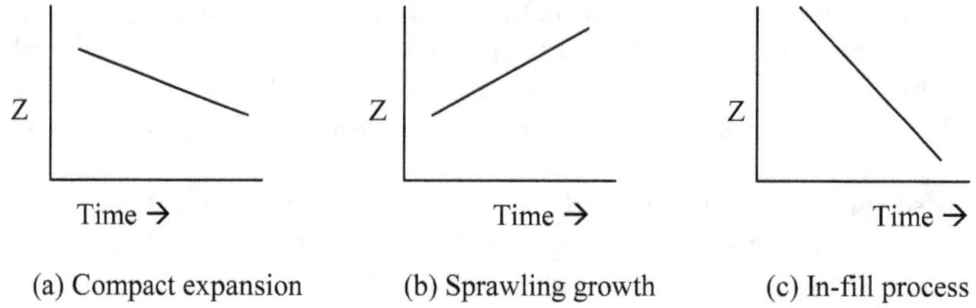

Figure 3. Three temporal trends of residential development.

In Figure 3a, a gentle decrease of Z scores over time gives the temporal trend of a compact expansion as measured by Join-Count Statistics.

Next, a *sprawling growth* is a development pattern that newly built residential lands occurred in a leap-frogging, haphazard manner. In this pattern, the developers typically look for lands that are the most economical, efficient to build, opting for the most profit. The sprawling pattern is also the results of a fast-pace growth and, to some degree, a growth pattern that is not under tight control or according to a defined plan. In a sprawling growth, it is often the case that many small, hard-to-use vacant lands are created by the leap-frogging developmental pattern. In quantitative terms as shown in Figure 3b, a sprawling growth is a pattern where Z scores of Join-Count Statistics increase over time. The rate of increase in Z scores depends on the rate of urban sprawl.

Finally, an *in-fill process* would occur after sprawling growth begins to slow down and the growth of the urbanized area expands into filling up the vacant lands left behind by urban sprawl. Using the Z scores of Join-Count Statistics as shown in Figure 3c, the in-fill process is characterized by steep slope for fast decrease of Z scores over time. This is because the number of O_{BV} is quickly reduced with each new residential development that fills in a hole created by the surrounding built parcels. Compared Figure 3c with Figure 3a, the slope of an in-fill trend is much steeper than that of a compact expansion.

In spatial terms, Figure 4 shows the three temporal trends by adding new residential developments. Figure 4a shows that the three new residential developments (shown in gray) impact on the Z scores of Join-Count Statistics only modestly since O_{BV} only decreases slightly. The number of observed BV joints decreases slightly from 9 to 8.

With also three new residential developments (shown in gray), all detached from existing developments (shown in black), Figure 4b shows that O_{BV} increases dramatically in a leap-frogging development. In this case, O_{BV} increases significantly from 9 to 19. In turn, this increase will be translated to a significant increase in the Z scores of Join-Count Statistics.

The last of the three trends is shown in Figure 4c. Also being added with three new residential developments, the number of BV joints decreases from 9 to 7. This is a decrease that is faster than that of a compact expansion. It is because for every infill land parcel, there are three or four BV joints converted

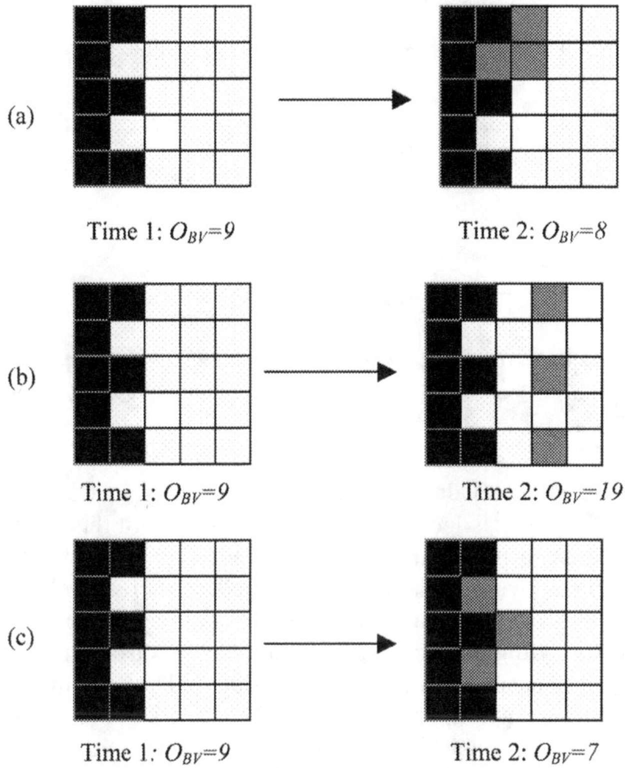

Figure 4. Three temporal trends. (a) a compact expansion, (b) a sprawling growth, (c) an in-fill process.

to either *BB* or *VV* joints. Thus, we would expect a faster decrease of O_{BV} if new land parcels are being developed during the infill stage.

It should be pointed out that it is possible to combine the three temporal trends to model long-term urbanization process. For example, Figure 5 shows that a population settlement started out as a small village going through a slow growth of compact expansion (segment a), then following up with a fast growth before a urban sprawl (segment b), growing in a sprawling manner (segment c), and finally ending up with a slower growth when sprawling is over (segment d).

In summary, the changes in *Z* scores of Join-Count Statistics as calculated from geographical patterns of residential lands at different time period allow us to construct a curve with the vertical axis being the *Z* scores and the horizontal axis being the time to show temporal trends. The directions of changes and their slopes help to define both the geographical patterns and their temporal trends. With these models, it is now possible to examine how they perform with real world data sets.

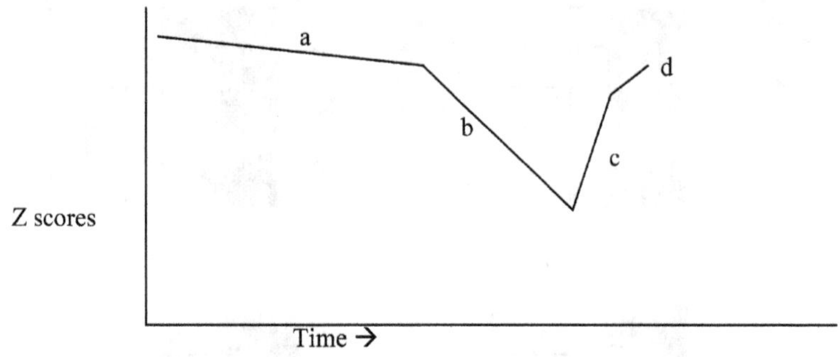

Figure 5. Long-term trend of urbanization modeled by Z scores from Join-Count Statistics.

4. CASE STUDY

We use the parcel level data set of Geauga County in Ohio of the United States. Geauga County is located in Northeast Ohio with the greater Cleveland metropolitan area to its west. Due to the process of sub-urbanization occurring in Cleveland over the last several decades, there has been significant urban sprawl observed in Geauga and nearby counties at the same time.

While the sprawling built-out in Geauga County is by no means an isolated phenomenon, the reasons and patterns of residential development displayed in it are similar to many other regions in the country, and perhaps in other countries as well. As such, it is suitable to be used as an example here for demonstrating the quantitative models established earlier for the study of urbanization processes. We expect that similar application of these models to other geographic regions is possible if data are available in similar forms.

In Figure 6, land parcels that were built earlier are shown in darker colors while parcels built recently are shown in lighter color. The four categories are separating parcels into: pre–1900, 1900–1950, 1950–1970, and 1970–2000.

As shown in Figure 6, Geauga County has seen much of its lands been developed within just the last several decades. The sprawling of developed parcels is certainly to a great extent being influenced by the growth of the greater Cleveland metropolitan from the west. However, it is difficult to gain a detailed understanding just by visually examining this map. A quantitative measure that describes the spatial pattern of how the parcel development proceeded will allow us to fully explore the trend of land development.

For Geauga County, we obtained from the county's Auditor's Office a data set that contains boundaries of all land parcels, their assessed land values, the years parcels were built (or developed), land use types, and many other administrative attributes. For the purpose of this study, we use only the attribute that contains the years when the residential parcels were built. With this attribute information, we are able to construct, dating back to pre-1900 time, the spatial patterns of developed lands in the county for any year between 1800 and 2000, or any duration of the time periods.

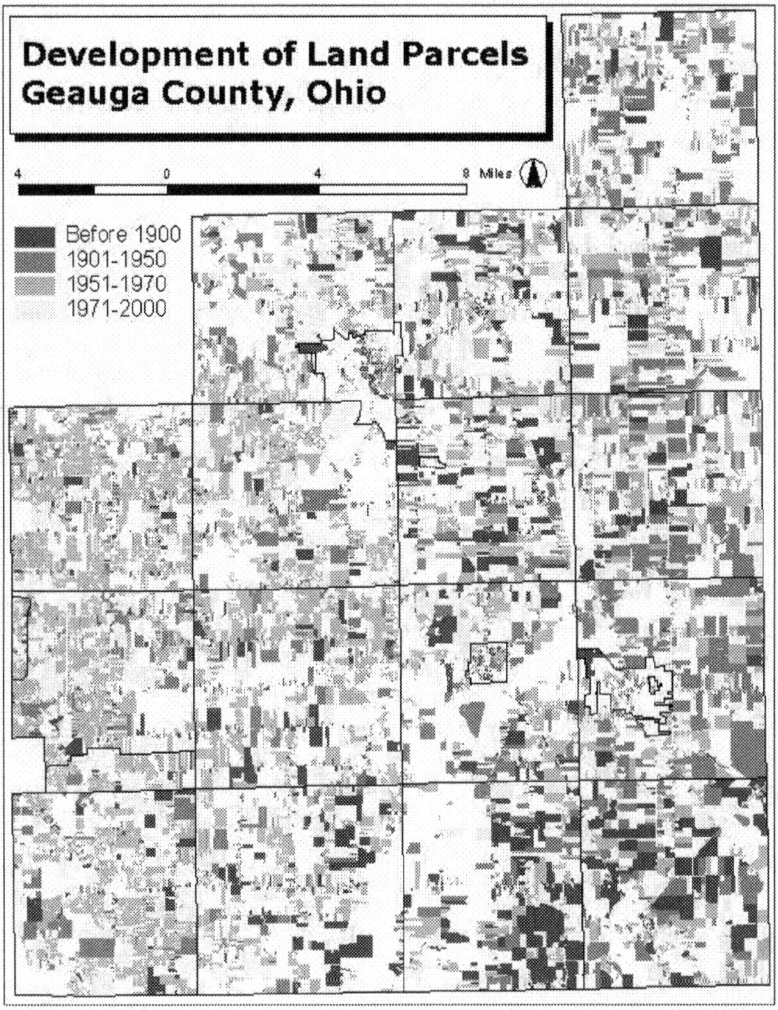

Figure 6. Development of land parcels in Geauga County, Ohio, USA.

With this study, we show that Join-Count Statistics can be used to measure the geographical and temporal trends of residential development. This method is appropriate because the land parcels are represented by polygons with a built/vacant binary form of attribute information. What we are measuring is, for any given year, the statistical significance of its geographic pattern being different from a random pattern. With this, we will also examine any significant difference regarding its direction and strength as indicated by the Z score of the Join-Count Statistics.

There are a total of 23 townships and villages in Geauga County. Built/Vacant data are available from 1800 to 2000. For reasons of annexation and simplifying data processing, 23 townships/villages were integrated into 21 areal units. Along the temporal axis, yearly data were integrated to every 10th year, starting with pre-1900, then 1910, 1920, . . . , and 2000.

Specifically, we performed the following steps when calculating the Z scores for the Join-Count Statistics:

For each township/village,
1. Count the actual number of shared boundaries (*BV* joints) between neighboring built and vacant parcels (polygons), O_{BV}.
2. Calculate the expected number of built-vacant joints for a random pattern, given the number and the structure of polygons, E_{BV}.
3. Calculate the standard deviation of the number of built-vacant joints, (σ_{BV}).
4. Calculate the Z score using the observed number of joints, expected number of joints, and the standard deviation of joints ($Z = [(O_{BV} - E_{BV})/\sigma_{BV})$.

In this way, the value of *Z* score is an indication of:

- If *Z* approximates 0, it means that the observed pattern is not significantly different from a random pattern of the same polygonal structure.
- If $Z > 0$, it means that the observed pattern has more built-vacant joints than a random pattern, implying that the observed pattern is likely a dispersed pattern. For this study, a high positive *Z* score would mean an urban sprawl pattern because very few newly built parcels would be adjacent to existing built parcels.
- If $Z < 0$, it means that the observed pattern has less built-vacant joints than a random pattern, implying that the observed pattern is likely a clustered pattern. For our purpose here, a strong negative *Z* score may mean that the built-out pattern to be similar to a compact expansion where newly built parcels are adjacent to existing built parcels.

For the issue of statistical significance level, we used the conventional $\sigma = 0.05$, which translates to a critical value of ±1.96 for *Z* scores:

- If $Z < -1.96$, it means that the observed pattern is a more clustered pattern than a random pattern. Typically, a more contagious expansion of residential development will yield such *Z* values. The more negative the *Z* value is, the more clustered the observed pattern would be.
- If $Z > 1.96$, it means that the observed pattern is a more dispersed pattern than a random pattern. Similarly, a more leap-frogging development will yield such *Z* values. The larger the *Z* value is, the more severe the dispersion is.

With the consolidated 21 townships, Figure 7 shows the relative position of these townships and villages. Note that Geauga County is located east of Cleveland. Therefore, the influence by Cleveland is apparently coming from the west (left) side of the map.

With Geauga County divided into 21 areal units and the year-built attribute information integrated to every 10th year, we have compiled the following table (Figure 8) that shows the *Z* scores for testing the significance level of

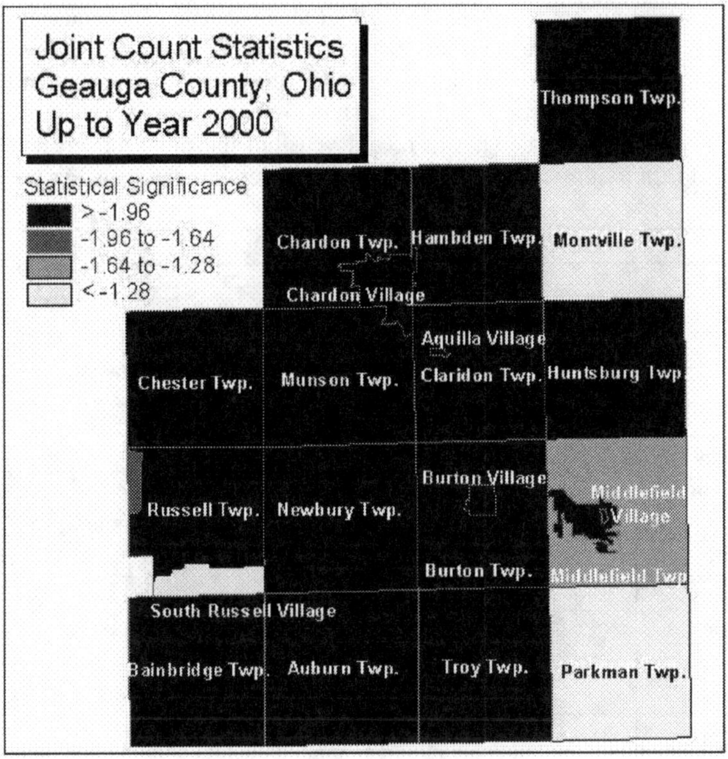

Figure 7. Z scores of Join-Count Statistics for townships and villages in Geauga County, Ohio, USA.

computed Join-Count Statistics. In the case when Z scores are greater than +1.96, they are listed in bold font. For Z scores that are smaller than –1.96, they are listed in italic font.

Among the 21 area units, it can be seen that they display different geographic patterns of development of residential parcels over time. To simplify the description, the townships and villages are grouped to the following categories:

Old, intensively developed areas – This group includes Chardon Village, Burton Village, and Middlefield Village. The three villages were developed the earliest in the county and were urbanized the earliest. Their Z scores are statistically significant dating back to pre-1900 era. This indicates that these villages are matured urbanized areas with few vacant parcels for new development.

Mature, developed areas – This group includes Montville Township, Humbden Township, Huntsburg Township, Middlefield Township, and Troy Township. They are mostly in the eastern part of the county. These areas were sparsely developed in the earlier part of the century but began to develop since 1950 with influence from the old, intensely developed areas.

Young, developing areas – This group includes the rest of the county. As can be seen in Figure 7 and Figure 8, these townships did not display any significant spatial pattern until 1950's when they began receiving influences from Cleveland metropolitan area. Due to space limit, the rate of growth has slowed down somewhat in recent year but the dramatic growth experienced in the county is apparent as the *Z* scores indicated in Figure 8.

	~1910	~1920	~1930	~1940	~1950	~1960	~1970	~1980	~1990	~2000
Thompson Twp.	-0.07	0.42	0.30	0.29	0.01	-0.23	-1.29	-3.07	-3.13	-4.08
Montville Twp.	**2.55**	**2.76**	**2.65**	1.69	0.87	-0.43	-2.40	-3.93	-5.05	-5.67
Hambden Twp	**2.81**	**2.36**	**2.26**	1.69	0.82	-1.19	-5.20	-7.82	-10.50	-9.32
Chardon Twp.	1.23	1.06	1.04	-0.01	0.05	-1.64	-12.10	-16.83	-18.51	-15.12
Chardon Village	-5.83	-9.84	-11.53	-14.86	-15.79	-18.23	-25.03	-28.72	-32.50	-29.85
Huntsburg Twp.	**2.39**	**2.71**	**3.21**	**3.40**	**2.98**	**3.00**	0.04	-1.68	-4.13	-5.61
Claridon Twp.	0.67	0.53	0.67	-1.10	-1.24	-2.15	-5.10	-6.04	-7.21	-6.00
Munson Twp.	-0.65	-0.89	-0.44	-0.74	-1.35	-2.46	-7.99	-13.43	-17.66	-19.76
Chester Twp.	0.76	0.22	0.00	-0.20	-0.76	-3.45	-21.03	-34.64	-22.04	-14.78
Aquilla Vlg.	0.67	0.53	0.67	-1.10	-1.24	-2.15	-5.10	-6.04	-7.21	-6.00
Middlefield Twp.	**4.12**	**2.69**	**2.63**	**5.11**	**4.68**	1.13	-0.75	-3.54	-5.60	-3.86
Burton Twp.	-0.08	-0.19	-0.19	-0.97	-0.93	-2.04	-7.69	-10.03	-15.35	-15.20
Newbury Twp.	0.38	-0.06	-0.31	-1.49	-1.71	-2.43	-9.01	-9.41	-17.44	-19.54
Russell Twp.	-0.06	-0.36	-0.38	-0.56	-1.38	-4.95	-15.14	-22.03	-16.72	-9.90
Hunting Valley V.	-1.15	-1.45	-0.98	-1.03	-1.26	-1.94	-0.99	-2.36	-2.63	-1.71
Burton Vlg.	-4.82	-5.77	-6.95	-8.91	-10.14	-9.50	-9.82	-9.67	-8.18	-7.18
Middlefield Vlg.	-1.33	-2.71	-3.41	-9.28	-9.83	-11.74	-14.70	-15.36	-17.60	-18.55
South Russell Vlg.	0.78	0.13	0.14	-0.54	-0.80	-1.46	-8.73	-27.74	-30.27	-16.51
Parkman Twp.	1.85	1.63	1.41	1.51	1.15	-0.88	-2.78	-3.58	-5.72	-5.94
Troy Twp.	**3.22**	**2.65**	**2.81**	**1.90**	**2.02**	-0.10	-2.83	-4.38	-5.75	-5.96
Auburn Twp.	-0.03	0.07	-0.01	-0.53	-1.45	-3.86	-6.68	-7.77	-14.93	-22.78
Bainbridge Twp.	-0.65	-0.89	-1.14	-1.49	-1.74	-3.15	-11.62	-20.95	-37.35	-47.49

Figure 8. Z scores of Join-Count Statistics for townships and villages in Geauga County, Ohio, USA.

To examine the temporal trends of how *Z* scores change over time, they were translated to trend lines in Figure 9. In graphic form as shown in Figure 9, most townships and villages showed decreasing *Z* scores until the 1950's. After that time, townships and villages developed in very different rates because the slopes of the trend lines show very different steepness. This is because different townships/villages developed with different degrees of compactness. Different townships and villages arrived at the lowest *Z* scores at different times that were also the times they began to experience the rebound of *Z* scores.

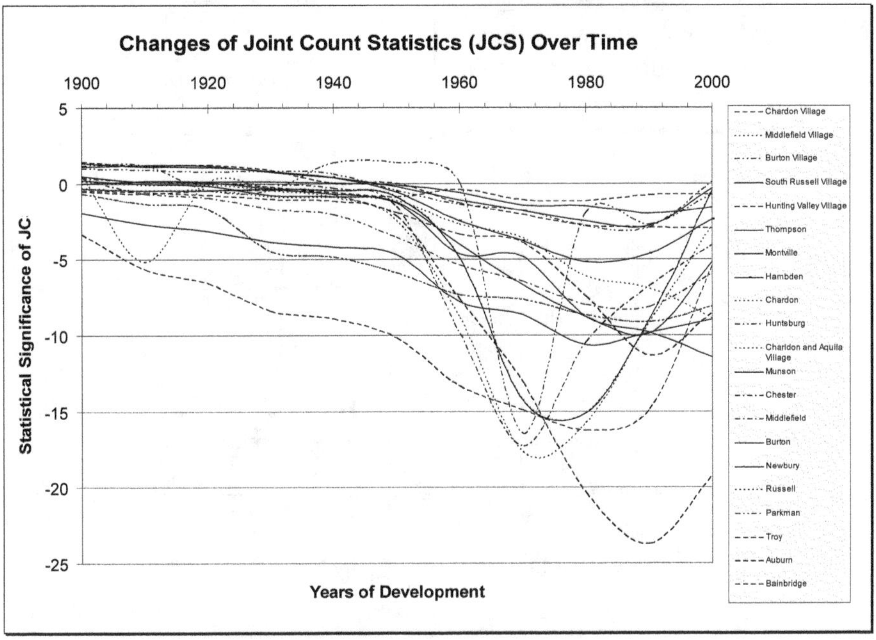

Figure 9. Temporal trend lines of Z scores.

Note that the upward trend of *Z* scores shows fast growth of the township or village, often to a sprawling pattern.

Finally, in graphic form, the maps in Figure 10 show a temporal account of the changes in *Z* scores by the townships and villages in Geauga County. It can be seen that the earlier development in the county was mostly generated from regional urban cores within the county but the recent development was clearly influenced by Cleveland to the west (left).

Based on the temporal trends showed in Figure 9, we chose to present the changes of *Z* scores by townships and villages for up to 1910, 1940, 1960, 1970, 1990, and 2000. These years were chosen to reflect the shape of trend lines in Figure 9. As can be seen in Figure 10, a wave-like shift of strong negative *Z* scores go through the county from west to east. In the early years up to 1910, most townships and villages showed only weak negative *Z* scores, then the whole county showed stronger *Z* scores. In the most recent years, some old intensively developed area began to show changes of *Z* scores from strong negative to weak negative.

5. DISCUSSION AND CONCLUDING REMARKS

Study of urban form and the process of urbanization is by no means a new topic. As early as in 1939, Hoyt examined the geographic patterns of the growth of the built-up areas in five American cities (Hoyt, 1939). With vivid graphics

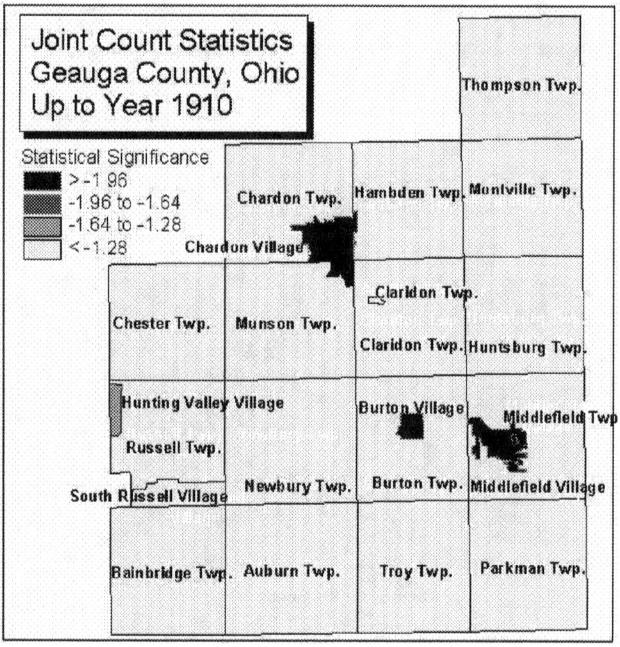

Figure 10a. Statistical significance of Join-Count Statistics, up to 1910.

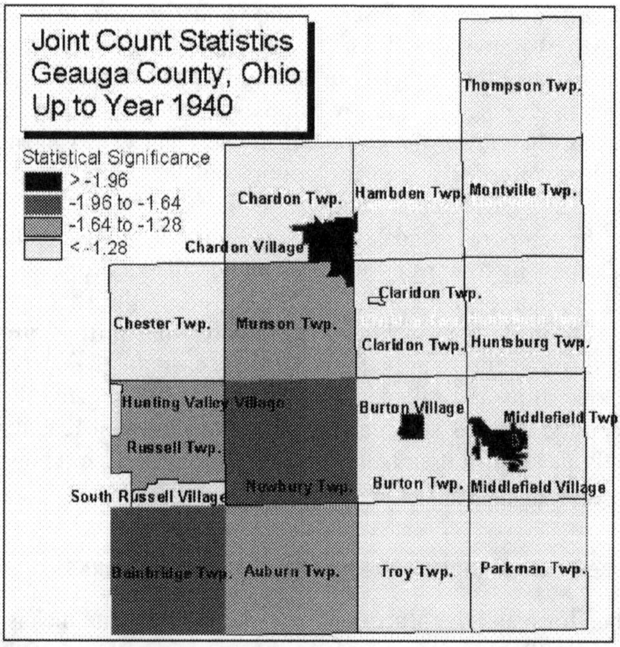

Figure 10b. Statistical significance of Join-Count Statistics, up to 1940.

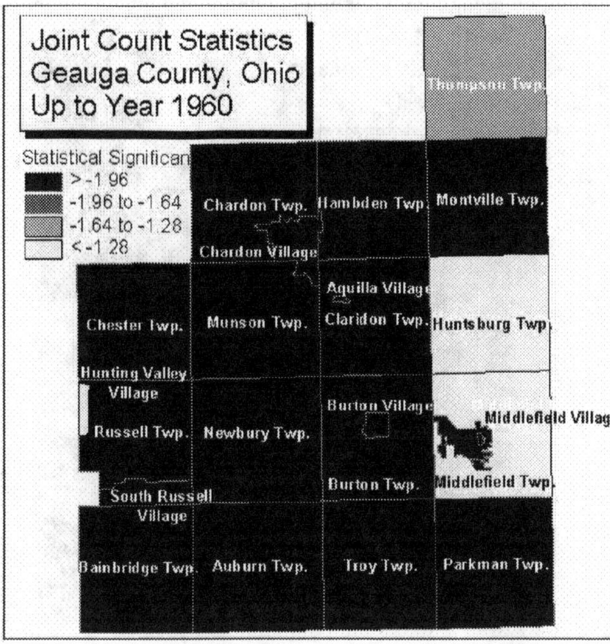

Figure 10c. Statistical significance of Join-Count Statistics, up to 1960.

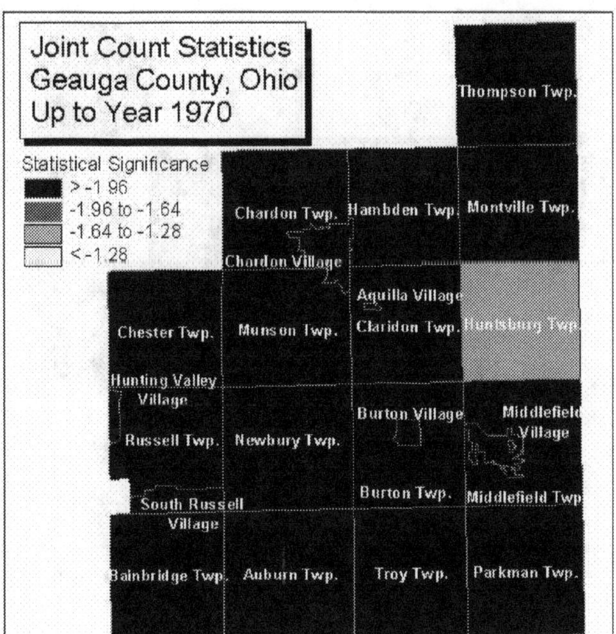

Figure 10d. Statistical significance of Join-Count Statistics, up to 1970.

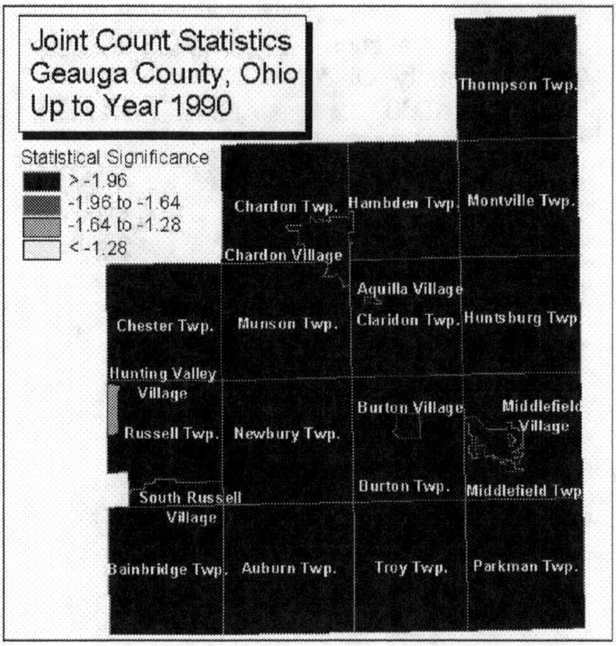

Figure 10e. Statistical significance of Join-Count Statistics, up to 1990.

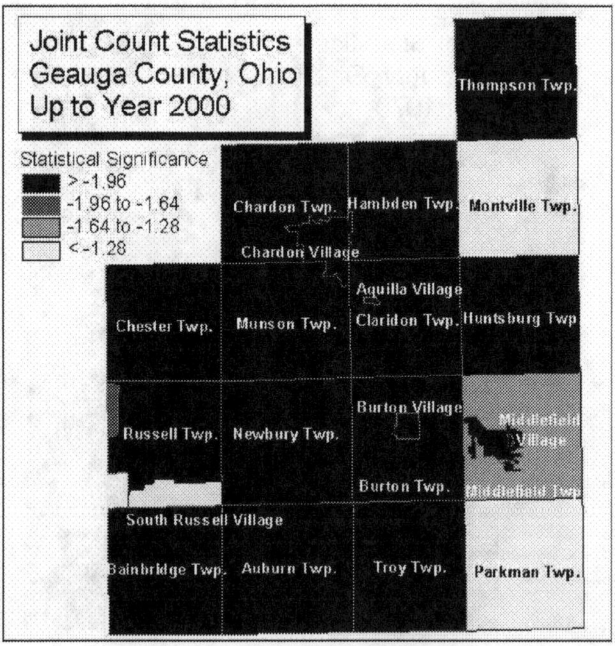

Figure 10f. Statistical significance of Join-Count Statistics, up to 2000.

showing how residential developments expanded over time for San Francisco, California, New Orleans, Louisiana, Baltimore, Maryland, Salt Lake City, Utah, and Dallas, Texas, He showed how the industrial revolution prompted the dramatic growth in population and territorial extent.

Following that, there had been models proposed to describe internal structure of urban land use patterns. These include the well known concentric zone land use model originated from von Thuüen's land rent theory, Hoyt's own sectoral model of urban structure (Hoyt, 1939), and the more recent multiple-nuclei model of urban form by Harris and Ullman (1945) – all helped to explain some aspects of how urban land use changed geographically with various variables and over time.

Specific to the urban core, the core-frame concept of central business district (CBD) by Horwood and Boyce (1959) helped to establish the relationship between urban core and its surrounding areas. For sub-urbanization, Erickson (1983) proposed the three-stage process of spillover and specialization, dispersal and diversification and infilling and multinucleation. Together, these models provided the literature of urban geography a solid foundation pointing clear directions for more in-depth understanding of the urbanization and suburbanization processes.

With the proliferation of geographic information systems (GIS), researchers began to use them to examine changes of urban form with an integrated and operational approach to understand urbanization process. With parcel data for the city of Buffalo, New York, Batty (1997) showed ways to simulate urban expansion for practical applications. In similar way, Clarke *et al.* (1997) integrated raster data layers to examine urban sprawl for large geographic regions. It should be noted that contemporary efforts in studying geographic forms of urban growth have geared toward dynamically simulating urban built-out scenarios by incorporating various geographic factors, planning practices, growth management strategies, and newly available technology in GIS. Examples for this may be found in Landis' work for California future land use modeling (Landis and Zhang, 1998), Lee's GIS-based simulation of built-out scenarios for Northeast Ohio counties (Lee *et al.*, 1999, also the Urban Growth Simulator at http://empact.geog.kent.edu) and many others.

In light of proposing an effective and applicable quantitative approach to studying geographic patterns of residential developments, we presented the method of using the Join-Count Statistics whose Z scores may be used to formulate various models to describe different urban growth patterns. The models presented here suggest a possible taxonomy of three growth patterns to help explain how residential temporal developments evolved in different localities. These three growth patterns, to be used to characterize localities with different developmental patterns, are: old intensely developed areas, mature developed areas, and young developing areas.

In many ways, this is a practical approach to understanding how any given locality developed. This is because the approach allows for the incorporation of the established urban land use models and the sub-urbanization processes of spillover, sprawling growth, and infill. As a result, urbanization and suburbanization can be better understood both conceptually and operationally.

On the technical side, the studies of urban form have been conducted mostly using raster data structure that really did not capture the spatial configuration of land parcels in a city. Examples for raster-based studies on urban sprawl/form can be found in Couclelis (1997), Clarke *et al.* (1997), Batty (1997), and Landis and Zhang (1998). The cell size in a raster-based analysis of this kind is almost always difficult to define or to be justified. Analyzing land parcels in their original form of vector data structure is the only way to allow researchers to take full accounts of the geographic relationship among land parcels. With the methods and models described here, this geographic relationship can finally be measured and tested. Of course, this is assuming that the parcel-level data and the temporal data of when parcels were built are available to researchers – an assumption sometimes does not hold in places where such data are not available. Also, the success and usefulness of the analysis greatly depend on the types of data available. For example, if parcel data record the history of land ownerships and their transactions, it will be possible to extend studies to examine how such process impacts spatial patterns over time.

Finally, the many spatial analytical methods that geographers developed have not been incorporated into modern GIS software packages. The use of these spatial analytical methods is very much limited to only academia and perhaps a small circle of researchers who are knowledgeable in spatial statistics. Although some scattered efforts have attempted to widen the usage of spatial statistics for more practical applications, the few software packages developed did not gain much popularity because of their long learning curves and the requirements of user's computer literacy. To this end, Lee and Wong (2001) developed and bundled a suite of computer scripts that work directly within ArcView GIS, a widely used GIS system, to allow even novice users to take full advantage of the powerful tools in spatial statistics. With these scripts, the calculation of Join-Count Statistics and the associated Z scores can be easily carried out for parcel level data in any geographical locations.

ACKNOWLEDGEMENT

This study was funded by the US Environmental Protection Agency under Grant #985989-01-0. The funding and support from US EPA Chicago, Illinois and Westlake, Ohio offices are greatly appreciated.

Department of Geography,
Kent State University,
Kent, Ohio 44242-0001, USA

REFERENCES

Batty, M. (1997), 'Cellular automata and urban form: a primer.' *Journal of the American Planning Association*, **63**(2), 216–274.

Clarke, W. (1997), 'A self-modifying cellular automaton model of historical urbanization in the San Francisco Bay area.' *Environment and Planning B*, **24**, 237–261.

Cliff, A. and J. Ord (1981), *Spatial Processes: Models and Applications*. Pion, London.

Couclelis, H. (1997), 'From cellular automata to urban models: new principles for model development and implementation.' *Environment and Planning B*, **24**, 165–174.

Donovan, T. and M. Neiman (1993), 'Community social status, suburban growth, and local-government restriction on residential development.' *Urban Affair Review*, **28**, 323–336.

Fader, S. (2000), *Density by Design: New Directions in Residential Development.* Urban Land Institute, Washington DC.

Fulford, C. (1996), 'The compact city and the market: the case of residential development.' In M. Jenks, E. Burton and K. Williams (eds.), *The Compact City: A Sustainable Urban Form.* E & FN Spon, London, pp. 122–134.

Erickson, R.A. (1983), 'The evolution of the suburban space economy.' *Urban Geography*, **4**, 96–108.

Goodchild, M. (1986), *Spatial Autocorrelation.* CATMOG 47. Geobooks, Norwich.

Griffith, D. (1987), *Spatial Autocorrelation: A Primer*. Washington, DC: Association of American Geographers.

Harris, C.D. and E.L. Ullman (1945), 'The nature of cities.' *Annals, American Academy of Political and Social Science*, **CCXLII**, 7–17.

Hitt, K. (1994), *Refining 1970's Land-Use with 1990 Population to Iindicate New Residential Development.* U.S. Geological Survey, Reston, VA:

Horwood, E.M. and R.R. Boyce (1959), *Studies of the Central Business District and Urban Freeway Development.* University of Washington Press, Seattle.

Hoyt, H. (1939), *The Structure and Growth of Residential Neighborhoods in American Cities.* Federal Housing Administration, Washington, DC.

Landis, J. and M. Zhang (1998), 'The second generation of the California urban future model. Part I: Model logic and theory.' *Environment and Planning B*, **25**, 651–666.

Lee, J. and D.W.S. Wong (2001), *Statistical Analysis with ArcView GIS.* Wiley, New York.

Lee, J., L. Erickson, T. Kulikowski and L. Tian (1999) 'Managing urban sprawl with geographic information systems.' *Environment and Planning B*, **25**, 865–879.

Levia, D. and D. Page (2000), 'The use of cluster Aanalysis in distinguishing farmland prone to residential development: a case study of sterling.' *Massachusetts. Environmental Management*, **25**, 541–548.

Morcombe, K. (1984), *The Residential Development Process: Housing Policy and Theory.* Gower, Brookfield, VT.

Upton, G. and B. Fingleton (1985), *Spatial Data Analysis by Examples*. Wiley, New York.

Barry Boots

Using Local Statistics for Boundary Characterization

Geographic boundary analysis involves the definition and evaluation of boundaries in spatial patterns (see Jacquez *et al.*, 2000, for a review). In this paper, we are concerned with the evaluation of polygon boundaries which are recorded in choropleth maps of thematic properties. In such situations, the polygons may have been determined independently of the mapped phenomena (Mark and Csillag, 1989). Further, the same set of polygons may be used to display a number of different variables. For example, in forestry, the same set of photo-interpreted polygons may be used to display mean dominant height, stand density, species composition, etc. Thus, when we view a single map of a given variable, it is possible that adjacent polygons may have the same value (see Figure 1 for an example). In such circumstances, we may wish to determine how distinctive any given boundary is relative to other boundaries in the mapped pattern. We characterize this as the strength (certainty) of the cartographic boundary.

Although there has been little work on this topic so far (see De Groeve and Lowell, 2000), there are parallels in techniques involved with boundary detection (see Fortin and Edwards, 2000; Jacquez *et al.*, 2000). Experience here suggests that the success of our endeavours will depend on the type of boundary, i.e., the nature of the difference in the real world which it represents. For example, edge detection methods appear better for identifying boundaries in transition zones, while clustering methods are better when there are spatially abrupt changes in the underlying data (Fortin and Edwards, 2000). This finding indicates that we may well experience problems if both kinds of boundary are present in the pattern. Further, there is clear evidence that the results of boundary detection are influenced by the scale at which the pattern is examined (Fortin and Edwards, 2000; Fortin *et al.*, 2000). For example, some boundaries are easier to detect locally while others can only be detected when large regions are considered. Recognizing these concerns, we develop an exploratory spatial data analysis (ESDA) procedure for characterizing boundary strength which can be used for boundaries of different types and at different geographic scales.

The procedure uses local statistics. Since such statistics focus on the identification of variations within a study region (Fotheringham, 1997, 1999; Fotheringham and Brunsdon, 1999), they are consistent with the notion of boundaries representing locations of change. We define a neighbourhood (or kernel) on each side of the boundary of interest (see Figure 2) and calculate a

B. Boots et al. (eds.), Modelling Geographical Systems, 33–44.

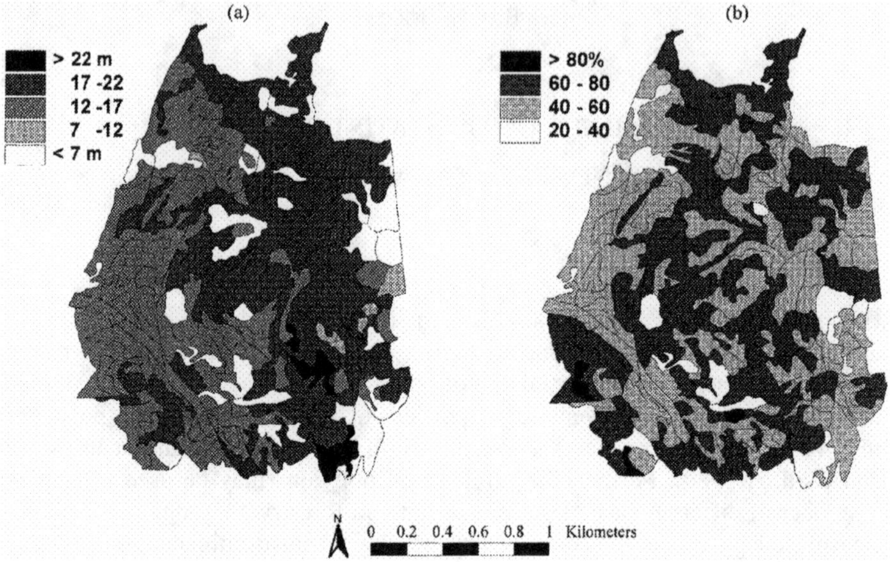

Figure 1. Forest stand height (in metres) and density for a sample area of the Monmorency Forest, Quebéc, 1992. (Source: Centre de Recherche en Geomatique, Université Laval).

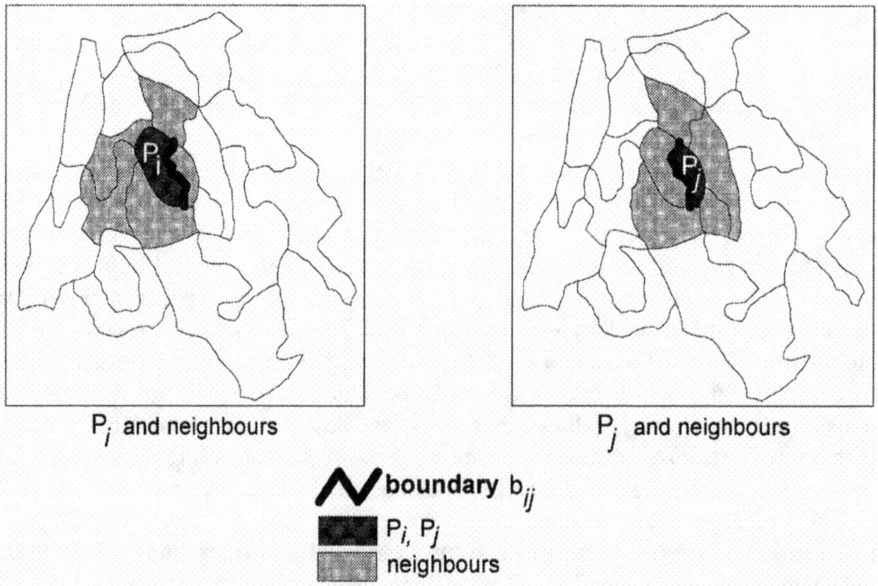

Figure 2. Definition (using topologic distance) of neighbourhoods associated with the boundary between polygons i and j.

local (or neighbourhood) measure (Getis and Ord, 1996; Fotheringham, 1997; 2000) for each neighbourhood. The values of the local measures for the neighbourhoods on either side of the boundary are then compared and a measure of their of their difference is assigned to the boundary as a measure of boundary strength. The relationship of scale to the strength of the boundary is explored by examining the behaviour of the difference measure as the size of the kernel is expanded. The procedure is described formally in the next section.

In general, if a boundary separates two very different areas, we would expect the difference between the two neighbourhood measures to be large and *vice versa*. Further, if a boundary is only important at the local scale, the difference between the two neighbourhood measures should be most extreme at this scale. In contrast, if the boundary separates two distinctive spatial regimes (regions), the difference should remain extreme as the size of the kernel increases. Using our procedure, for any given boundary between two polygons, we can:

1. at any given scale, identify the strength of the boundary relative to other boundaries;
2. identify the scale(s) at which it is strongest;

while for a given pattern we can:

3. at any given scale, identify the strongest boundaries;
4. identify the strongest boundaries over all scales.

In addition, the relative measures of boundary strength so derived can be used in any further analyses involving the boundaries instead of assigning them all equal weight.

Finally, we note that our procedure also has some parallels with the "moving window" analysis of gridded data (Turner *et al.*, 1991) which involves sliding a window along a transect. At each sampled location the window is divided into halves and a measure of dissimilarity between the two is computed. The choice of window size and separation between sample sites can be manipulated to examine the effects of scale on dissimilarity.

In the next section we describe the procedure more formally, as well as considering various operational variants, while in Section 3 we report on its application to both "real" and "artificial" data sets. The final section presents our conclusions and discusses further development of the procedure.

2. METHODOLOGY

Here we describe the procedure in formal terms. Let R be the study region and $P = \{p_1, p_2, \ldots, p_n\}$ be a set of n non-overlapping polygons covering R (i.e., $p_1 \cup p_2 \cup \ldots \cup p_n = R$). Let a_i be the value of some quantitative attribute for polygon p_i, and b_{ij} be the boundary between polygons p_i, p_j. In general terms our procedure involves the following steps.

1. Define a neighbourhood $D(p_i)$ for p_i consisting of all polygons within distance δ of p_i. Thus, if the distance between p_i, p_j is $\|p_i - p_j\|$, $D(p_i) = \{p_j: \|p_i - p_j\| \leq \delta; p_j \in P\}$.
2. Define a weight w_{ij} ($w_{ij} \geq 0$) representing the spatial association between p_i and every other polygon in $D(p_i)$, including p_i.
3. Define a local statistic m_i for $D(p_i)$.
4. Define a difference statistic $d_{ij} = f(m_i, m_j)$.
5. Attach the value of d_{ij} to b_{ij} and express as a quantile measure q_{ij}.
6. Use q_{ij} as a relative measure of the distinctiveness of the boundary at the scale of the analysis.
7. Repeat steps 1–6 using increasing values of δ.

Note that steps 1–4 allow for a number of alternative definitions. For example, in step 1 δ may be measured using either topological (as in Figure 2) or metric distances (Unwin and Unwin, 1998). In step 2, the weights can range from simple, symmetric, binary forms to ones which take into account various characteristics of b_{ij} (e.g., its length relative to the length of the perimeter of p_i). There are various possibilities for m_i. One general form is $m_i = \sum_{k=1}^{n_D} w_{ik} a_k$, $p_k \in D(p_i)$, where n_D is the number of polygons in $D(p_i)$, i.e., m_i is the weighted sum of a_i in $D(p_i)$ (note $D(p_i)$ may be defined to either include or exclude p_i). Another possibility is to use a local indicator of spatial association (LISA) such as local Moran's I_i or the Getis statistics G_i and G_i^* (Ord and Getis, 1995). Further, since $D(p_i)$ defines a subregion R_i of R, landscape ecological indices (LEIs) (Hargis *et al.*, 1997, 1998; Haynes-Young and Chopping, 1996) can be calculated for R_i and used in place of m_i. Similarly, d_{ij} can take a number of forms, the most obvious of which are $d_{ij} = (m_i - m_j)^2$ and $d_{ij} = |m_i - m_j|$.

3. APPLICATIONS

In order to evaluate the procedure we have by applied it to both data from artificial and real world environments. The former involve landscapes created by mapping the values of the elements of eigenvectors of an adjusted binary contiguity matrix of a tessellation of regular hexagons (Boots and Tiefelsdorf, 2000). More precisely, let $\mathbf{C}$ be the contiguity matrix with elements $c_{ij} = 1$, if cells i and j share a common boundary, and $c_{ij} = 0$, otherwise, and $\mathbf{M}_{(1)} \equiv \mathbf{I} - \mathbf{1}\,(\mathbf{1}^T\mathbf{1})^{-1}\,\mathbf{1}^T$ denote the idempotent projection matrix. If $\{\lambda_1, \lambda_2, \ldots, \lambda_n\}$ ($\lambda_i \leq \lambda_{i+1}$) and $\{\mathbf{e}_1, \mathbf{e}_2, \ldots, \mathbf{e}_n\}$, are the sets of eigenvalues and eigenvectors, respectively, of $\mathbf{M}_{(1)} \cdot \mathbf{C} \cdot \mathbf{M}_{(1)}$, then

$$1 = \frac{\mathbf{e}_i^T \cdot \mathbf{M}_{(1)} \cdot \mathbf{C} \cdot \mathbf{M}_{(1)} \cdot \mathbf{e}_i}{\mathbf{e}_i^T \cdot \mathbf{M}_{(1)} \cdot \mathbf{e}_i} = \mathbf{e}_i^T \cdot \mathbf{C} \cdot \mathbf{e}_i \forall_i \tag{1}$$

Griffith (1996) (see also Griffith and Amrhein, 1997, pp. 66–76) shows that the map pattern displayed by $\mathbf{e}_i$ has a value of Moran's spatial autocorrelation coefficient I equal to λ_i. Collectively, the set of maps associated with $\{\mathbf{e}_1, \mathbf{e}_2, \ldots, \mathbf{e}_n\}$ identify all the mutually distinct patterns of spatial autocorrelation which can be realized for the tessellation. Thus, we are able to create a spectrum of landscapes which extends from those dominated by first-order effects, through those with no spatial effects, to those dominated by second-order effects. So far, we have concentrated on examining the first type of artificial landscape since these involve a higher proportion of adjacent cells with similar values (i.e., the landscapes are characterized by zones of transition rather than abrupt changes) (see Figure 3).

The real environments are maps of mean dominant height and stand density for parts of the Montmorency Forest, the research and teaching forest of Laval University, Quebec City, in 1992. These maps were created from human interpretation of infrared aerial photographs (De Groeve and Lowell, 2000). Here there are no obvious transition zones, although some boundaries separate polygons with disparate values (see Figure 1).

Although we have begun to explore many of the variants of the procedure described in the previous section, here we report on the results using only one set of choices. This involves using topologic distance δ, symmetric binary weights w_{ij}, the standardized form of the Getis statistic G_i^* for m_i, $d_{ij} = (m_i - m_j)^2$, with the q_{ij} values being grouped into quartiles. Thus,

$$m_i = G_i^* = \frac{\sum_j^{n_D} w_{ij}a_j - W_i\bar{a}}{s\sqrt{[W_i(n - W_i)/(n - 1)]}} \tag{2}$$

where $\bar{a} = \sum_{j=1}^{n} a_j/n$; $s^2 = \sum_{j=1}^{n} a_j^2/n - (\bar{a})^2$; $W_i = \sum_j^{n_D} w_{ij}$ with $w_{ii} = 1$, and all other symbols are as described in Section 2. Choice of a Getis type statistic (Ord

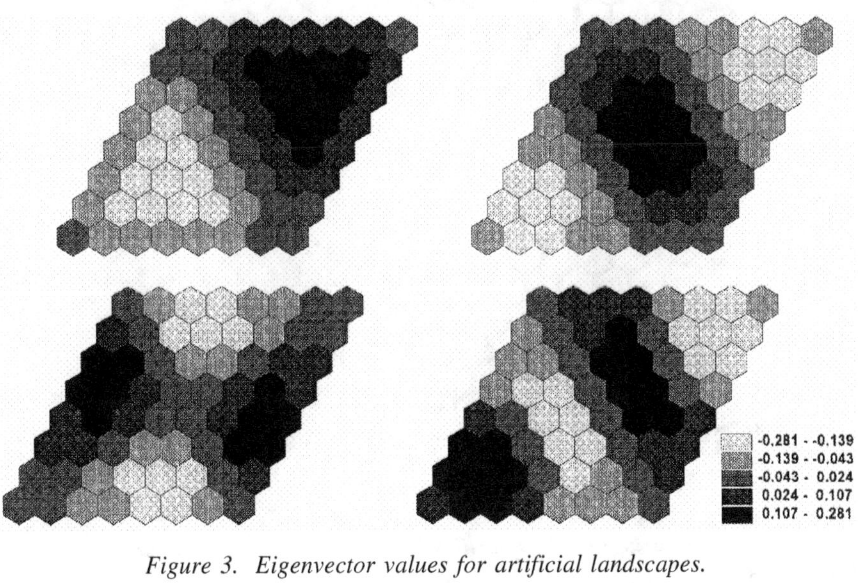

Figure 3. Eigenvector values for artificial landscapes.

and Getis, 1995) was favoured because such statistics have already been shown to be useful in identifying boundaries in remotely sensed images (Derksen *et al.*, 1998; Wulder and Boots, 1998, 2001) and to have the ability to capture scale effects (Amrhein and Reynolds 1996, 1997). Since δ is defined toplogically, we consider three sizes of neighbourhood, those consisting of first-order neighbours, first and second-order neighbours, and first, second, and third-order neighbours. For convenience, these are referred to as lags 1, 2, and 3, respectively.

Figure 4 shows the results for three increasing sizes of neighbourhoods for

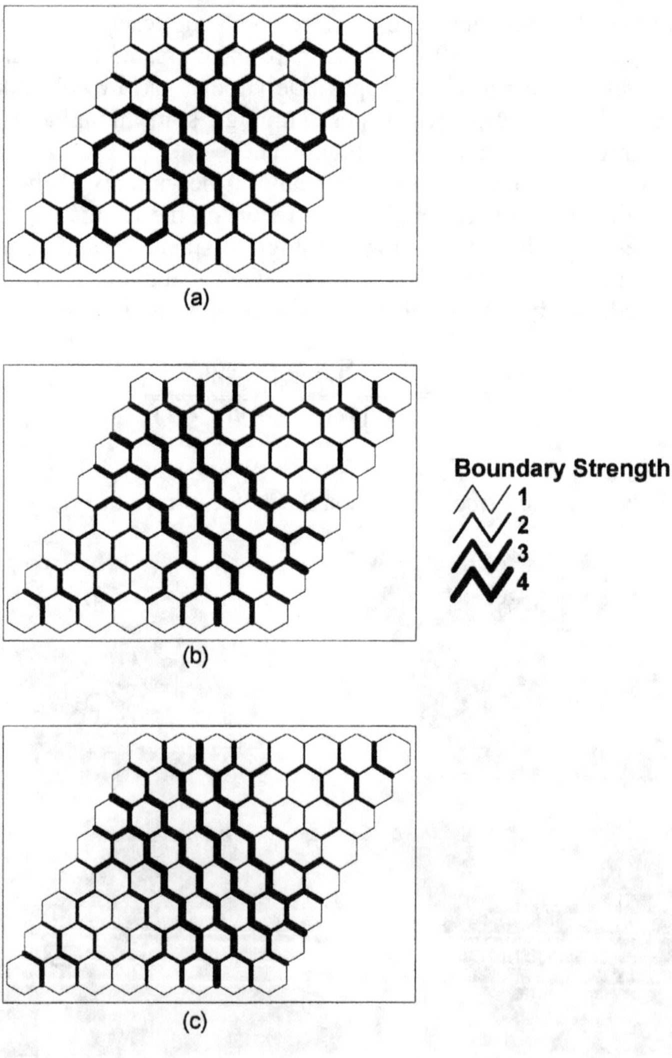

Figure 4. Boundary strength by quartiles for artificial landscapes: (a) lag 1; (b) lag 2; (c) lag 3.

the artificial landscape with the highest degree of positive spatial autocorrelation (I = 1.01) (see Figure 3) realizable for this particular tessellation. This is a landscape where there are only first-order effects (transition zones). The maps in Figure 4 enable us to consider items 1 and 3 from the Introduction. Note that the boundaries in the transition zone separating the clusters of high and low values are dominant at all scales examined. In contrast, the limited number of strong boundaries (those in the top quartile) which occur within the two contrasting clusters at the local scale lose their strength as we move towards a more global scale. These findings are echoed in Figures 5 and 6 which address items 2 and 4, respectively from the Introduction. In particular, Figure 6 illustrates how the procedure identifies the boundaries in the transition zone as both strong and persistent over all scales examined.

Figures 7, 8, and 9 show the corresponding results for one of the real landscapes (Figure 1a). Since this landscape contains a mixture of boundary types, these results are more complex visually than those for the artificial landscape examined. Nevertheless, a number of trends can be discerned. Looking at Figure 7 we see that strong boundaries (those in the top quartile) are most

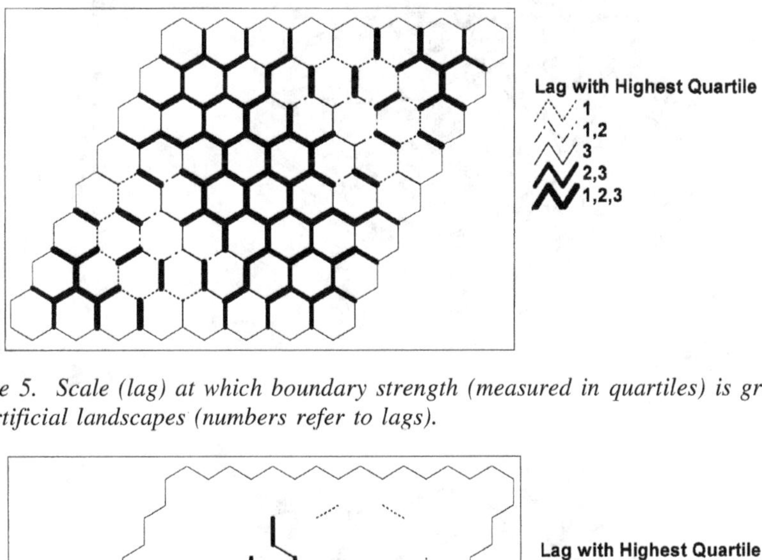

Figure 5. Scale (lag) at which boundary strength (measured in quartiles) is greatest for artificial landscapes (numbers refer to lags).

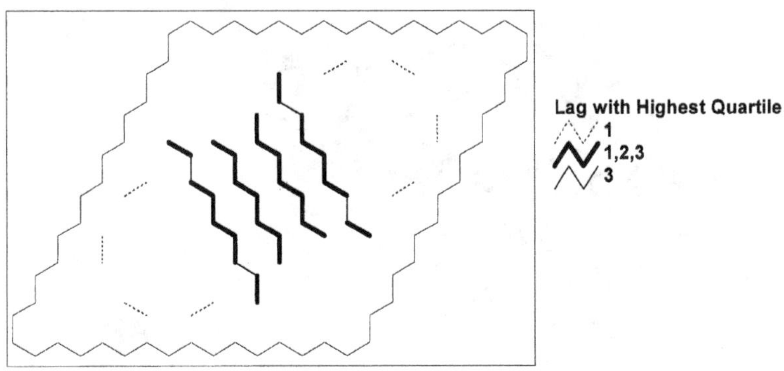

Figure 6. Strongest boundaries (those in the top quartile) at all scales (lags) for artificial landscapes (numbers refer to lags).

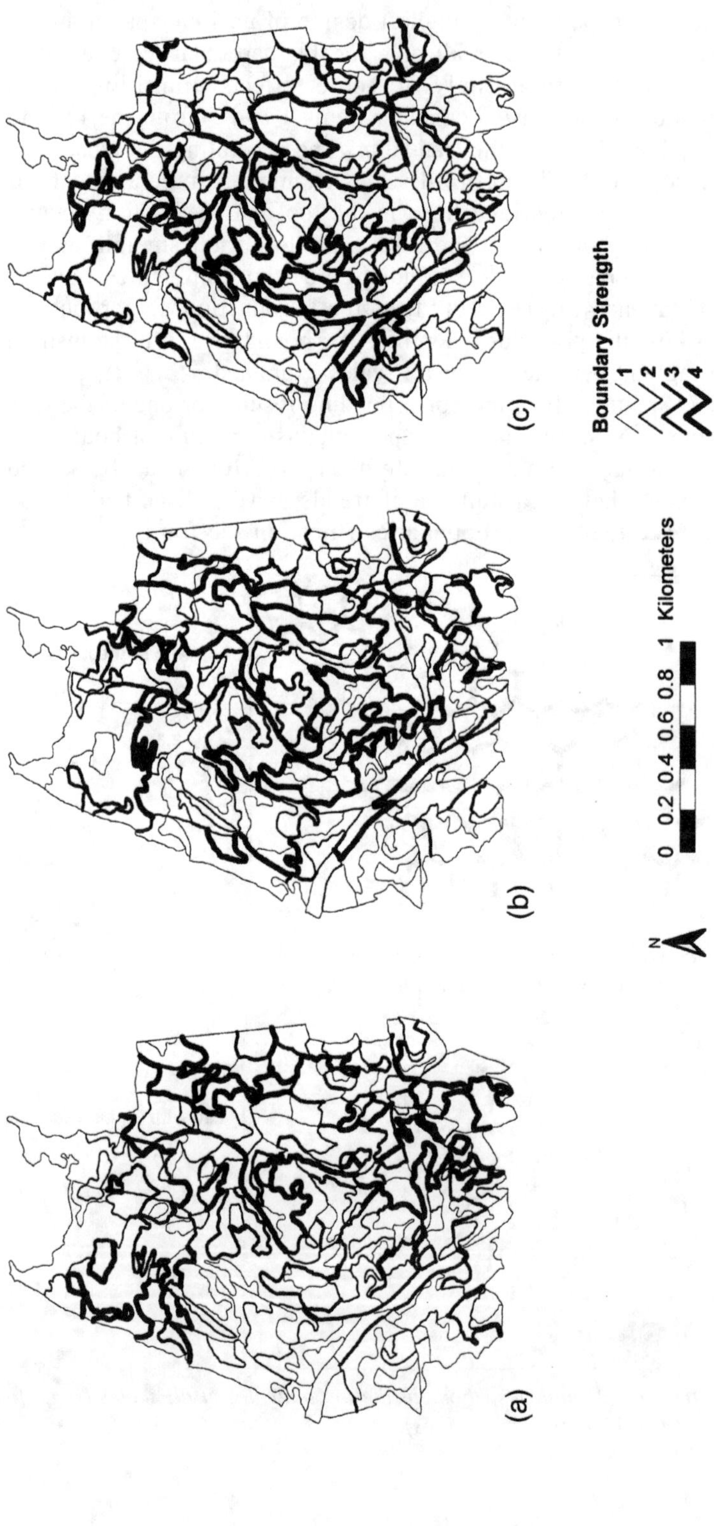

Figure 7. Boundary strength by quartiles for real landscape: (a) lag 1; (b) lag 2; (c) lag 3.

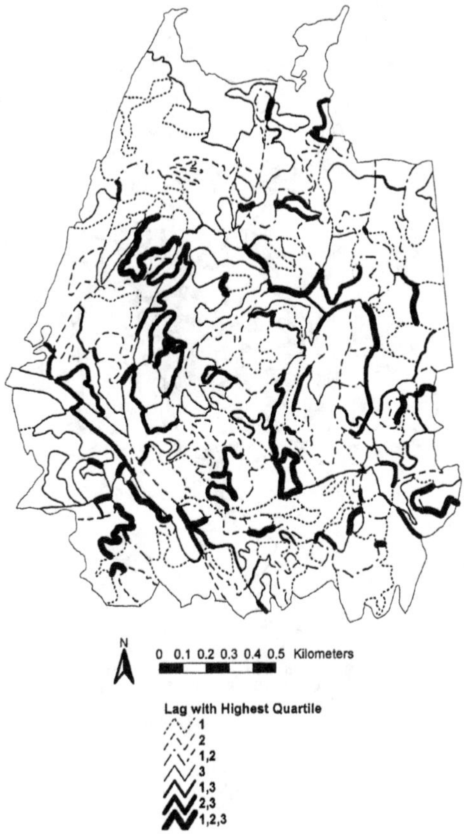

Figure 8. *Scale (lag) at which boundary strength (measured in quartiles) is greatest for real landscape (numbers refer to lags).*

dispersed geographically at lag 1, the local scale, becoming more concentrated as the lag (scale) increases. Paralleling this trend, the incidence of isolated strong boundaries is greatest at the local scale. As scale increases, there is a tendency for individual boundaries to be linked into longer boundary segments. Similarly, while closed polygons formed by strong boundaries occur at all scales, the size of such polygons increases with scale. Thus, while the closed polygons correspond with individual map polygons at lag 1, they are composed of more than one original map polygon at lag 3. Figure 8 suggests the presence of some spatial trends in boundary strength. Boundaries in the eastern and north-western parts of the study region tend to be strongest at the local scales (lags 1 and 2), while those in the south-west are generally stronger at the regional scale (lag 3). However, Figure 9 reveals that, relatively, there are few strong boundaries at any lag in the western, north-western, and south-western parts of the study area. In contrast, the strongest relative boundaries are found in the centre of the region.

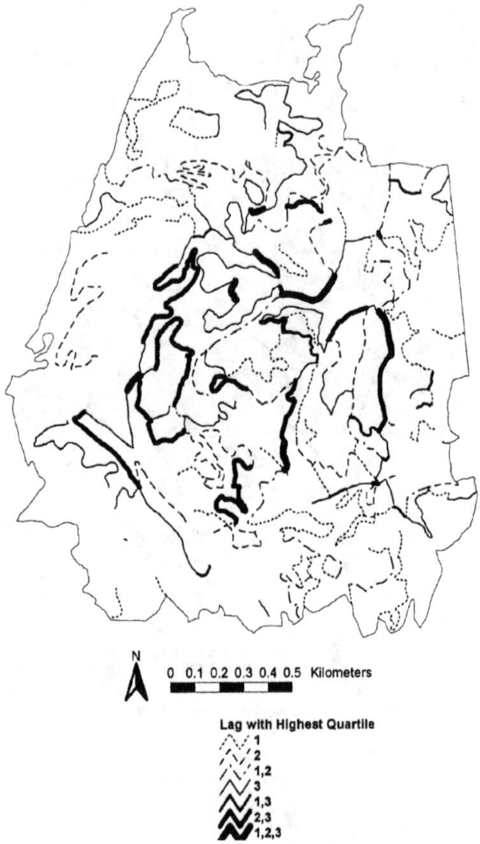

Figure 9. Strongest boundaries (those in the top quartile) at all scales (lags) for real landscape (numbers refer to lags).

4. Discussion and Conclusions

The results presented in Section 3, together with those obtained using other variants of the procedure, suggest that it meets the objectives described in Section 1. In particular, the procedure appears to be able to identify both sharp and transitional boundaries, as well as enabling us to explore how boundary strength responds to changes in the scale of analysis. However, much remains to be done.

First of all, we need to apply the procedure to a greater variety of landscapes, both real and artificial, in order to evaluate the sensitivity of the approach to landscape complexity. Then, for any given map, we need to determine how alternative ways of calculating m_i impact the definition of boundary strength (i.e., the impact of selecting different choices in steps 1–4 in section 2). We also need to undertake a rigorous examination of the impact of edge effects on our procedure regardless of how m_i is defined. So far we have considered two ways of dealing with edge effects. The first is to use all polygons in R and adjust the weights w_{ij} to allow for the reduction in the size of $D(p_i)$ for polygons close the boundary of R (e.g., $w_{ij} = 1/\sum_j w_{ij}$). An alternative is to ignore those polygons for which $D(p_i)$ includes part of the boundary of R, although this is only realistic for large data sets.

We are also exploring ways of adapting the procedure to deal with qualitative (categorical) data and multivariate data (i.e., if there are r attributes of p_i, replacing a_i with a vector $\mathbf{A}_i = \{a_i\!: i = 1, \ldots, r\}$). One way of accommodating the latter would be to compute d_{ij} using the average of the r values of m_i for the individual variables for each p_i. Another way would be to use the average of the values of q_{ij} computed for each of the r attributes. However, it would be better if we can develop procedures which make use of the variance/covariance structure of the r variables.

Finally, although the procedure is not designed as a boundary detection technique, it is equivalent to such techniques if both use the same set of spatial relationships. This would occur for regular lattices, such as the pixels of a remotely-sensed image, if a topologic measure is used to define δ. The same is true for irregular lattice data if we first define a set of spatial relationships between the data sites (e.g., by defining the Delaunay triangulation of the data sites). Thus, we can undertake comparative analyses of the performance of our procedure relative to the different kinds of edge detection procedures currently in use (Fortin, 1994; Fortin and Drapeau, 1995).

ACKNOWLEDGEMENTS

The author gratefully acknowledges financial support from the GEOIDE Network of Centres of Excellence (ENV #4) and the contribution of maps and ground-based data from personnel of the Montmorency Forest, Department of Wood and Forest Sciences, UniversitéLaval.

Department of Geography and Environmental Studies,
Wilfrid Laurier University,
Waterloo, Ontario N2L 3C5, Canada

REFERENCES

Amrhein, C.G. and H. Reynolds (1996), 'Using spatial statistics to assess aggregation effects.' *Geographical Systems*, **2**, 143–158.

Amrhein, C. and H. Reynolds (1997), 'Using the Getis statistic to explore aggregation effects in Metropolitan Toronto Census data.' *The Canadian Geographer*, **41**(2), 137–149.

Boots, B. and M. Tiefelsdorf (2000), 'Global and local spatial autocorrelation in bounded regular tessellations.' *Journal of Geographical Systems*, **2**(4), 319–348.

De Groeve, T. and K. Lowell (2000), 'Boundary uncertainty assessment from a single forest type map.' *Photogrammetric Engineering and Remote Sensing*, **67**, 717–726.

Derksen, C., M. Wulder, E. LeDrew and B. Goodison (1998), 'Associations between spatially derived autocorrelation patterns of SSM/I-derived prairie snow cover and atmospheric circulation.' *Hydrological Processes*, **12**, 2307–2316.

Fortin, M-J. (1994), 'Edge detection algorithms for two-dimensional ecological data.' *Ecology*, **75**, 956–965.

Fortin, M-J. and P. Drapeau (1995), 'Delineation of ecological boundaries: comparison of approaches and significance tests.' *Oikos*, **72**, 323–332.

Fortin, M-J. and G. Edwards (2001), 'Accuracy issues related to the delineation of vegetation boundaries.' In C. Hunsaker, M. Goodchild, M. Friedl and T. Case (eds.), *Spatial Uncertainty in Ecology*. Springer-Verlag, Berlin, pp. 158–174.

Fortin, M-J., R.J. Olson, S. Ferson, L. Iverson, C. Hunsaker, G. Edwards, D. Levine, K. Butera and V. Klemas (2000), 'Issues related to the detection of boundaries.' *Landscape Ecology*, **15**, 453–466.

Fotheringham, A.S. (1997), 'Trends in quantitative methods I: Stressing the local.' *Progress in Human Geography*, **21**(1), 88–96.

Fotheringham, A.S. (1999), 'Guest editorial: local modelling.' *Geographical & Environmental Modelling*, **3**(1), 5–7.

Fotheringham, A.S. (2000), 'Context-dependent spatial analysis: a role for GIS?' *Journal of Geographical Systems*, **2**(1), 71–76.

Fotheringham, A.S. and C. Brunsdon (1999), 'Local forms of spatial analysis.' *Geographical Analysis*, **31**(4), 340–358.

Getis, A. and J.K. Ord (1996), 'Local spatial statistics: an overview.' In P. Longley and M. Batty (eds.), *Spatial Analysis: Modelling in a GIS Environment.* GeoInformation International, Cambridge, pp. 261–277.

Griffith, D.A. (1996), 'Spatial autocorrelation and eigenfunctions of the geographic weights matrix accompanying geo-referenced data.' *The Canadian Geographer*, **40**, 351–367.

Griffith, D.A. and C.G. Amrhein (1997), *Multivariate Statistical Analysis for Geographers*. Prentice Hall, Upper Saddle River, N.J.

Haines-Young, R. and M. Chopping (1996), 'Quantifying landscape structure: a review of landscape indices and their application to forested landscapes.' *Progress in Physical Geography*, **20**(4), 418–445.

Hargis, C.D., J.A. Bissonette and J.L. David (1997), 'Understanding measures of landscape pattern.' In J.A. Bissonette (ed.), *Wildlife and Landscape Ecology: Effects of Pattern and Scale*. Springer-Verlag, New York, pp. 231–261.

Hargis, C.D., J.A. Bisonette and J.L. David (1998), 'The behavior of landscape metrics commonly used in the study of habitat fragmentation.' *Landscape Ecology*, **13**, 167–186.

Jacquez, G.M., S. Maruca and M-J. Fortin (2000), 'From fields to objects: a review of geographic boundary analysis.' *Journal of Geographical Systems*, **2**, 221–241.

Mark, D.M. and F. Csillag (1989), 'The nature of boundaries on 'area-class' maps.' *Cartographica*, **26**, 65–78.

Ord, J.K. and Getis, A. (1995), Local spatial autocorrelation statistics: distributional issues and an application.' *Geographical Analysis*, **27**, 286–306.

Turner, S.J., R.V. O'Neill, W. Conley, M.R. Conley and H.C. Humphries (1991), 'Pattern and scale: statistics for landscape ecology.' In M.J. Turner and R.H. Gardner (eds.), *Quantitative Methods in Landscape Ecology: The Analysis and Interpretation of Landscape Heterogeneity*. Springer-Verlag, New York, pp. 17–76.

Unwin, A. and Unwin, D. (1998), 'Exploratory spatial data analysis with local statistics.' *Journal of the Royal Statistical Society: Series D (The Statistician)*, **47**(3), 415–421.

Wulder, M. and B. Boots (1998), 'Local spatial autocorrelation characteristics of remotely sensed imagery assessed with the Getis statistic.' *International Journal of Remote Sensing*, **19**(11), 2223–2231.

Wulder, M. and B. Boots (2001), 'Local spatial autocorrelation characteristics of Landsat TM imagery of a managed forest.' *Canadian Journal of Remote Sensing*, **27**(1), 67–75.

Tomoki Nakaya

Local Spatial Interaction Modelling based on the Geographically weighted regression Approach

1. Introduction

One of the recent major trends in spatial analysis is local modelling by which spatial analysts examine local properties in geographical phenomena (Fotheringham, 1997). Indeed, spatial processes tend to vary over space due to different geographical contexts so that spatial non-stationarity emerges (Jones III and Hanham, 1995). In such cases, global models that postulate universally acceptable properties fail to capture the real phenomena under study. We could say that inferences of local incidence rates in disease mapping are the simplest form of local modelling (Openshaw *et al.*, 1987, Nakaya, 2000). As for more complicated association analyses, Casetti's (1972) expansion method is popular to model explicitly the property of non-stationarity in regression analysis (e.g. Casetti, 1990). According to the method, we can specify geographical drifts of regression parameters by polynomial or harmonic expansion series of locational variables. Recently, the Newcastle school (Brunsdon *et al.*, 1996, Fotheringham *et al.*, 1998) has developed a more generalised local regression methodology, called geographically weighted regression (GWR). The approach estimates local regression coefficients with a moving weighting kernel.

The aim of this paper is to develop the way in which the GWR approach is applied to spatial interaction modelling and to present an empirical example applied to the migration flows in Japan during the latter half of 1980. Although spatial interaction models are regarded as basic components of a lot of spatial mathematical models, their poor ability to describe the observed flows has been pointed out (Openshaw, 1976, 1979). There are many considerations indicating that the non-stationarities in the spatial interaction process contribute to this problem. Origin or destination specific models are classic approaches that assume the effect of the explanatory variables in gravity models may differ between origins or destinations. Lovett and Flowerdew (1989) show an example to incorporate many origin or destination specific explanatory variables in Poisson spatial interaction modelling. Fitting models separately to relative subsets of flows is an effective way to infer parameter drifts depending on different origin and destination pairs reflecting different underlying processes. For example, longer and shorter distance migrations would be influenced mainly by employment factors and by housing factors, respectively. Gordon (1991) advocates multi-stream modelling in which gravity models are separately calibrated for each subset of migration data grouped by distances between origins

B. Boots et al. (eds.), Modelling Geographical Systems, 45–69.

and destinations. Boyle (1993) extends the idea enabling us to divide migration data utilising more variables. Openshaw and Connolly (1977) insist that functional forms measuring distance decay should also vary regionally. They then build up a heuristic algorithm to allocate each flow to subsets with different functions. These approaches dramatically improve the model performance.

Instead of dividing data sets, the GWR approach employs all data weighted by a geographical kernel to infer local variations of parameters. The scheme leads to desirable properties of geographical local modelling. Firstly, we do not need to predefine the geographical categories that are necessary for the data dividing approach. Secondly, we need not predefine the functional form of local variations, such as the power or harmonic series specifications in the expansion method. Thirdly, the weighting is more flexible to fit data locally rather than the expansion series specifications in the expansion method. Namely, we can expect that GWR attains satisfactory goodness of fits in most cases and identifies the spatial trends and local anomalies of parameter drifts. Rather than residual maps by ordinal regression models, that is, global models, the parameter drift map derived from GWR directly visualises local displacements of regularities in comparison with global regularities. Fourthly, because GWR utilises the ordinal statistical estimation process, we can extend conventional statistical indicators measuring model performance. Although spatial adaptive filtering (SAF) (Foster and Gorr, 1986; Gorr and Olligschlaeger, 1994) is an alternative for modelling geographically varying parameters, its ad hoc manner to estimate local parameters by using damped negative feedback of predictive errors makes statistical inferences difficult (Brunsdon *et al.*, 1996).

So far GWR has been applied to several regression models including ordinal multi-regression models and spatially autoregression models (Brunsdon *et al.*, 1996, 1997; Fotheringham *et al.*, 1997, 1998). In addition to non-flow modelling, these properties of GWR are expected to be useful for exploring the non-stationarity of spatial interaction processes with least *a priori* assumptions. Applying the GWR approach to spatial interaction modelling has never been implemented, though Berglund and Karlström (1999) suggested its possibility. Fotheringham and Pitts (1995) estimated directional drifts of distance decay parameters in spatial interaction models by the expansion method. In this paper we will develop a more generalised local model for spatial interaction adapting to not only directional but also other kinds of geographical parameter drifts. We could regard the new model as an origin-destination specific or doubly specific spatial interaction models.

The remainder of this paper is organised in seven sections. In the next section we discuss the general framework of the localised or doubly specific spatial interaction models. We then focus on an approach extending origin-specific models and outline the details on how to apply the GWR approach to spatial interaction modelling in aspects of its estimation (section 3), the model specification (section 4) and the statistical evaluation (section 5). An empirical application to the migration flows in Japan in the late 1980s is made in section 6 and conclusions are drawn in section 7.

2. ORIGIN – DESTINATION SPECIFIC MODELS

A spatial interaction model can be considered as a specialised regression model for spatial flows. Its general form is described as below:

$$Y_{ij} = f(V_i^{(1)}, V_i^{(2)}, \ldots, X_j^{(1)}, X_j^{(2)}, \ldots, D_{ij}^{(1)}, D_{ij}^{(2)}, \ldots) \cdot \varepsilon_{ij}, \tag{1}$$

where the dependent variable, Y_{ij}, is the flow from the origin i to the destination j and the independent variables are categorised into three groups; the origin specific variables $V_i^{(1)}, V_i^{(2)}, \ldots$; the destination specific variables $X_j^{(1)}, X_j^{(2)}, \ldots$; and the interrelationship variables between origins and destinations $D_{ij}^{(1)}, D_{ij}^{(2)}, \ldots$; and ε_{ij} is a multiplicative error term. There are a lot of variations in the details of the model specification.

In order to explain the approach in this study, let us begin with the simplest model, the distance decay model:

$$Y_{ij} = \exp(c - \beta \log D_{ij}) \cdot \varepsilon_{ij}, \tag{2}$$

where D_{ij} is the distance between i and j; c and β are parameters to be estimated. As stated earlier, the distance decay parameter β tends to vary depending on origin-destination pairs. One way to estimate local estimates of β is assigning different parameter sets to the model for each subset of the data. The most popular approach is the origin specific parameterisation:

$$Y_{ij} = \exp(c_i - \beta_i \log D_{ij}) \cdot \varepsilon_{ij}, \tag{3}$$

where the parameters have the index of the origin i and are calibrated using flows from each origin separately. It indicates that the conventional approach of the spatial interaction modelling can be used for estimating local variations of parameters partially, but not fully. The fully parameterised distance decay model is shown as follows:

$$Y_{ij} = \exp(c_{ij} - \beta_{ij} \log D_{ij}) \cdot \varepsilon_{ij}, \tag{4}$$

where each parameter has two indices: the origin and the destination. We could then call this an origin-destination specific model or doubly specific model. This model is expected to capture the local variation of the flow regularities depending on the combination of origin and destination. We cannot, however, calibrate the model because of the lack of a degree-of-freedom: we can pool only one flow data for estimating an origin-destination specific parameter.

Instead of pooling the data, GWR specifies the parameter based on the location. To clarify the idea let us respecify the origin-destination specific model:

$$Y_{ij} = \exp(c_{(ij)} - \beta_{(ij)} \log D_{ij}) \cdot \varepsilon_{ij}, \tag{5}$$

The parameter $c_{(ij)}$ and $\beta_{(ij)}$ are parameters of flows *around* the location of the

ij flow. In order to overcome the shortage of the degree-of-freedom, GWR employs all the flow data weighted gradually according to the distance from the highlighted location of the flow ij in the calibration process.

To construct weighting functions in GWR, we should define the degree of spatial separation between flows. Several ideas that measure it have been proposed for incorporating spatial autocorrelations in flow modelling (Berglund and Karlström, 1999; Black, 1992; Bolduc, 1992; Bolduc *et al.*, 1995). There is, however, high complexity due to the four dimensional nature of measuring distances between two flows which contain two geographical locations each.

For simplicity, in the remainder of this paper we focus on extending the origin specific model. The restriction enables us to consider the application of geographical weighting only for destination locations. It also contributes to reducing computational tasks. We should rewrite the origin-destination specific model again:

$$Y_{ij} = \exp(c_{i(j)} - \beta_{i(j)} \log D_{ij}) \cdot \varepsilon_{ij}. \tag{6}$$

If the amount of flows from each origin were large enough, the extension from the origin-specific model would be adequate to estimate local parameters depending on not only origins but also destinations.

3. An Estimation Scheme for Local Parameters

Letting l_{ij} be the log-likelihood of the flow from i to j, we can estimate parameters based on the maximum likelihood principle. In the case of the origin-specific distance decay model, the problem is written as:

$$\text{Max } L_i = \textstyle\sum_i l_{ik}(c_i, \beta_i \mid Y_{ik}, D_{ik}). \tag{7}$$

In order to estimate the local parameters around each destination, the maximisation problem is extended to use geographical weights:

$$\underset{c_{i(j)},\, \beta_{i(j)}}{\text{Max}} \; L_{i(j)} = \textstyle\sum_i l_{ik}(c_{i(j)}, \beta_{i(j)} \mid Y_{ik}, D_{ik}) w_{jk}(D_{jk}), \tag{8}$$

where $w_{jk}(D_{jk})$ is a weighting function which has a peak on the location of destination k. The geographical weight designates the degree of spatial separation of the flow k relative to the focused location j. This calibration procedure can be called the weighted maximum likelihood principle, which is the essence of GWR.

Regarding the specification of the weighting function, there are many options (Fotheringham *et al.*, 1997). In this study we apply the squared Cauchy function for weighting, which is the function of t distribution with three degrees of freedom:

$$w_{jk}(D_{jk}) = \frac{1}{(1 + D_{jk}^2/G^2)^2}, \tag{9}$$

where G is the size parameter called bandwidth. The function makes a bell-shaped kernel similar to the Gaussian kernel. Figure 1 shows the shape of the squared Cauchy. Inside the bandwidth, the function is similar to the Gaussian curve but the function is characterised by long tail. Owing to the properties of this function, if there is little data inside the bandwidth of the kernel, the localised model fits the data set outside the bandwidth. Using this function we can expect that the estimators of local parameters shrink to global values if there is little data inside the kernel like Bayesian estimators in disease mapping (Clayton and Kaldor, 1987; Lawson, 2001) so that the localised estimator should be robust. In general, the spatial interaction matrices are highly skewed and sparse; it means that most elements, origin-destination pairs, have very few or no flows. Therefore the squared Cauchy function would be suitable for GWR applications to flow datasets.

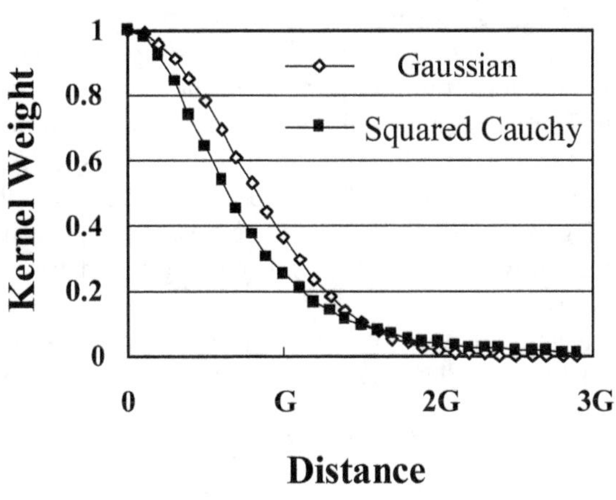

Figure 1. A comparison between Gaussian and Squared Cauchy kernel function.

Moving the weighting kernel (Figure 2) provides the estimates of local parameters for each origin-destination pair. In comparison with the conventional geographically weighted regression, the localised spatial interaction model has higher complexities of the estimated results. The reason is that since the parameters are shown as origin-destination matrices, it is not easy to map the result as with previous studies. Specialised visualisation of origin-destination values is needed.

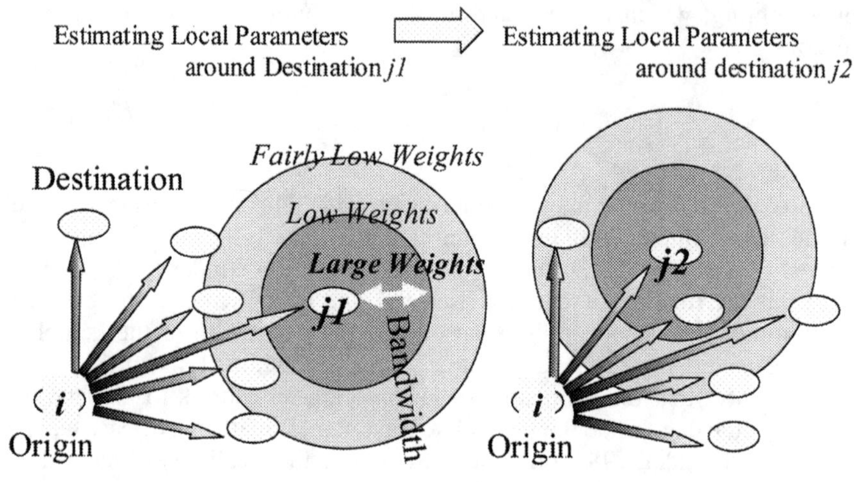

Figure 2. Moving kernel weighting flow data.

4. SPECIFICATION OF A LOCALISED CONSTRAINED SPATIAL INTERACTION MODEL

Spatial analysts usually make use of more complex models, such as singly and doubly constrained models, rather than the simple distance decay model. Singly constrained models are quite popular in the context of modelling spatial choice behaviour. The Poisson spatial interaction models assuming that the error term is following the Poisson distribution facilitate the specification and calibration of such constrained models (Baxter, 1984; Flowerdew, 1991). Therefore, the application of geographically weighted regression approach to the Poisson regression is a natural way to specify localised origin-constrained spatial interaction models.

We can write a general form of the origin-specific Poisson spatial interaction model with a liner predictor $\psi_{ij} = \boldsymbol{\theta}_i \boldsymbol{X}_{ij}$ by using the above notations:

$$Y_{ij} = \exp(\boldsymbol{\theta}_i \boldsymbol{X}_{ij}) \cdot \varepsilon_{ij} \tag{10}$$

$$= \exp(\psi_{ij}) \cdot \varepsilon_{ij}, \tag{11}$$

where $\boldsymbol{\theta}_i$ and $\boldsymbol{X}_{ij}$ are a vector of parameters and explanatory variables for the flow from i to j, respectively; and ε_{ij} is the Poisson error term.

Let us think of the competing destination model as a typical origin-constrained model for migration, consumer behaviour, and several other spatial choice models (Fotheringham, 1983, 1988; Ishikawa, 1987; Yano *et al.*, 2000):

$$Y_{ij} = \exp(c_i + \alpha_i \log S_j - \beta_i \log D_{ij} + \gamma_i \log A_j) \cdot \varepsilon_{ij} \tag{12}$$

$$= \frac{C_i S_j^{\alpha i} A_j^{\gamma i}}{D_{ij}^{\beta i}} \cdot \varepsilon_{ij}, \tag{13}$$

where $C_i = \exp(c_i)$; S_j is the population size of destination j; and A_j is the accessibility of destination j defined by:

$$A_j = \sum_{k \neq j} S_k/D_{kj}. \tag{14}$$

Here, α_i is size sensitivity, β_i is distance decay trend, and γ_i is agglomerating/competing effect of destination for the flows from the origin i. Although the theoretical meaning of the accessibility term is originally represented by the two-stage decision-making theory, the term is useful to describe directional tendency of flows toward or away from pivotal locations, such as the large metropolitan areas, as shown by Yano *et al.* (2000).

The conditional equation for the estimators of the parameters is derived from the first partial derivative of the likelihood function with the parameters as below:

$$\frac{\partial L_i}{\partial \theta_{m,\,i}} = \frac{\partial \sum_k l_{ik}(\boldsymbol{\theta}_i \mid Y_{ik}, \boldsymbol{X}_{ik})}{\partial \theta_{m,\,i}} = 0 \qquad \forall m, \tag{15}$$

where

$$\boldsymbol{\theta}_i = \{\theta_{m,\,i}\} = \{c_i, \alpha_i, \beta_i, \gamma_i\}, \tag{16}$$

$$\boldsymbol{X}_{ik} = \{X_{m,\,k}\} = \{1, \log S_i, \log D_{ik}, \log A_k\}. \tag{17}$$

The total amount of the flows from an origin should be the same for the observed and estimated one arising from the condition of the most likelihood estimator of the origin specific constant term, c_i:

$$\frac{\partial L_i}{\partial c_i} = \frac{\partial \sum_k \{-\hat{Y}_{ik} + Y_{ik} \ln \hat{Y}_{ik}\}}{\partial c_i} = 0, \tag{18}$$

where $\hat{Y}_{ik}$ is the estimated amount of the flows from i to k. Solving the above equation, we can obtain the constrainted equation as:

$$\exp(c_i) = \frac{O_i}{\sum_k S_k^{\alpha i} A_k^{\gamma i} / D_{ik}^{\beta i}}, \tag{19}$$

Then we can see that the model is the origin-constrained model in a familiar multiplicative form:

$$Y_{ij} = O_i \frac{S_j^{\alpha i} A_j^{\gamma i} / D_{ij}^{\beta i}}{\sum_k S_k^{\alpha i} A_k^{\gamma i} / D_{ik}^{\beta i}} \varepsilon_{ij}, \tag{20}$$

where O_i is the total amount of the flows from the origin i.

The localised competing destination model extending an origin-specific model is written as:

$$Y_{ij} = \exp(c_{i(j)} + \alpha_{i(j)} \log S_j - \beta_{i(j)} \log D_{ij} + \gamma_{i(j)} \log A_j) \cdot \varepsilon_{ij} \tag{21}$$

$$= \frac{C_{i(j)} S_j^{\alpha i(j)} A_j^{\gamma i(j)}}{D_{ij}^{\beta i(j)}} \cdot \varepsilon_{ij}, \tag{22}$$

We can define the localised log-likelihood, $L_{i(j)}$, and conditional equations for maximising it:

$$\frac{\partial L_{i(j)}}{\partial \theta_{m,\, i(j)}} = \frac{\partial \sum_k l_{ik}(\boldsymbol{\theta}_{i(j)} \mid Y_{ik}, \boldsymbol{X}_{ik}) w_{jk}(D_{jk})}{\partial \theta_{m,\, i(j)}} \tag{23}$$

$$= \frac{\partial \sum_k \{-\hat{Y}_{ik} + Y_{ik} \ln \hat{Y}_{ik}\} w_{jk}(D_{jk})}{\partial \theta_{m,\, i(j)}} = 0 \quad \forall m, \tag{24}$$

where

$$\boldsymbol{\theta}_{i(j)} = \{\theta_{m,\, i(j)}\} = \{c_{i(j)}, \alpha_{i(j)}, \beta_{i(j)}, \gamma_{i(j)}\}. \tag{25}$$

A constraint for the localised model is deduced as:

$$\sum_k Y_{ik} w_{jk} = \sum_k \hat{Y}_{ik} w_{jk}. \tag{26}$$

The condition indicates that the weighted sum of the flows from the origin is the same for observed and estimated one. Therefore, letting $O'_{i(j)}$ denote the weighted sum, we obtain the origin-specific locally constrained model based on GWR as below:

$$Y_{ij} = O'_{i(j)} \frac{S_j^{\alpha i(j)} A_j^{\gamma i(j)} / D_{ij}^{\beta i(j)}}{\sum_k w_{jk} S_k^{\alpha i(k)} A_k^{\gamma i(k)} / D_{ik}^{\beta i(k)}} \varepsilon_{ij}. \tag{27}$$

We can understand that the model is a kind of spatial choice model in which choice alternatives are emphasised and restricted to the destinations around the interested destination j. Conventional spatial choice models often assume homogeneous decision makers and choice sets. However, there might be heterogeneous groups of movers having different choice sets. If we expect that the choice sets are composed of neighbourhood areas around the destination that movers finally decide, we can then infer the local rules of spatial choice by using this locally constrained model.

5. GOODNESS OF FITS AND BANDWIDTH SELECTION FOR THE LOCALISED SPATIAL INTERACTION MODEL

Loader (1999) shows that goodness of fits and model selection indicators for localised models can be developed in a similar manner in generalised linear modelling by McCullagh and Nelder (1989). As a goodness of fit indicator, the deviance for flows from the origin i is defined by:

$$DEV_i = \sum_j 2 \left[\sup_{\theta_{i(j)}} l_{ij}(\boldsymbol{\theta}_{i(j)} \mid Y_{ij}, \boldsymbol{X}_{ij}) - l_{ij}(\hat{\boldsymbol{\theta}}_{i(j)}, Y_{ij}, \boldsymbol{X}_{ij}) \right], \tag{28}$$

where $\sup_{\theta_{i(j)}} l_{ij}(\boldsymbol{\theta}_{i(j)} \mid Y_{ij}, \boldsymbol{X}_{ij})$ and $l_{ij}(\hat{\boldsymbol{\theta}}_{i(j)}, Y_{ij}, \boldsymbol{X}_{ij})$ are the log-likelihood of the full model and the calibrated model for the flow from i to j, respectively. On the assumption of the Poisson errors, the formula for the deviance is:

$$DEV_i = 2\sum_j Y_{ij} \ln \frac{l_{ij}}{\hat{Y}_{ij}(\hat{\boldsymbol{\theta}}_{i(j)}, \boldsymbol{X}_{ij})} . \tag{29}$$

The lower the deviance is, the better the model fits the observed flows. If the calibrated model is correct, the distribution of the deviance is asymptotically chi-squared, $\chi^2_{n_i - p_i}$, with $(N_i - p_i)$ degrees of freedom, and the average of the deviance converges to the degree of freedom as the sample size increases:

$$DEV_i \sim \chi^2_{n_i - p_i}, \tag{30}$$

$$\mathrm{E}[DEV_i] \cong N_i - p_i \tag{31}$$

where N_i is the number of observations and p_i is the number of free parameters of the model for the flows from the origin i.

For the localised model, p_i can be approximated by:

$$p_i = -\sum_j W(0) X_{ij}^T \boldsymbol{J}_{i(j)}^{-1} \boldsymbol{X}_{ij} \frac{\partial^2 l_{ij}(Y_{ij}, \hat{\psi}_{i(j)})}{\partial \hat{\psi}_{i(j)}^{\,2}} \tag{32}$$

$$= -\sum_j \boldsymbol{X}_{ij}^T \boldsymbol{J}_{i(j)}^{-1} \boldsymbol{X}_{ij} \hat{Y}_{ij}, \tag{33}$$

where $\hat{\psi}_{i(j)} = \hat{\boldsymbol{\theta}}_{i(j)} \boldsymbol{X}_{ij}$ and $\boldsymbol{J}_{i(j)}$ is the Jacobian matrix of the local likelihood:

$$\boldsymbol{J}_{i(j)} = -\sum_k W(D_{kj}) \boldsymbol{X}_{ij}^T \boldsymbol{X}_{ij} \frac{\partial^2 l_{ik}(Y_{ik}, \hat{\psi}_{i(k)})}{\partial \hat{\psi}_{i(k)}^{\,2}} \tag{34}$$

$$= -\sum_k W(D_{kj}) \boldsymbol{X}_{ij}^T \boldsymbol{X}_{ij} \hat{Y}_{ij}, \tag{35}$$

For the bandwidth selection, we can use indicators for model selection, such as AIC, Akaike Information Criterion (Akaike, 1973):

$$AIC_i = DEV_i + 2p_i. \tag{36}$$

The model with the lowest AIC should be chosen as the best model on the basis of information statistics theory. We can also develop the approximate cross-validation (CV) indicator for the local model (Loarder, 1999). However, as the number of observations increase, the best model based on AIC and CV is attained for a more complex model than the true one. It means that these indicators lead to a smaller bandwidth size for the local modelling. As alternatives, the criterion given by BIC, Bayesian Information Criterion (Schwarz, 1978) and MDL, Minimum Description Length (Rissanen, 1983, 1986) is proposed to give a consistent estimator by introducing the increased penalty of the complexities weighted by the number of observations. Although the estimator selected by BIC/MDL is more biased than the one selected by AIC, the bias could be negligible in large samples. The formula of BIC/MDL is given by:

$$BIC_i = DEV_i + p_i \ln(N_i). \tag{37}$$

The criterion can be used only if the assumed probability distribution is correct. For the Poisson spatial interaction modelling, however, we often encounter the problem of overdispersion. Although in Poisson regression the expected amount and variance of flows should be the same, $\mathrm{E}[\hat{Y}_{ij}] = \mathrm{Var}[\hat{Y}_{ij}]$, the observed variance is often far larger than the expected variance based on the observed flows. In order to overcome this problem, the quasi-likelihood method developed by Wedderburn (1974) is useful (Davies and Guy, 1987). In quasi-likelihood modelling, we assume simply:

$$\mathrm{Var}[\hat{Y}_{ij}] = \sigma^2\, \mathrm{E}[\hat{Y}_{ij}], \tag{38}$$

where σ^2 is called the dispersion parameter to be estimated. It should be noted that the parameter estimation of the quasi likelihood model is identical to the ordinal likelihood model except for the dispersion parameter.

The dispersion parameter can be estimated by:

$$\hat{\sigma}_i^2 = \frac{N_i}{N_i - p_i} \frac{\sum_j \left[\dfrac{\partial l_{ij}(Y_{ij}, \hat{\psi}_{i(j)})}{\partial \hat{\psi}_{i(j)}} \right]^2}{\sum_j \left[\dfrac{\partial^2 l_{ij}(Y_{ij}, \hat{\psi}_{i(j)})}{\partial \hat{\psi}_{i(j)}{}^2} \right]}. \tag{39}$$

We can now define scaled BIC (SBIC) that takes the overdispersion into account as:

$$SBIC_i = DEV_i/\hat{\sigma}_i^2 + p_i \ln(N_i), \tag{40}$$

where $DEV_i/\hat{\sigma}_i^2$ is the scaled deviance.

For bandwidth selection, the dispersion parameter should be the same for all the models with the different bandwidth sizes under consideration. The estimator derived from a fit at the smallest bandwidth under study is desirable to use for assuming that bias is negligible like Mallows' (1973) CP criterion for selection of regression models.

6. AN APPLICATION TO MIGRATION FLOWS IN JAPAN

6.1. *Data*

The aim of this section is to present an application of the locally constrained spatial interaction model developed previously. The data under study is the Japanese migration between 727 zones (Figure 3) during 1985–1990 without intra-zone flows ($N_i = 727 - 1 = 726$ for all i). The original data is taken from the 1990 Japanese census report, which has the migration table between the 3372 municipalities. The migration table is constructed based on the difference of residence between the census date (1st October, 1990) and five years ago. For the details, see Yano *et al.* (2000). For reducing computational tasks, the migration data between 3372 units is aggregated into the table between 727 functional zones. The 727 zones are defined in *Minryoku*, a database for marketing made by a Japanese newspaper company, *Asahi-Shinbun*. The locally constrained model is calibrated where S_j is the population size of zone i and D_{ij} is the great circle distance between zones i and j in equation (21).

6.2. *Global Parameters*

Before carrying out the local modelling, we calibrate the conventional origin-specific origin-constrained spatial interaction models as global models. The estimated global parameters are shown in Figure 4. The patterns almost match with the result of the previous study (Yano *et al.*, 2000) in which origin-specific competing destination models are calibrated to the migration flows between 3372 municipalities based on the same statistics. There are distinctive parameter variations for all the origin-specific parameters.

The attractiveness parameter α_i is smaller and less than 1.0 in the metropolitan areas, such as the Tokyo and Osaka metropolitan areas. The distance-decay parameter β_i is less negative along the major transportation networks through these metropolitan areas. The accessibility parameters γ_i in the metropolitan areas are significantly negative. The movers from these origins tend to choose less accessible destinations. The tendency indicates two kinds of

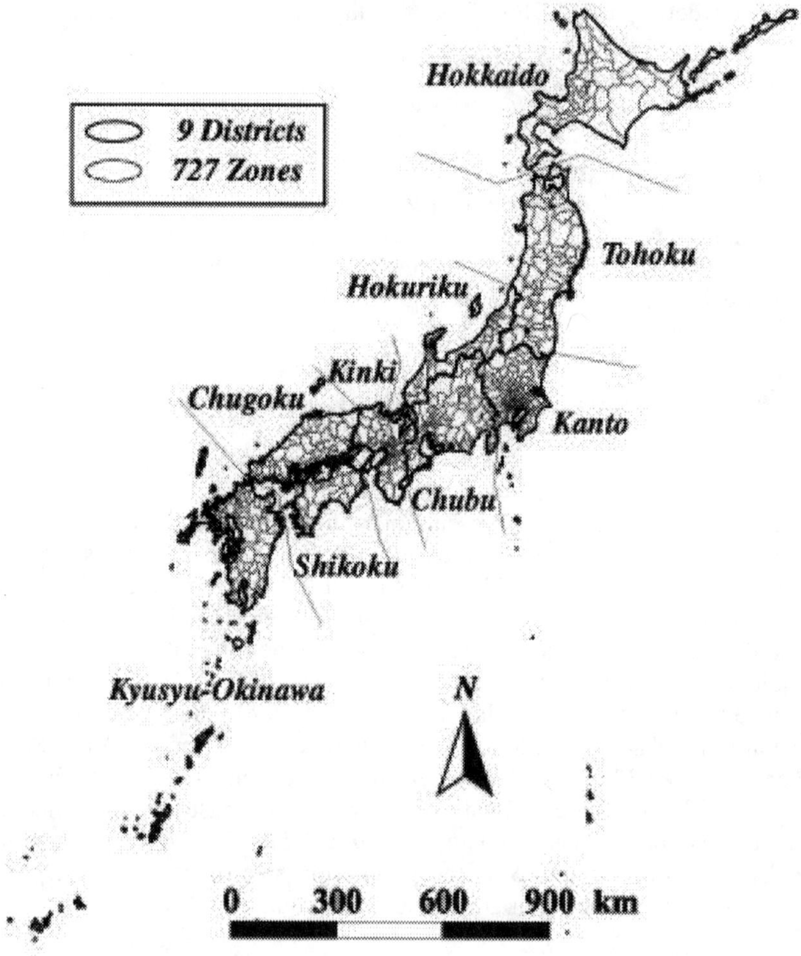

Figure 3. The 727 zones and the 9 districts in Japan.

movements: first, the outer movement from the inner parts to suburban parts inside metropolitan areas; second, the return migration to their birthplaces in more peripheral districts.

Most of peripheral parts, such as Tohoku and Kyusyu-Okinawa districts, have higher attractiveness and positive accessibility parameters. The movers from these origins tend to prefer the accessible and populous destinations, that is, the major metropolitan areas. This movement trend from periphery to core was prominent in the late 1980s, in particular, leading the large net flows into the Tokyo (Keihin) metropolitan area with the prosperous economic condition called "bubble boom"(Ishikawa and Fielding, 1998).

The highest degrees of negative distance-decay are shown in the mountainous parts where the attractiveness and the accessibility parameters are positively

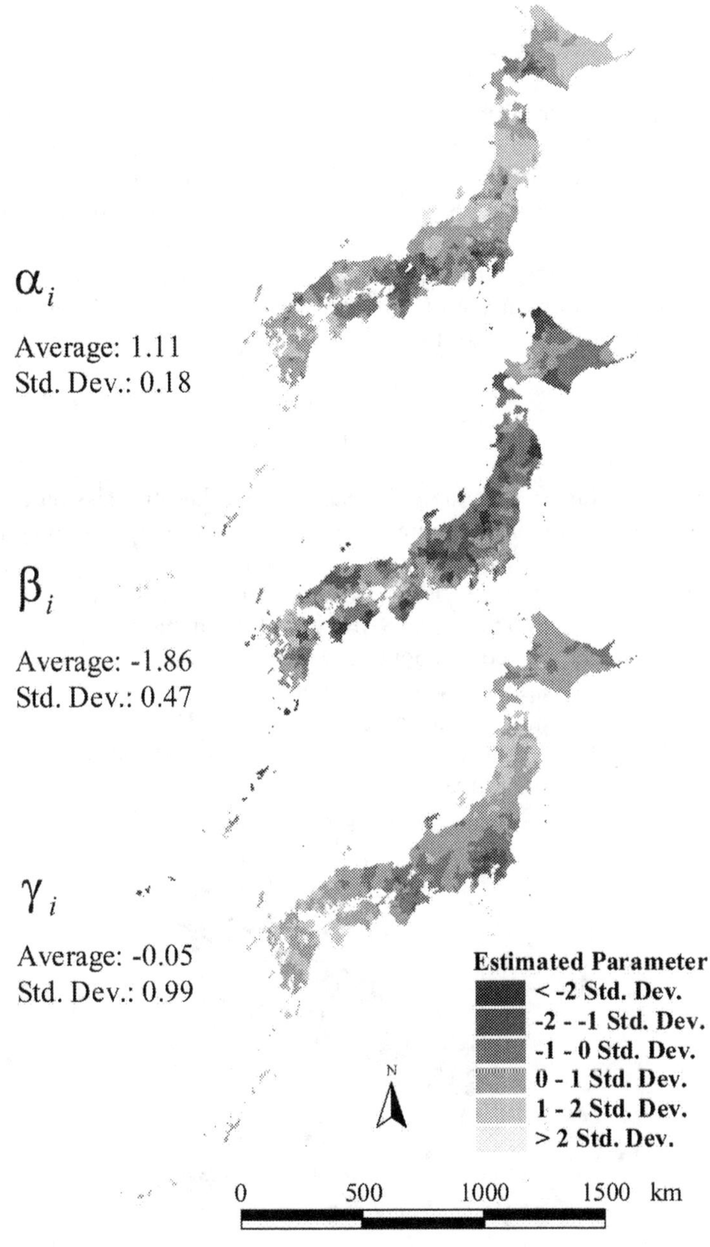

Figure 4. Geographical variations of the estimated global parameters by the conventional origin-constrained models.

high values. It reflects that the dominant destinations for the movers from these areas are to nearest local capitals.

6.3. *Bandwidth Selection*

The localised origin-specific origin-constrained competing destination models are calibrated with bandwidths from 50 km to 250 km by 50 km steps. In order to estimate the dispersion parameter, the deviances with a 50 km bandwidth are analysed. Figure 5 clearly shows that the estimated dispersion parameter (called scaled value in the Figure 5) of an origin, $\hat{\sigma}_i$, depends on the total number of flows from an origin. Therefore we can expect a probalistic process that the variance of the observations becomes larger as the size of flows is larger. The relationship is described well by a power function (R^2 = 0.71):

$$\hat{\sigma}_i^2 = 0.0079(\textstyle\sum_j Y_{ij})^{0.6798}. \tag{41}$$

Instead of estimating $\hat{\sigma}_i$ separately for each origin, these regression estimators are used for the bandwidth selection in order to ignore the random component in the variation of $\hat{\sigma}_i$.

Table 1 shows the average SBIC (see Equation 40) with the selected bandwidth size under consideration. The SBIC with 100 km bandwidth is the lowest and the deviance is reduced to about 55% of the global model deviance. The effective number of parameters p_i = 43 means that the degree of freedom of the model is almost equivalent to the case of 11 (≈ 43/4) conventional models separately fitting 11 data subsets. The bandwidth size almost follows with the size of major metropolitan areas or districts.

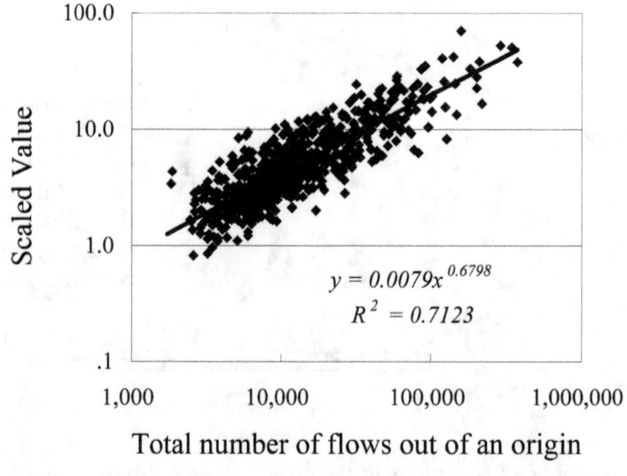

Figure 5. *Dependency of scaled values on the flow size.*

Table 1. Performance indicators of the models with different bandwidth sizes.

Bandwidth G	Average p	Average of Scaled Deviance	Average of Scaled BIC
Global	4.0	1647.8	1674.2
250 km	15.2	1166.6	1267.1
200 km	19.3	1109.0	1236.9
150 km	26.6	1031.2	1207.7
100 km	43.1	910.2	1195.3
50 km	100.1	678.4	1337.7

Although we can separately decide the best bandwidth for each origin, we employ the same bandwidth size for all origins for interpreting the results under the same condition. The two-stage destination choice theory by Fotheringham (1983, 1984) might be applicable to the interpretation of the bandwidth size: migrants might firstly select a fuzzy cluster of destinations approximately covered by a kernel with the bandwidth size and then choose a specific destination with a local choice rule. Ishikawa (1990) argues that the district scale clusters of destinations would be recognised as the choice sets of the first stage destination choice for Japanese migrants by using a nested-logit model.

6.4. Visualisation of Spatial Variation of Parameters

For demonstrating the origin-destination specific local parameters, Figure 6 shows the estimated parameters for migrations from Setagaya Ward located in the central part of the Tokyo metropolitan area. It can be seen that the parameters vary regionally with distinctive patterns. It should be emphasised that the variances of the parameters based on the flows from only one origin almost match with the ones estimated by the origin-specific models. The migrants from one origin have significantly diverse destination choice rules.

In and around the Tokyo metropolitan area, the destination choices are characterised by low attractive sensitivity, negatively high distance decay trends and negatively high accessibility parameters meaning preference for less accessible areas. The features are, however, not valid for other destinations. For example, the distance decay parameters are less negative in the migration from the origin to the peripheral areas. It could be reasonable that in the destination choice sets distant from an origin, a difference of the distances from the origin is not significant for migrants. Although the tendency is partly described by the power function of distance decay (Wilson, 1970), the function is not adequate for the overall destination choices for the migrants from Setagaya Ward.

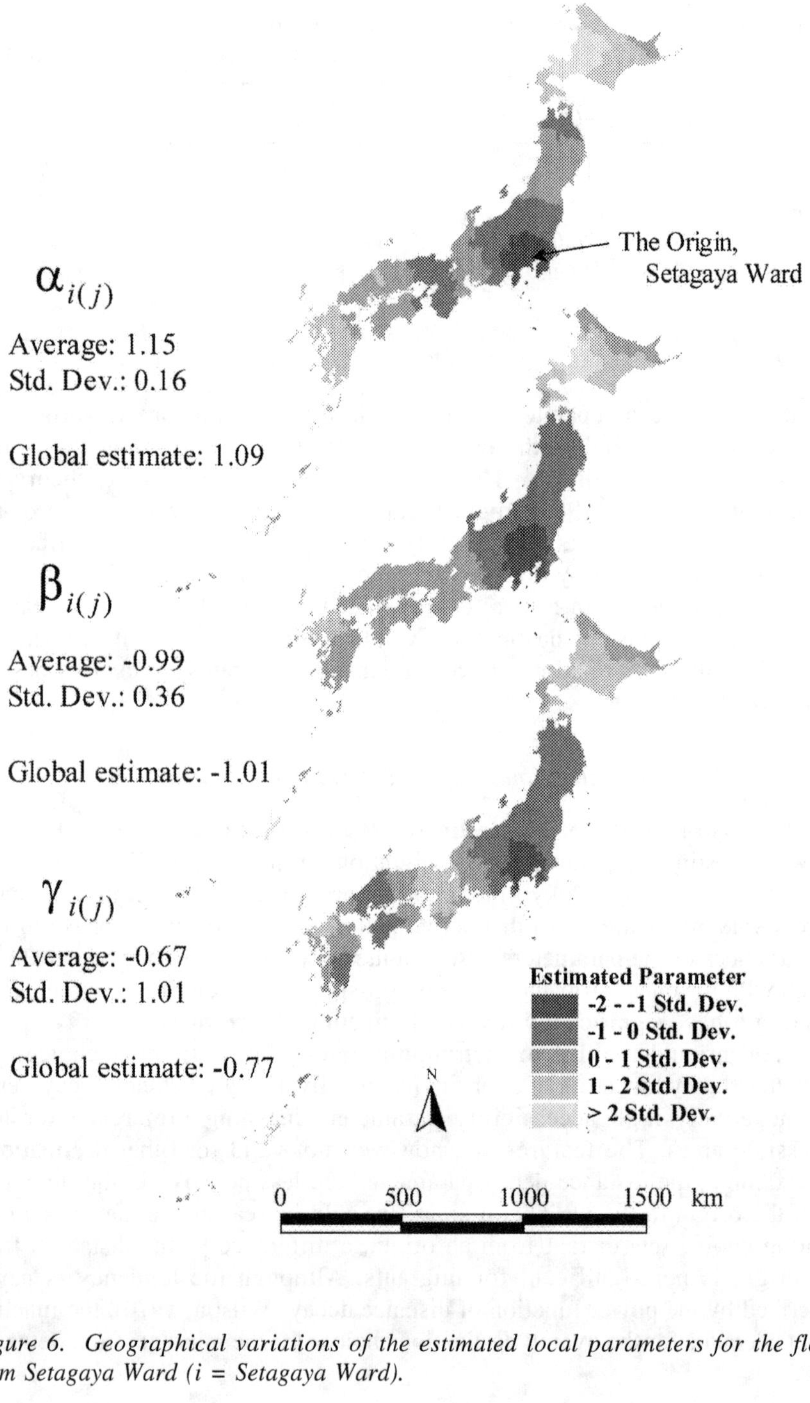

Figure 6. Geographical variations of the estimated local parameters for the flows from Setagaya Ward (i = Setagaya Ward).

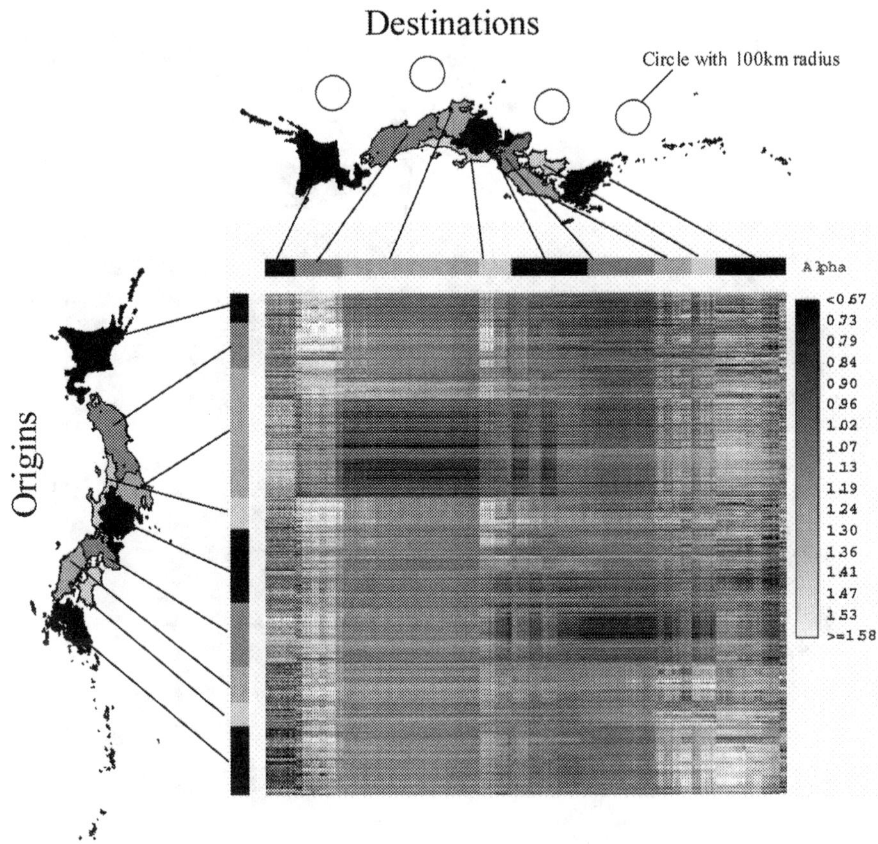

Figure 7. The OD matrix of estimated local attractiveness parameters.

Although the mapping of the local parameters may be the best way to interpret the parameter drifts, maps for all the 727 origins are difficult to handle. Instead of such thematic mapping, 'images' of OD matrices (Marble *et al.*, 1997) are adopted for visualising all the parameter variations (Figures 7–9). In these images (727 by 727), each row and column corresponds with each origin and destination, respectively. The rows and columns are arrayed based on the major 9 districts. The two elongated index bars in the left and top side of the images have each nine parts corresponding to the nine districts in Japan.

These OD images reveal the distinctive parameter variations depending on both origin and destination locations. It should be noted that the conventional origin-specific models show parameter variations depending only on origin locations. For example, the flows from any origin to peripheral districts, such as the Tohoku district, show higher attractive parameter. In the peripheral area the destinations with larger population size have prominent attractiveness for the immigrants. The weakest population sensitivity is seen in the intra-move-

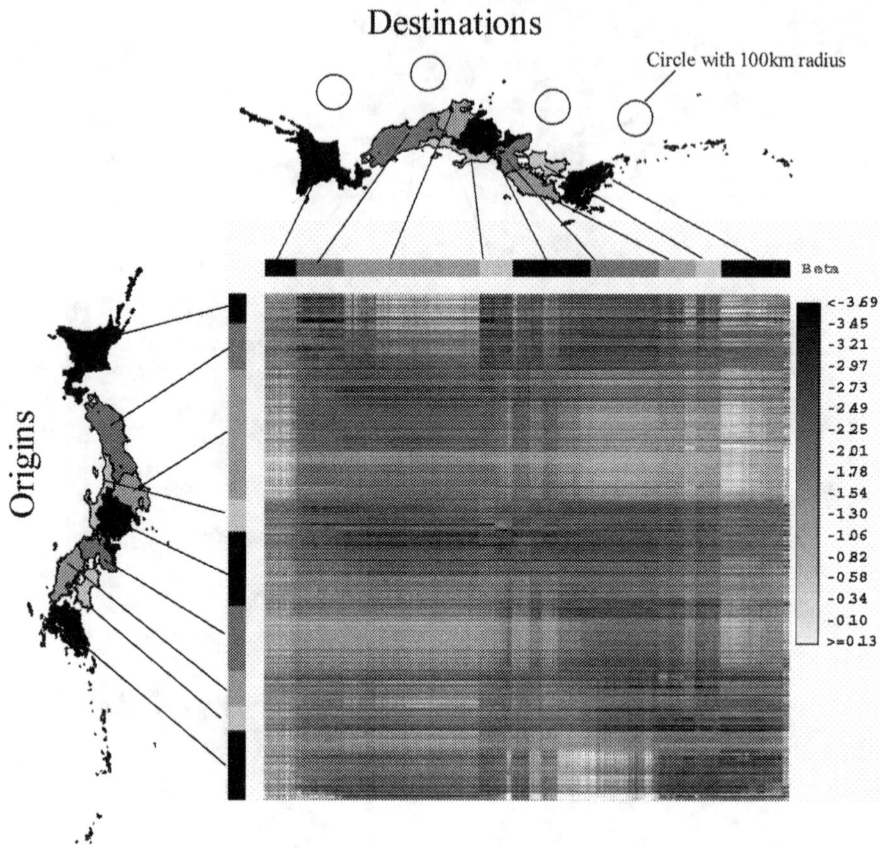

Figure 8. The OD matrix of estimated distance decay parameters.

ments of the major metropolitan areas, that is, the Kanto, Chubu and Kinki districts. Such movements would be mainly caused by housing relocations and we could expect that population size is insufficient for estimating housing attractiveness. On the other hand, for the migrants moving toward destinations in peripheral areas, job and educational purposes would be the main cause of the movements. In non-metropolitan areas, job and educational opportunities concentrate on local major cities with high population so that the population should be a good indicator of migration attractiveness.

6.5. *Clustering of Local Parameters*

We can make clusters of the origin-destination pairs based on the similarities of the estimated parameters. Such clustering contributes to identifying the overall pattern of spatial variations in local spatial choice rules concisely. Because the OD matrices can be handled as images, we can easily implement an unsu-

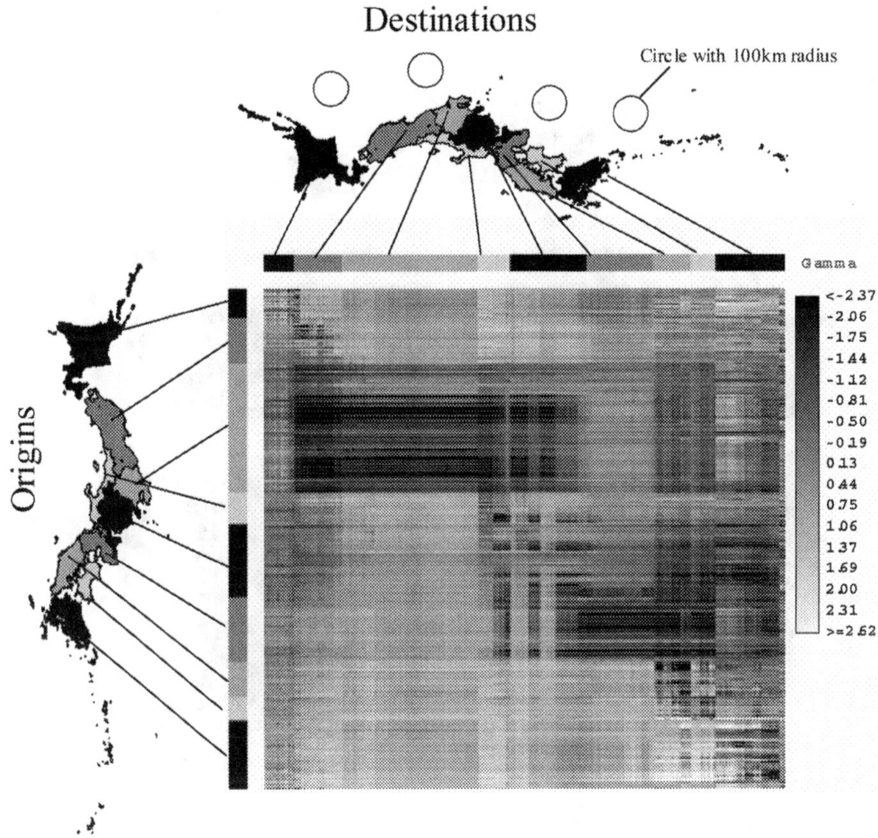

Figure 9. The OD matrix of estimated local accessibility parameters.

pervised classification technique by using any raster GIS with image processing functions.

At first, the three images of $\alpha_{i(j)}$, $\beta_{i(j)}$, and $\gamma_{i(j)}$ are standardised. If the standardised values are higher than two or lower than minus two, the values are replaced by the two or minus two, respectively, in order to avoid small clusters with outliers. Then images with 256 classes are made from the linear stretch of the standardised image. The ISOCLUST function in IDRISHI (a raster GIS developed by Clark University) is applied to the stretched images. The routine of the function is almost the same as the famous non-hierarchical cluster analysis, such as ISODATA, H-means and K-means (Anderberg, 1973). Among several trials of the clustering with the number of clusters from 2 to 10, the result of 3 clusters is the most tractable (Figure 10). This image of clusters reveals a generally symmetrical structure that mirrors the correspondence between the eastern and western Japan; the migration characteristics of local destination choices are partly determined by the relative locations of the origin and desti-

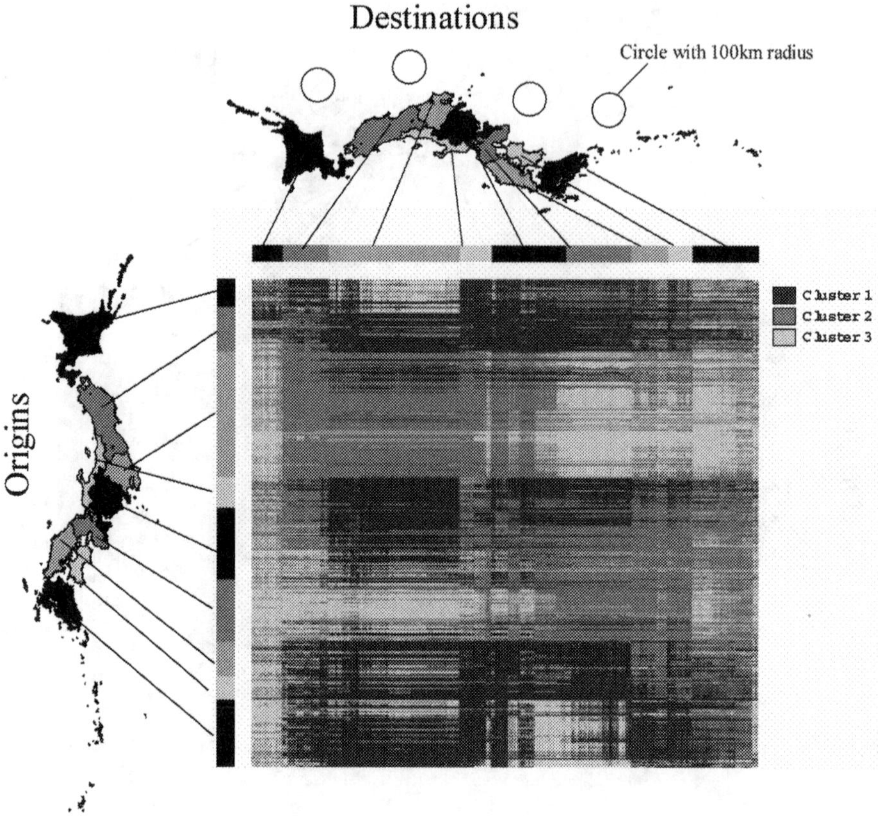

Figure 10. The result of the non-hierarchical OD clustering based on the local parameter similarities.

nation in the core-periphery structure of Japan. Table 2 shows the average and standard deviation of each estimated parameter for each cluster.

Cluster 1 is mainly composed of the OD pairs from periphery. The more negative distance-decay parameters imply that the opportunities of migration, in particular, about information of job and education, in such peripheral places are restricted for use by migrants. The highest population and accessibility sensitivities show that the migrants from the periphery prefer agglomerated destinations. In the major metropolitan areas, such as the Kanto and Kinki districts, the agglomerated areas are the central parts of the metropolitan areas. It is well known that young people (around 18-years-old) move from peripheral parts to the core of the metropolitan area in order to find jobs and educational opportunities and then move later to suburban areas after their marriage and birth of children.

Cluster 2 captures such later movements inside the metropolitan area. The negative accessibility parameters reflect the direction of movements from the

Table 2. Summary Statistics of Estimated Parameters for each OD Cluster.

		$\alpha_{i(j)}$	$\beta_{i(j)}$	$\gamma_{i(j)}$
Cluster 1	Average	1.16	–2.21	1.02
	Std. Dev.	0.16	0.58	0.57
Cluster 2	Average	1.10	–1.95	–1.23
	Std. Dev.	0.27	0.40	1.05
Cluster 3	Average	1.12	–0.90	0.39
	Std. Dev.	0.21	1.12	0.63

centre of the metropolitan areas to suburban areas within the same metropolitan area (short-distance movement). The strong distance decay of the cluster is reasonable because the cluster includes the short distance movements mainly caused by housing-relocation needs. At the same time, the negative accessibility parameter also reflects the return-movements from these metropolitan areas to their birthplaces. The directions of such movements are biased to the north in the migrations from the Kanto district including Tokyo. Yano *et al.* (2000) show that the hinterland of migration to the Tokyo metropolitan area contains northern Japan, such as the Tohoku and Hokkaido districts. The migrations of cluster 2 from the Kinki district including Osaka have similar directional biases to the south.

Another implication of the negative accessibility parameters is the existence of the hierarchical destination choice in migrant's decision making (Fotheringham, 1983). According to Fotheringham (1991), the larger the number of potential destinations, the more apparent the hierarchical choice becomes, because migrants consciously or unconsciously reduce their task of making comparisons between possible destinations by introducing the hierarchical decision making. It is plausible that for migrants from a major metropolitan area, the area and its hinterland have more potential destinations competing with each other than other areas.

Cluster 3 is mainly composed of the OD pairs from the major metropolitan areas to other distant areas (e.g. from the Kanto to the Kinki and west districts) and from some peripheries to their nearest major metropolitan areas (e.g. from the Kyusyu to the Kinki district). The OD pairs are mainly characterized by their long-distance between origin and destination, and the weak distance-decay, but the signs of the average estimated parameters are similar to those of cluster 1; the migrants prefer populous and agglomerated destinations. The migration purpose of cluster 3 would be the same as that of cluster 1, that is, mainly job-educational. The less negative distance-decay points out that the distance from the origin does not greatly restrict the opportunities for the migration.

When we compare the origin-destination specific parameters by the local models with the origin-specific parameters by the global models, it is clear that the origin-specific parameters are valid only for some parts of the total OD pairs. The local parameters are highly dependent on the combination of origin and destination; we can call their nature doubly dependency of the geographical parameter drifts. As suggested in the multi-stream migration modelling, the spatial variation could be interpreted as the difference between migration groups in which migrants have different motivations and opportunities to fulfil their purpose.

7. CONCLUDING REMARKS

This research has considered the specification of a localised spatial interaction model by applying the GWR approach to an origin-specific origin-constrained spatial interaction model. The model being proposed here takes account of parameter variations depending on origin and destination pairs. Therefore, the model could be called an origin-destination specific or doubly specific spatial interaction model. So far origin-specific or destination-specific models have shown considerable regional variations of parameters, in particular, focusing on distance decay parameters (e.g. Southworth, 1979, Stillwell, 1991). The empirical result of the localised spatial interaction model in the previous section is evidence for parameter variations depending on the combinations of origins and destinations in a migration system.

Fotheringham (1983, 1984) proposes the competing destination model and explains how a spatial structure (a configuration of origins and destinations) biases the estimates of distance decay parameters according to the two stage spatial decision-making theory. The spatial structure effects could certainly be a reason for the variation (Fotheringham *et al.*, 2001). However, even if we use the competing destination model to take into consideration the spatial structure effect, there still remain significant geographical variations of parameters (Ishikawa, 1987; Yano *et al.*, 2000). The empirical result of this research also shows the significant geographical variation of the accessibility parameters. These results indicate that the manner of the hierarchical destination choice process could change based on the geographical context for migrants. Moreover, as Gordon (1985) explains, there are numerous explanations for the parameter variations in migration modelling which could cause the spatial non-stationarity in observed migration systems. It is then reasonable to expect that the spatial non-stationarity appears in every kind of spatial interaction.

Visualising local variations of parameters is helpful to explore the spatial non-stationarity in spatial interaction in order to examine reasonable explanations for underlying processes that cause the spatial non-stationarity. The localised spatial interaction modelling based on the GWR approach might be the most generalised and sophisticated method to attain this purpose at present.

This is not to say that the method proposed in this paper is complete and unique in applying local modelling to spatial interactions. Congdon (2000) shows a Bayesian spatial interaction model regarding parameters as random variables. Although the modelling does not assume the spatial dependency of parameter variations, we can specify the dependency based on Bayesian approach to disease mapping (Clayton and Kaldor, 1987; Lawson, 2001). Neural network would be another flexible modelling technique to infer systematic parameter drifts with complex interaction terms between explanatory variables of origins and destinations (Nakaya, 1995, 1996). The modelling can also incorporate locational variables to model regionally different sensitivities of explanatory variables (Murnion, 1999). As Fotheringham (1999) edited a series of comparison between several local modelling approaches for non-flow data in *Geographical and Environmental Modelling Vol.* 3 (1), the comparison of localised spatial interaction modelling derived from different methodologies would be a potential issue to be explored.

There are also several areas where we should pursue future developments in the GWR approach for spatial interaction modelling. Theoretically speaking, we can specify other formulas of origin-destination specific models. The weighting of not only destinations but also origins would be desirable to get statistically stable predictions. Especially, unconstrained conventional gravity models having elastic flow generations should use the weighting origins and destinations simultaneously. However, performing such doubly weighting might require too heavy a computational load. Regarding visualisation, the doubly specification of parameters makes it difficult to see the result. This paper proposes a method of using OD images and non-hierarchical clustering. The better visualisation techniques of OD data sets would be considered necessary. With these improvements the localised spatial interaction modelling would encourage the spatial explorative data analysis for flow data well.

ACKNOWLEDGEMENTS

The author would like to thank Professor Stewart Fotheringham, the editors (Professors Atsuyuki Okabe and Richard Thomas), and the anonymous referee for their useful help to improve this paper. This paper is supported by a Grant-in-Aid for Encouragement of Young Scientists of Japan Society for the Promotion of Science.

Department of Geography,
Ritsumeikan University,
56-1 Tojiin-kita-machi, Kita-ku, Kyoto, 603-8577, Japan

REFERENCES

Akaike, H. (1973), 'Information theory and an extension of the maximum likelihood principle.' In B.N. Petrov and F. Csaki (eds.), *2nd International Symposium on Information Theory*. Akademiai Kiado, Budapest, pp. 267–281.

Anderberg, M.R. (1973), *Cluster Analysis for Applications*. Academic Press, New York.

Baxter, M. (1984), 'A note on the estimation of a nonlinear migration model using GLIM.' *Geographical Analysis*, **16**, 282–286.

Berglund, S. and A. Karlström (1999), 'Identifying local spatial association in flow data.' *Journal of Geographical Systems*, **1**, 219–236.
Black, W.R. (1992), 'Network autocorrelation in transport network and flow systems.' *Geographical Analysis*, **24**, 207–222.
Bolduc, D. (1992), 'Spatial autoregressive error components in travel flow models.' *Regional Science and Urban Economics*, **22**, 371–385.
Bolduc, D., R. Laferriere and G. Santarossa (1995), 'Spatial autoregressive error components in travel flow models.' In L. Anselin and R.J.G.M. Floarax (eds.), *New Directions in Spatial Econometrics*. Springer, Berlin, pp. 96–108.
Boyle, P. (1993), 'A three-dimensional approach to multistream migration modelling: local-level flows in Hereford and Worcester.' *Environment and Planning A*, **25**, 1279–1293.
Brunsdon, C.F., A.S. Fotheringham and M. Charlton (1996), 'Geographically weighted regression: a method for exploring spatial non-stationarity.' *Geographical Analysis*, **28**, 281–298.
Casetti, E. (1972), 'Generating models by the expansion method: applications to geographical research.' *Geographical Analysis*, **4**, 81–91.
Casetti, E. (1990), 'The investigation of parameter drift by expanded regressions: generalities, and a "family-planning" example.' *Environment and Planning A*, **23**, 1045–1061.
Clayton, D.G. and J. Kaldor (1987), 'Empirical Bayes estimates of age-standardised relative risks for use in disease mapping.' *Biometrics*, **43**, 671–691.
Congdon, P. (2000), 'A Bayesian approach to prediction using the gravity model, with an application to patient flow modelling.' *Geographical Analysis*, **32**, 205–224.
Davies, R.B. and C.M. Guy (1987), 'The statistical modeling of flow data when the Poisson assumption is violated.' *Geographical Analysis*, **19**, 300–314.
Flowerdew, R. (1991), 'Poisson regression modelling of migration.' In J. Stillwell and P. Congdon (eds.), *Migration Models: Macro and Micro Approaches*. Belhaven Press, London, pp. 92–112.
Foster, S. and W. Gorr (1986), 'An adaptive filter for estimating spatially varying parameters: application to modelling police hours spent in response to calls for service.' *Management Science*, **32**, 878–889.
Fotheringham, A.S. (1983), 'A new set of spatial interaction models: the theory of competing destinations.' *Environment and Planning A*, **15**, 15–36.
Fotheringham, A.S. (1984), 'Spatial flows and spatial patterns.' *Environment and Planning A*, **16**, 529–543.
Fotheringham, A.S. (1988), 'Consumer store choice and choice set definition.' *Marketing Science*, **7**, 299–310.
Fotheringham, A.S. (1991), 'Migration and spatial structure: the development of the competing destinations model.' In J. Stillwell and P. Congdon (eds.), *Migration Models: Macro and Micro Approaches*. Belhaven Press, London, pp. 57–72.
Fotheringham, A.S. (1997), 'Trends in quantitative methods I. Stressing the local.' *Progress in Human Geography*, **21**, 88–96.
Fotherigham, A.S. and T.C. Pitts (1995), 'Directional variation in distance decay.' *Environment and Planning A*, **27**, 715–729.
Fotheringham, A.S., C. Brunsdon and M.E. Charlton (1998), 'Geographically weighted regression: a natural evolution of the expansion method.' *Environment and Planning A*, **30**, 1905–1927.
Fotheringham, A.S., M.E. Charlton and C. Brunsdon (1997), 'Measuring spatial variations in relationships with geographically weighted regression.' In M.M. Fisher and A. Getis (eds.), *Recent Developments in Spatial Analysis*. Springer, Berlin, pp. 60–82.
Fotheringham, A.S., T. Nakaya, K. Yano, S. Openshaw and Y. Ishikawa (2001), 'Hierarchical destination choice and spatial interaction modelling: a simulation experiment.' *Environment and Planning A*, **33**, 901–920.
Gordon, I. (1985), 'Economic explanations of spatial variation in distance deterrence.' *Environment and Planning A*, **20**, 907–924.
Gordon, I. (1991), 'Multi-stream migration modelling.' In J. Stillwell and P. Congdon (eds.), *Migration Models: Macro and Micro Approaches*. Belhaven Press, London, pp. 73–91.

Gorr, W.L. and A.M. Olligschlaeger (1994), 'Weighted spatial adaptive filtering: Monte Carlo studies and application to illicit drug market modelling.' *Geographical Analysis*, **26**, 67–87.

Ishikawa, Y. (1987), 'An empirical study of the competing destinations model using Japanese interaction data.' *Environment and Planning A*, **19**, 1359–1373.

Ishikawa, Y. (1990), 'Explorations into the two-stage destination choice.' *Geographical Review of Japan*, **62B**, 75–85.

Ishikawa, Y. and A.J. Fielding (1998), 'Explaining the recent migration trends of the Tokyo metropolitan area.' *Environment and Planning A*, **30**, 1797–1814.

Jones III, J.P. and R.Q. Hanham (1995), 'Contingency, realism, and the expansion method.' *Geographical Analysis*, **27**, 185–207.

Lawson, A.B. (2001), *Statistical Methods in Spatial Epidemiology*. Wiley, Chichester.

Loader, C. (1999), *Local Regression and Likelihood*. Springer, New York.

Lovett, A. and R. Flowerdew (1989), 'Analysis of count data using Poisson regression.' *Professional Geographer*, **41**, 190–198.

Mallows, C.L. (1973), 'Some comments on Cp.' *Technometrics*, **15**, 661–675.

Marble, D., Gou, Z., Lin, L. and Saunders, J. (1997), 'Recent advances in the exploratory analysis of interregional flows in space and time.' In Z. Kemp (ed.), *Innovations in GIS 4*. Taylor and Francis, London, pp. 75–88.

McCullagh, P. and J.A. Nelder (1989), *Generalized Linear Models*. Chapman and Hall, London.

Murnion, S. (1999), 'Exploring spatial non-stationarity with radial basis function neural networks.' *Geographical and Environmental Modelling*, **3**, 35–45.

Nakaya, T. (1995), 'Spatial interaction modelling using the PR-Perceptron: a case study of migration in Japan.' *Jinbun-Chiri (Human Geography)*, **47**, 521–540 (in Japanese with English abstract).

Nakaya, T. (1996), 'A comparison of the gravity model and the perceptron: an analytical method for perceptrons by partial differential coefficients.' *Geographical Reports of Tokyo Metropolitan University*, **31**, 173–182.

Nakaya, T. (2000), 'An information statistical approach to the modifiable areal unit problem in incidence rate maps.' *Environment and Planning A*, **32**, 91–109.

Openshaw, S. (1976), 'An empirical study of some spatial interaction models.' *Environment and Planning A*, **8**, 23–41.

Openshaw, S. (1979), 'A methodology for using models for planning purposes.' *Environment and Planning A*, **11**, 879–896.

Openshaw, S. and C.J. Connolly (1977), 'Empirically derived deterrence functions for maximum performance spatial interaction models.' *Environment and Planning A*, **9**, 1067–1079.

Openshaw, S., M.E. Charlton, C. Wymer and A.W. Craft (1987), 'A mark 1 geographical analysis machine for the automated analysis of point data sets.' *International Journal of Geographical Information Systems*, **4**, 279–311.

Rissanen, J. (1983), 'A universal prior for integers and estimation by minimum description length.' *The Annals of Statistics*, **11**, 416–431.

Rissanen, J. (1986), 'Stochastic complexity and modelling.' *The Annals of Statistics*, **14**, 1080–1100.

Schwarz, G. (1978), 'Estimating the dimension of a model.' *The Annals of Statistics*, **6**, 461–464.

Southworth, F. (1979), 'Spatial structure and parameter disaggregation in trip distribution models.' *Regional Studies*, **13**, 381–394.

Stillwell, J. (1991), 'Spatial interaction models and the propensity to migrate over distance.' In J. Stillwell and P. Congdon (eds.), *Migration Models: Macro and Micro Approaches*. Belhaven Press, London, pp. 34–56.

Wedderburn, R.W.M. (1974), 'Quasilikelihood functions, generalized linear models and the Gauss-Newton method.' *Biometrika*, **61**, 439–447.

Wilson, A.G. (1970), *Entropy in Urban and Regional Modelling*. Pion, London.

Yano, K., T. Nakaya and Y. Ishikawa (2000), 'An analysis of inter-municipal migration flows in Japan using GIS and spatial interaction modelling.' *Geographical Review of Japan*, **73B**, 165–177.

SECTION B

SPACE-TIME ANALYSIS

CHANG-HYEON JOH, THEO A. ARENTZE AND HARRY J.P. TIMMERMANS

UNDERSTANDING ACTIVITY SCHEDULING AND RESCHEDULING BEHAVIOUR: THEORY AND NUMERICAL ILLUSTRATION

1. INTRODUCTION

Inspired by Hagerstrand's time geography, the activity-based modelling approach has become an active area of research since the mid 90s in transportation research (Ettema and Timmermans, 1997). As a consequence, it has regained interest in geography as well. The central concept underlying the activity-based approach is that travel is derived from the participation in activities instead of being pursued for its own sake, and therefore, the understanding, analysis and forecasting of travel behaviour should be based on the understanding of activities (Burnett and Hanson, 1982). Individuals try to meet their personal and family needs by participating in activities in everyday life, subject to a set of constraints. In the process of organising activities in time and space, travel is derived as a by-product to overcome the distance between activity locations. Any direct causation of socio-demographic characteristics and the physical environment to travel behaviour without explicitly considering the choice of activity participation may, therefore, be incorrect or at least theoretically inappropriate.

The last two decades have witnessed an enormous amount of literature that illustrates a variety of research directions in this paradigm. Two distinctive approaches should be mentioned here. First, the utility maximisation approach assumes that individuals choose the best solution through the search of all possible alternatives. Key studies in this direction include Bowman and Ben-Akiva (2000), Fujii *et al.* (1998), McNally (1997), Recker *et al.* (1985) and Recker (1995). As pointed out in Gärling *et al.* (1994), however, while the utility maximisation principle might explain which factors affect the final choice, it does not account for the process of making decisions that also impact on outcomes. If the goal is to forecast travel demand, this may not be an urgent research question. It is however important for a better understanding of travel behaviour.

Secondly, to overcome this limitation, the computational process approach, which assumes that individuals pursue satisfactory, possibly sub-optimal, solutions through the search of a subset of the universe of alternatives, has been introduced. This approach focuses on the decision making process and offers more insight into how people process information and arrive at the observed

B. Boots et al. (eds.), Modelling Geographical Systems, 73–95.

activity-travel pattern. Because of its focus on the decision process, this approach potentially allows one to better evaluate the implications of various transportation policy measures (e.g. Bhat and Koppelman, 2000). Examples include Arentze *et al.* (2000), Ettema *et al.* (1993) and Gärling *et al.* (1998). Among these, Arentze *et al.* (2000) presented the latest and most comprehensive operational model to date, named *Albatross*, a rule based system, which derives choice heuristics from diary data and predicts activity-travel patterns.

Yet, research to date has paid little attention to the decision making process underlying activity schedule changes under time pressure and unexpected events. Everyday life is full of uncertainty, and frequently, individuals are forced to reconsider previously scheduled activities. In fact, scheduling and rescheduling are a never-ending process over the entire lifetime of individuals and are complementary to each other. Scheduling is an incremental process that gradually increases the level of detail of the sequence of activities as activities are executed in a given time horizon. Doherty and Miller (2000) found in a week-long survey of activity scheduling in Canada that 60% of modifications and 41% of cancellations of activities were made *during* the execution of the schedule.

Although Gärling *et al.* (1999) have undertaken pioneering work on this topic and provided a principle of schedule change where people delete or postpone particular activities under time pressure, they only considered some selected aspects of rescheduling behaviour. The authors, therefore have developed a more comprehensive model of scheduling and rescheduling behaviour. It includes the theoretical foundation of a short-term dynamics model of activity-travel behaviour (Timmermans *et al.*, 2001), the elaborated utility functions (Joh *et al.*, 2001), and the complementary model component representing decision heuristics (Joh *et al.*, 2002). As a continuation of this modelling work, the current research aims to examine whether the model exhibits the envisioned activity behaviour as a function of some important model parameters. To this end, the current chapter simulates initial scheduling and time pressure-induced rescheduling behaviour as conducted by different hypothetical individuals characterised by different parameters.

This article is organised into the following three sections. Section 2 introduces the conceptual framework underlying the model and provides a formal description of the model components. Section 3 then examines the behaviour of the model depending on the variation in some selected parameters. Section 4 concludes the chapter with an agenda for future research.

2. THE MODEL

Our theory of individuals' scheduling and rescheduling behaviour involves the following conceptual considerations. First, individuals execute activities to meet a variety of needs. Fulfilling activities returns satisfaction or utility as a reward for meeting the needs. The list of activities to conduct is determined by an individual's personal desires, responsibility for family and work contract (Damm and Lerman, 1981; Bhat and Koppelman, 1993).

Secondly, a set of circumstantial conditions limits the extent to which individuals can increase the utility. These conditions include individuals' physical condition, their role in society and the physical environment surrounding them. These conditions encourage individuals and society to develop particular ways or means of fulfilling activities. Activities of particular purposes are then organised in space and time (Pred, 1981; Thrift, 1983).

Thirdly, uncertainty should also be taken into account. Given the needs and conditions, individuals identify and evaluate activities in terms of their anticipated utilities for possible implementation. Uncertainty, however, affects the evaluation due to the fact that the activities in the later positions of the planned schedule may involve a larger amount of uncertainty. The evaluation results differ between individuals faced with uncertainty, dependent on their personalities or decision styles in dealing with uncertainty, which in turn affect the schedule.

Fourthly, individuals are assumed to use heuristics in looking for alternatives instead of becoming involved in an exhaustive search, due to the fact that their rationality is bounded. Individuals usually have numerous alternative ways of planning a schedule given a time horizon, each of which may result in a different level of utility. Cognitive constraints 'however' prevent individuals from identifying and evaluating every single one of a universe of alternatives. Individuals therefore use a set of heuristics to reduce the burden of search and to pursue cost effectiveness. A typical example is habitual behaviour (Gärling, 1998) that does not concern any other alternatives than a routinised alternative, which is far from optimising behaviour but is frequently the case. Heuristic behaviour may result in sub-optimal, satisfying choices.

Finally, an activity schedule is tentative and may be changed at any time. Every moment in time, there may be the need for changing the schedule of remaining not-yet-completed activities. An individual may be forced to change the schedule due to time pressure or may actively decide to change to improve the existing schedule. Any (sub-optimal) decision is enforced until a further need to reschedule the activities arises.

Based on the above discussion, we formulate a conceptual framework of individuals' scheduling and rescheduling behaviour as illustrated in Figure 1. Initially, a tentative schedule is given. The set of activities included in the current schedule is a subset of the activity program. The individual evaluates the utility of activities for possible implementation as well as non-implementation. When an individual with a certain decision style evaluates the utility of alternative activities under a set of constraints, he/she examines whether some change of the schedule is necessary. More specifically, the individual examines whether there is any time pressure or any improvement of utility level is possible by changing the existing schedule. Because it is impossible to identify and evaluate all possible alternatives due to cognitive constraints, individuals adopt certain heuristic strategies to effectively and efficiently reduce the search space to reach

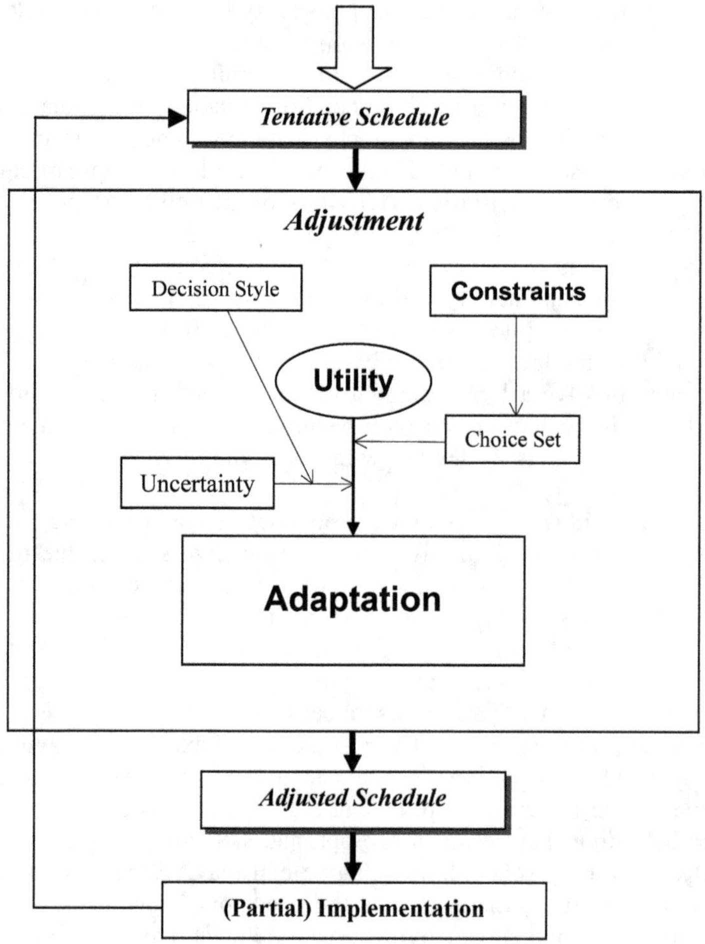

Figure 1. Conceptual framework.

reasonably good solutions in real time. The adjusted activity schedule then is implemented. The utility of the remaining schedule is again subject to unexpected events causing increased or reduced time pressure. Therefore, the schedule will often be only partially implemented, and the adjusted schedule remains tentative.

In the following sub-sections we provide a formal description of two major components of the suggested model: the utility functions and the heuristic solution methods.

2.1. The Utility Function

Individuals derive a certain level of utility directly or indirectly from participating in activities. The utility function associates some selected characteristics of activities with particular numerical values that are assumed to correspond to particular utility levels. The utility also varies with characteristics of individuals and households. We assume that the utility of an activity is a function of the amount of time spent on the activity, where longer duration provides a higher level of utility.

Microeconomic theory assumes an increasing utility function with a diminishing marginal utility over the entire range of input variable values. The utility functions of activities used in activity-based approaches adopting time-use microeconomic theory (Becker, 1965) almost invariably rely on the same assumption of an ever-diminishing marginal utility over the entire range of activity duration (Kitamura *et al.*, 1996), although many other functional forms are conceivable, as argued by Supernak (1988). We introduced such an alternative utility function (Timmermans *et al.*, 2001), paying particular attention to the possibility of increasing marginal utility due to a negative saturation effect in the beginning phase of the implementation of certain activities. Many leisure activities and some initial information-acquisition activities such as shopping activity possibly display this property.

Here, we reintroduce the proposed functional form of the utility function and provide a detailed formal description of its properties. Although utility is primarily a function of the time spent on the concerned activity, many other choice facets such as location, transport mode, etc. also affect the utility that individuals derive from becoming involved in an activity. Furthermore, as mentioned earlier, uncertainty impacts utility, dependent on decision styles and the process of schedule adjustment. In this sub-section, therefore, we do not concern ourselves with this variety of choice facets and uncertainty-related behaviour, but focus only on the assumed utility function under certainty. (Section 3 focuses on the impact of utility parameter values on scheduling and rescheduling behaviour as a function of the amount of change in activity duration.) Readers interested in other choice facets and uncertainty are referred to Joh *et al.* (2001).

Let AP refer to the given activity program, $AP = \{1, \dots, A\}$. The proposed utility function can be written as:

$$U_a = U_a^{\min} + \frac{U_a^{\max} - U_a^{\min}}{\{1 + \exp[-\beta_a(v_a - \alpha_a)]\}^{\gamma a}} \tag{1}$$

where:

α is an index of activities ($a \in$ AP);

v is the duration ($v \geq 0$);

$U^{\max}$ is the upper asymptote of the curve ($U^{\max} > 0$);

$U^{\min}$ is the lower asymptote of the curve ($U^{\min} \leq 0$);

α is parameter of the location of the curve ($\alpha > 0$);
β is parameter of the slope of the curve ($\beta > 0$);
γ is parameter of the inflection point of the curve ($\gamma > 0$).

Equation (1) dictates that the utility of an activity is a function of duration with a set of activity-specific parameters. In particular, the utility function in Equation (1) has the following properties. First, it is a monotonically increasing function of duration since:

$$\frac{dU}{dv} = \frac{\gamma\beta(U^{\max} - U^{\min})}{\exp[\beta(v - \alpha)]\ \{1 + \exp[-\beta(v - \alpha)]\}^{\gamma + 1}} > 0 \qquad (2)$$

This implies that the utility of an activity is primarily determined by the duration spent on the activity and that a longer duration gives a higher utility.

Secondly, utility is however not infinitely increasing but converges to a maximum level as:

$$\lim_{v \to \infty} U = U^{\max} \qquad (3)$$

The maximum level of utility differs between activities.

Thirdly, the minimum utility or zero-duration utility is given by:

$$U_{v=0} = U^{\min} + \frac{U^{\max} - U^{\min}}{(1 + \exp[\alpha\beta])^{\gamma}} \qquad (4)$$

The size of $U_{v=0}$ may differ between activities. A zero or approximately zero value is considered the normal case. However, for some activities the value may be negative. For example, not conducting a leisure activity may produce fatigue or other intrinsic negative effects that reduces total utility for the schedule as a whole. A large negative utility would facilitate the inclusion of the concerned activity. On the other hand, a positive value would not be consistent with the theory and, therefore, would avoid imposing this constraint on the values the parameters can take.

Fourthly, the utility function is asymmetrically S-shaped, and the marginal utility increases up to a certain level of utility (inflection point) and decreases afterwards, i.e.:

$$\frac{d^2U}{dv^2} = \frac{\gamma\beta^2(U^{\max} - U^{\min})\{1 + \exp[-\beta(v - \alpha)]\}^{r}\{\gamma - \exp[\beta(v - \alpha)]\}}{\exp[2\beta(v - \alpha)]\{1 + \exp[-\beta(v - \alpha)]\}^{2\gamma + 2}} \gtreqless 0 \qquad (5)$$

and the inflection point is given as $U(v^*)$ where $v^* = (\alpha\beta + \ln\gamma)/\beta$. The utility function has a concave part as well as convex part. A larger γ causes a lower utility at the inflection point, $U(v^*) = U^{\min} + (U^{\max} - U^{\min})/(1 + \gamma^{-1})^{\gamma}$, as $\partial U(v^*)/\partial\gamma < 0$, and a smaller proportion of increasing marginal utility. The particular values of $U(v^*)$ may vary across activities and individuals, and can be captured by estimating the γ value.

Finally, the activation and the speed of utility change vary with parameters α and β, respectively. As implied by Equation (1), α and β determine the location of the utility function along the duration axis and the slope of the curve, respectively. The change in utility takes place after a certain amount of time, and the amount that activates the change varies across activities. An increasing α shifts the curve to the right, meaning that more time is needed to reach a particular utility level. The speed of utility change over time also varies across activities. The curve is steeper with a larger β, implying that the change in utility is more sensitive to the change in duration, keeping everything else constant.

Given the utility function for individual activities, the total function for an entire schedule should aggregate the utilities across activities. The aggregation equation can be manifold, and some multiplicative form may be desirable to reflect the mutual relationships between activities. Obviously, certain activities have complementary or substitution relationships. A business meeting, for example, has a complementary relationship with a follow-up social event. In-home and out-of-home leisure activities are likely to have a substitution relationship because of the fixed amount of time in the evening after work. To cope with these activity-specific relationships, the constituent activities should be grouped in a multiplicative term of an additive aggregate function. In the current study 'however' we simplify the problem and assume a simple additive aggregate utility function:

$$U = \sum_{a=1}^{A} U_a \tag{6}$$

with

$$\sum_{a=1}^{A} v_a = B \tag{7}$$

where B is the total duration ($B > 0$), representing the budget constraint that limits the total amount of time that can be assigned to activities at the moment of scheduling.

The additive form with the time budget constraint implies that activities have a general relationship with all others regarding their duration in the schedule. Increasing the duration of an activity means without exception that the duration of other activities is decreased, depending on their utility function. To increase total utility, therefore, the increase of utility for the extended activitie(s) must exceed the utility decrease of the reduced activitie(s). Othewise, no change is induced. Secondly, however, an activity does not have any activity-specific relationship with other activities.

2.2. *Decision Heuristics*

Models predicting the choice of a complete activity schedule (Jones *et al.*, 1983; Recker *et al.*, 1985) typically generate all feasible schedules and select the best one. In contrast, models of incremental activity scheduling behaviour (Gärling *et al.*, 1998; Ettema *et al.*, 1993) employ a set of operators and heuristic rules that control the application of those operators and induce incremental adjustments of the schedule. Incremental adjustment implies that the schedule is changed in a stepwise manner and the adjustment of the current step is made on and hence affected by the current state of the schedule.

We adopt the latter modelling approach. That is, we employ a set of operators including duration adjustment and insertion, deletion and substitution operators to implement incremental adjustments of the schedule. (We should mention again that other operators that might be related to choice facets such as location and transport-mode are not considered here.) The duration operator adjusts the duration of the scheduled activities such that the marginal utility becomes the same across the activities included in the schedule, and the system reaches an equilibrium state. Insertion, deletion and substitution operators change the list of scheduled activities such that unscheduled activities are added to the schedule, and scheduled activities are removed if this increases the total utility.

Among these operators, only the duration operator involves a fine-tuning of duration and realises the equalisation of the marginal utilities across scheduled activities. All three other operators involve discontinuous adjustments in which sudden larger changes in duration takes place, and the resulting schedule may well be out of equilibrium. Note that the operators are considered only if the adjustments meet the schedule constraints. The envisioned scheduling and rescheduling behaviour is therefore highly discontinuous due to constraints that limit feasible adjustments.

We assume the following decision heuristics. First, at each time of incremental adjustment, an individual mentally simulates the schedule change implied by each operator, and chooses the best operator that offers the highest increase in total utility. Secondly, the mental simulation of an operator requires an effort and discounts utility, the amount of which is constant and specific to each operator. This effort can be interpreted as operator-specific resistance to change. Likewise, any actual implementation of schedule change by the chosen operator counts as effort and reduces the total utility, the amount of which is an increasing function of the number of adjustment steps up to the current adjustment. This effort can be regarded as cumulative mental fatigue. Finally, these adjustment processes continue until no improvement of the schedule is possible. In other words, if the highest increase of total utility is negative, the system stops the adjustment process.

The evaluation is initiated at the beginning of a day for an initial schedule and resumes at any time of necessity including the completion of an activity and the occurrence of an unexpected event. Let S and R be the set of activities that are currently scheduled and not scheduled yet, respectively, implying $S \cup R = AP$ and $S \cap R = \varnothing$. Let O be a subset of operators, $O = \{O_A, O_I, O_D, O_S\}$, where O_A denotes duration adjustment, O_I insertion, O_D deletion and O_S substitution. Given the suggested utility function and heuristics, scheduling and rescheduling behaviour is represented as an algorithm below.

Thus, the system first calculates for each operator o the amount of improvement (ΔU^o_{t-1}), computed as the incremental total utility ($\tilde{U}^o_{t-1} - U_{t-1}$), discounted by resistance to change (θ_o) and mental fatigue cumulated up to the current adjustment step ($h(t)$). Next, the system selects the operator (o_t) that will return the highest utility increase at the current adjustment step. If the total utility after adjustment is higher than the existing schedule's utility, the system implements the adjustment using the chosen operator ($S_t \leftarrow AP^{o_t}_{t-1}$). Note that operators are applied not only on S_{t-1} but also on R_{t-1} because of insertion of activities into S from R and deletion of activities from S into R. The member activities of S and R may differ between AP_{t-1} and AP_t. Finally, the termination condition is expressed as the non-improvement of the existing schedule above the fatigue and resistance threshold.

```
do apply_O_A
t = 1
while U_t > U_{t-1} do
begin
   t = t + 1
   o_t = o(   max      [ΔU^o_{t-1}]     where ΔU^o_{t-1} = Ũ^o_{t-1} − U_{t-1} − θ_o − h(t)
          o ∈ O; o ≠ O_A
   U_t > U^{o_t}_{t-1} = U_{t-1} + ΔU^{o_t}_{t-1} = U_{t-1} + Ũ^{o_t}_{t-1} − U_{t-1} − θ_{o_t} − h(t)
                = Ũ^{o_t}_{t-1} − θ_{o_t} − h(t)
   if U_t > U_{t-1} then
   begin
      S_t ← AP^{o_t}_{t-1}
      do apply_O_A
   end
end.
```

where:

t is an index of adjustment steps ($t > 0$);
o is an index of operators;
S_t is S at adjustment step t, where S_0 implies the initial schedule;
U_{t-1} is total utility of AP_{t-1};
$\tilde{U}^o_{t-1}$ is non-discounted total utility when applying operator o on AP_{t-1}.
θ_o is the level of resistance to change by operator o ($\theta_o \geq 0$);
$h(t)$ is the cumulative mental fatigue at adjustment step t;
U^o_{t-1} is the increment of U_{t-1} discounted by θ_o and $h(t)$ when applying operator o on AP_{t-1}.

Throughout the adjustment processes, the duration adjustment operator plays an important role that supports other operators. The activities of the initial schedule are not likely to have an equal marginal utility, so that a fine-tuning of duration between activities to increase total utility is needed. Schedule adjustment by other operators thereafter likely disturbs this duration equilibrium. The duration operator is then again activated to get the schedule back into the equilibrium state.

3. AN EXAMINATION OF SCHEDULING AND RESCHEDULING BEHAVIOUR

We examine scheduling and rescheduling behaviour by varying some selected utility parameters representing individuals' characteristics. In particular, we simulate two different values for respectively U^{max}, β and γ of certain activities and investigate the impact of the different parameter values on scheduling and rescheduling behaviour. Note that the focus of the current chapter is to study the impact of utility parameters given the decision heuristics. No variation of heuristics parameters is introduced. The system keeps all resistance to change and mental fatigue parameters constant to zero ($\theta_o = 0\ \forall o$, and $h(t) = 0\ \forall t$) throughout the simulations. In the following subsections, we first describe the different parameter settings used in the simulations, then discuss expected results, and finally, describe the results of the simulation.

3.1. *Simulation Settings*

The simulation consists of two parts: initial scheduling and rescheduling induced by increased time pressure. Hypothetical full-time workers who are characterised by particular values of utility parameters are given an initial schedule that consists only of skeleton activities including sleep, work and in-home leisure. In the first part of the simulation, each individual gradually optimises the schedule. In the second part, a time pressure situation is imposed on the individuals that forces them to change their previously optimised schedules. The simulated time pressure situation is the consequence of congestion on the road to the first out-of-home activity in the morning. The duration of travel is increased by half an hour, and individuals clear the time pressure and further adjust the schedule.

Table 1 provides the list of activities of the activity program and the utility parameter values. A total of ten activities is included in the activity program. Out-of-home leisure, daily shopping and dinner activities respectively have two alternative values of U^{max}, β and γ. Each of the eight combinations of parameter values ($2 \times 2 \times 2$) characterises an individual. Table 2 shows the initial schedule where only the sequence and duration of the skeleton activities are specified. The current simulation does not consider the location attribute. Therefore, the classification of in-home and out-of-home activities follows an arbitrary definition. Table 3 represents the assumed time constraints in terms of facility opening hours and work contract hours.

3.2. *Individual Characteristics and Simulation Focus*

Table 1 identifies the following eight distinctive combinations of individual characteristics related to out-of-home leisure, daily shopping and dinner activities.

Hypothetical person 1: High Leis_O U^{max}, large Dshop β, and large Dinner γ.
Hypothetical person 2: High Leis_O U^{max}, large Dshop β, and small Dinner γ.
Hypothetical person 3: High Leis_O U^{max}, small Dshop β, and large Dinner γ.
Hypothetical person 4: High Leis_O U^{max}, small Dshop β, and small Dinner γ.
Hypothetical person 5: Low Leis_O U^{max}, large Dshop β, and large Dinner γ.
Hypothetical person 6: Low Leis_O U^{max}, large Dshop β, and small Dinner γ.
Hypothetical person 7: Low Leis_O U^{max}, small Dshop β, and large Dinner γ.
Hypothetical person 8: Low Leis_O U^{max}, small Dshop β, and small Dinner γ.

Table 1. Hypothetical Activity Programme and Parameters

Activity in AP	*ED*	U^{min}	U^{max}	α	β	γ
Sleep	360	–5	200	400	0.16	2.5
Pcare	10	0	130	55	0.5	5
Work1	180	–5	250	150	0.15	5
Lunch	30	0	135	35	0.15	5
Work2	300	–5	250	250	0.15	1
Dinner	30	0	160	55	0.1	5 1
Leis_I	360	0	60	100	0.1	1
Leis_O	60	0	210 60	100	0.1	1
Mcare	30	0	100	65	0.425	1
Dshop	10	0	125	35	0.5 0.01	5

Pcare, Work1, Work2, Leis_I, Leis_O, Mcare and Dshop respectively denote personal care, work in the morning, work in the afternoon, in-home leisure, out-of-home leisure, medical care and daily shopping. *ED* represents some expected duration of an activity that is assumed as given. The cells in grey denote the parameter variation in the current simulation.

Table 2. Initial Schedule

Activity	Duration	Begin time	End time	Activity	Duration	Begin time	End time
Sleep	480	0:00	8:00	Pcare	0	–	–
*	–	–	–	Dinner	0	–	–
Work1	180	9:00	12:00	Lunch	0	–	–
*	–	–	–	Leis_O	0	–	–
Work2	300	13:00	18:00	Mcare	0	–	–
*	–	–	–	Dshop	0	–	–
Leis_I	300	19:00	24:00				

In the activity column, '*' denotes that activity is not defined for the time slot yet. Activities having zero duration are those activities not included in the initial schedule. Activities in grey are assumed out-of-home, while other activities in-home.

Table 3. Institutional Constraints (Opening Hours and Work Contract)

Activity	Earliest possible begin time	Latest possible begin time	Earliest possible end time	Latest possible end time
Work1	–	09:30	11:30	12:30
Work2	12:30	13:30	17:30	–
Mcare	10:00	–	–	18:00
Dshop	08:00	–	–	20:00

The identification of eight hypothetical persons with different characteristics implies that the system will enforce eight times on the same initial schedule and will also implement eight times a congestion-induced rescheduling situation on the previously optimised schedules. Note that the initial schedule optimisation will be conducted on the same initial schedule across simulations of different parameter settings, while the rescheduling simulations will be conducted on the schedules that likely differ between parameter settings. In the following, we discuss the expected results of scheduling and rescheduling behaviour, dependent on these parameter settings.

Given the utility parameter settings provided in Table 1, we expect the following impacts. First, a high U^{max} of Leis_O implies that one has a higher utility when conducting leisure outside home (see Figure 2). This may prompt the person to spend more time on leisure activities outside home. Secondly, a large β of Dshop as reflected in Figure 3 implies that one reaches the maximum utility very quickly after some amount of time spent for daily shopping. Because in case of a large β only a small range of duration shows meaningful changes of utility, the duration of daily shopping is probably also limited across schedules. In case of a small β, on the other hand, the utility function is very flat

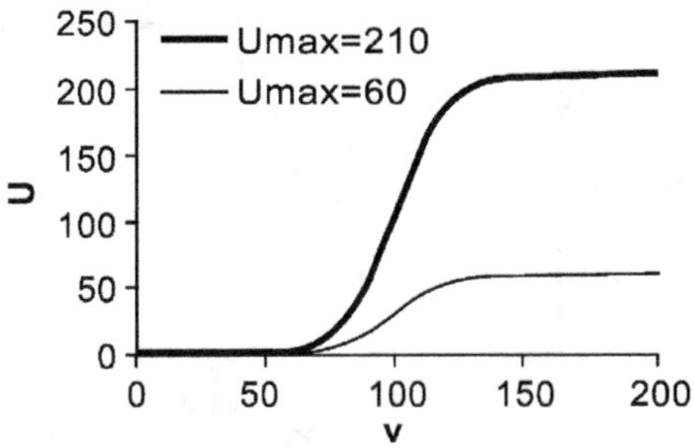

Figure 2. Utility of Leis_O activity varying with U^{max}.

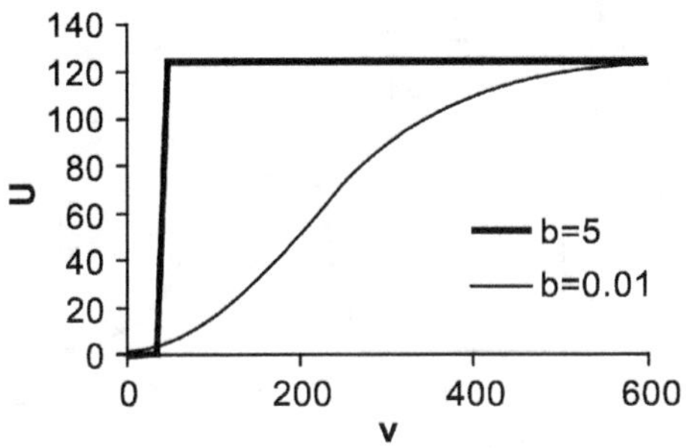

Figure 3. Utility of Dshop activity varying with β.

and shows smaller utility differences between durations, which probably induces a wide variation of duration across schedules. Furthermore, the curve with a small β provides a lower level of utility for most of the duration range. Finally, Figures 4 and 5 show that a large γ provides a much lower level of utility for the same duration, a very steep slope, while most part of utility function is increasing with diminishing marginality. This means that the relevant activity requires a longer duration to reach the near maximum utility. Once that level is reached, an individual is reluctant to reduce the duration of this activity because of the steeper slope and the larger loss of marginal utility.

Based on these expected impacts of each of the concerned parameter values, we postulate the following scheduling and rescheduling behaviour for each individual. Person 1 will spend more time on outside leisure activities and dinner and less time on daily shopping, while Person 2 would spend more time on outside leisure activities and less time on daily shopping and dinner. Person 3 is expected to spend more time on outside leisure activities and dinner, while the amount of time for daily shopping of flat-slope utility function would depend on the amount of time spent for other activities. Likewise, Person 4's amount of time spent on daily shopping will also depend on the amount of time for other activities, while spending more time on outside leisure activities and less time on dinner. Person 5 and 6 are all expected to spend less time on outside leisure

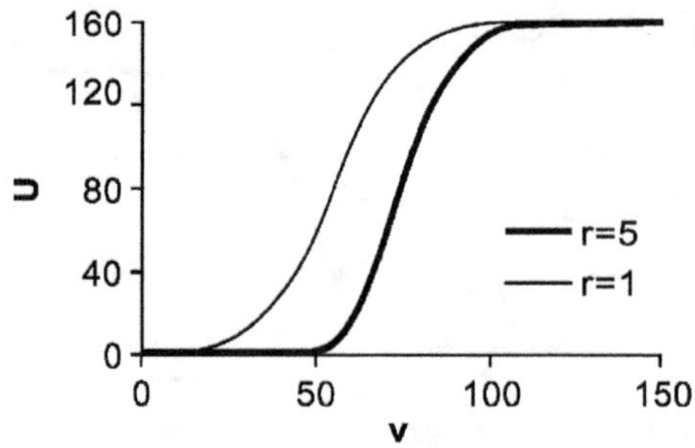

Figure 4. *Utility of Dinner activity varying with* γ.

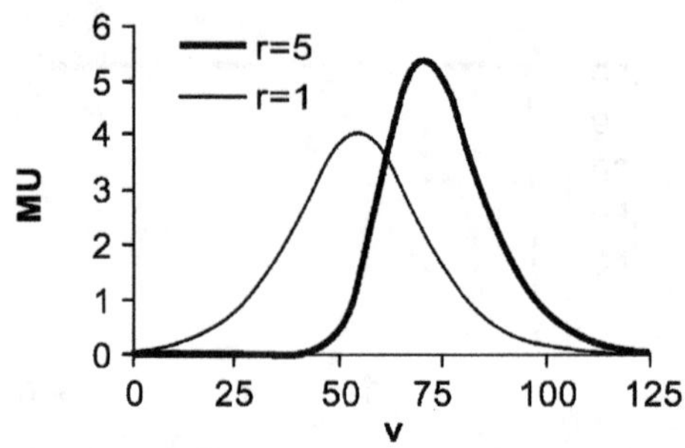

Figure 5. *Marginal utility of Dinner activity.*

activities and daily shopping, while more time on dinner for Person 5 and less time on dinner for Person 6. Person 7 may spend less time on leisure activities and more time on dinner, while Person 8 less time on dinner. The amount of time for daily shopping will be dependent upon the amount of time for other activities in case of Person 7 and 8.

3.3. RESULTS

Here, the simulated schedule adjustments will be compared for utility parameter settings with regard to the following aspects of activity-travel behaviour.

- The number of activities in the final schedule compared to the initial schedule
- The number of adjustment steps
- The list of operations used for the adjustments
- Whether or not the activities varying the parameters Umax, (and (are included in the final schedule
- The change in the duration distribution across activities before and after the adjustments
- The ratio of the duration of out-of-home to in-home activities in the final schedule
- The total utility of the final schedule.

3.3.1. *Initial Optimisation of the Schedule*

Overall, the application of operators (the sequence, number and list of operators) was almost the same across settings, and most adjustment processes involved duration and insertion operators. This is because the initial optimisation of the schedule with a large amount of open time slots naturally includes many insertions of activities with a certain duration and duration adjustment after each insertion. However, much diversity existed in the initial optimisation of the schedule, depending on the scheduling parameters of the activities (see Appendix I).

First, U^{max} of Leis_O had the following impacts. (1) Leis_O with a higher U^{max} was always included in the schedule across βs and γs, while a lower U^{max} failed to include Leis_O in most cases. Because Leis_O's duration for reaching U^{max} was relatively long, the inclusion of this out-of-home activity reduced the duration of other out-of-home activities. More specifically, when Leis_O was included, the duration of work was reduced. (2) Closely related to this result, a higher U^{max} induced a bigger ratio of the duration of out-of-home activities to the duration of in-home activities. This is because a lower U^{max} of outside leisure activities are in most cases substituted with inside leisure activities and dinner. (3) More importantly, the difference in utility is already very large between Leis_Os with higher and lower U^{max} as seen in Figure 2. As a result, a higher U^{max} resulted in a much higher level of total utility.

Secondly, the β parameter for the daily shopping activities had the following impacts. (1) Daily shopping activities with a larger β tend to have a fixed amount of time. This is because Dshop with a larger β shows a sharp increase of utility in the range of 30 to 50 duration units and, if included, should have a duration near 50 and should not be affected too much by other schedule circumstances (see Figure 3). Daily shopping activities with a smaller β, on the other hand, show a slow increase of utility over the range up to 600 and are more strongly affected by the state of other activities in the schedule, resulting in the deletion of daily shopping activities in several cases. (2) A daily shopping activity with a large β was included in all final schedules across the settings. This can be explained by the fact that a relatively short duration adds a reasonable amount of utility to the total utility without requiring other activities to be reduced much. (3) A larger (for daily shopping activity clearly induces a higher level of total utility. If β is large, the utility approximates its maximum; a smaller β provides a utility much lower than the maximum. Even if the duration is longer with for example Person 3 (53 units) than Person 1 (48 units), this episode is not long enough to approximate the maximum utility at all, as seen in Figure 3.

Finally, γ for dinner activity provided the following impacts. (1) Dinner was included in all final schedules across settings due to a high maximum utility with moderate duration. (2) The unitary value of the γ parameter (= 1) provides a typical symmetric S-shaped utility function. Amplifying γ has the effect of pushing the inflection point down, while keeping the duration with the maximum utility unchanged. As shown in Figure 4, the utility curve then has a smaller portion of increasing marginal utility and a larger portion of diminishing marginal utility from the viewpoint of utility axis (y-axis). Consequently, dinner activity with a large value of γ (> 1) comes closer to the conventional utility curve of diminishing marginal utility. Figure 5 shows that dinner activity with a unitary value of γ has a symmetric change of marginal utility with duration, while dinner activity with a larger γ has an asymmetric change that has a rapid increase up to the inflection point and a slow decrease afterwards. (3) A larger γ with everything else equal lowers the level of utility over the entire range of duration up to the point of maximum utility (see Figure 4). In other words, dinner activity with a larger γ requires a longer duration than dinner activity with a smaller γ to achieve the same level of utility. Consequently, a larger γ induced a longer duration of dinner activity, while providing a lower level of total utility. This is because a shorter duration of dinner activity with a smaller γ is already greater in utility than a longer duration of dinner activity with a larger γ. Furthermore, a shorter duration of dinner activity with a smaller γ enables other activities to have longer duration, which in turn increases total utility.

3.3.2. Rescheduling after Congestion

Overall, rescheduling involved a much smaller number of adjustments, compared to the initial scheduling, because the given schedules were previously optimised, and the adjustment was conducted mainly around the activity directly related to the time pressure scenario. Nevertheless, rescheduling behaviour clearly illustrated the amendments of previous scheduling decisions and involved subsequent deletion and insertion operations. Furthermore, besides the envisioned impacts of the simulated parameter values, the given initial states of the previously optimised schedules that differ between parameter settings provide a significant difference in rescheduling behaviour (see Appendix II).

The time pressure scenario is superimposed on a schedule that is in an (optimised) equilibrium state. It assumes that congestion, or any external force, will effectively reduce the optimised duration of a particular activity. If this reduction significantly disrupts the equilibrium state, the system will engage in a rescheduling procedure. In the current simulation, the time pressure scenario imposed congestion on the way to work in the morning, which caused 30 units longer travel time than usual.

The primary determinant of whether the imposed time pressure is significant enough to move the system in a rescheduling mode was the state of the activity that was directly related to the time pressure rather than the different settings of scheduling parameter values. This however, does not mean that the scheduling parameters representing personal characteristics of individuals are not important to the rescheduling of activities. The result of rescheduling (list of activities, activity sequence and duration) was very much dependent on the schedule that was given by the initial optimisation. Therefore, while the rescheduling process itself was not very different between settings, the final schedules were quite different.

Person 2 had the initial optimised schedule where daily shopping activity with a larger β was conducted on the way to work after sleep activity was conducted at home. Their rescheduling resulted in a significant increase of the duration of daily shopping activity, and consequently, a big increase in total utility. The explanation for this result is that the system first removed daily shopping activity of a too short duration on the way to work and inserted it again with a longer duration to another position after the second work activity. As a consequence of this increase in the duration of daily shopping activity, the near zero utility level of this activity increased to a near maximum utility level, which resulted in the increased total utility.

Likewise, when the first work episode was scheduled after in-home activities in the morning and its reduced duration was much smaller than 180 units, rescheduling involved a substantial increase in utility by increasing the duration of the first work activity, while changing the duration of other activities. Persons 4 and 8 exhibit this rescheduling pattern. This was, however, not the case when the reduced duration of the work activity was close to 180 units as in the cases of Persons 1, 3, 5, 6 and 7.

It follows from the above discussion that rescheduling choices strongly depended on the initial state of the optimised schedule. The impacts of different parameter settings can be summarised as follows. First, U^{max} had the following impact. (1) The increase in U^{max} enhances the possibility of the occurrence and long duration of the corresponding activity in the final schedule. (2) In case of an out-of-home activity, an increase in U^{max} increases the possibility of a higher ratio of out-of-home activity to in-home activity duration. (3) An increase in U^{max} generally results in a higher total utility value.

Secondly, β had the following impact. (1) A larger β for daily shopping activity results in the duration that is often the same across simulations, while a smaller β likely induces diversity in the duration. This outcome arises because most utility changes occur between the minimum utility and maximum utility duration, and this effective range is very small for an activity with a larger β.

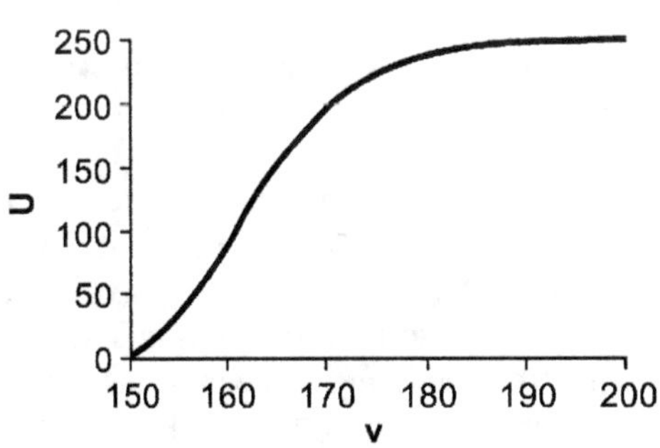

Figure 6. Utility of Work1.

This, therefore, explains an almost fixed amount of duration for an activity with a larger β. (2) An activity with a larger β had a much higher utility level than an activity with a smaller β over a wide range of duration, and the final schedule including an activity with a larger β, in general provides a higher level of total utility, keeping everything else equal. (3) If the duration of an activity implied by the maximum utility is short compared to other activities, the size of β does not show a clear impact on the ratio of the out-of-home duration to the in-home duration in the final schedule.

Finally, γ showed the following impacts. (1) Because, with everything else equal, an activity with a larger γ faces an earlier and sharper decrease of utility when the duration is reduced that of the maximum utility, an activity with a larger γ tends to have a longer duration to avoid this effect. (2) The final schedule including an activity with a smaller γ likely to have a higher level of total utility because a smaller γ has a higher level of utility than a bigger γ over the entire range of duration. This result in turn implies that, everything else equal, an activity with a smaller γ requires a shorter duration to achieve the same level of utility, which increases the possibility of longer duration of other activities in the schedule. (3) Interestingly, the difference in the ratio of the out-of-home duration between larger and smaller γs of an activity shows an interrelationship with the size of the U^{max} parameter. When U^{max} is large, outside leisure activities are included across the final schedules. A larger γ of dinner activity that results in a longer duration of the activity then implies that the schedule is not able to further include inside leisure activities. A smaller γ that results in a shorter duration of dinner activity on the other hand induces the inclusion of inside leisure activities in the schedule. Leisure activities have bigger units of duration than dinner activity, and consequently, a larger γ results in a bigger ratio of out-of-home duration compared to a larger γ, given a larger U^{max}. When U^{max} is small, outside leisure activities are mostly excluded, and dinner and inside leisure activities constitute the schedule after work. Given this situation, a larger γ of dinner activity that results in a longer duration of the activity directly induces a smaller ratio of out-of-home duration.

4. CONCLUSION

This article has described a model of activity scheduling and rescheduling behaviour under time pressure and examined the behaviour of the model in terms of a set of selected parameters. The model consists of two main components: a utility function and a set of decision heuristics. The assumed utility function of activity duration is S-shaped and able to represent a variety of changes in marginal utility over time. Decision heuristics involve a set of operators including duration adjustment, insertion, deletion and substitution. The heuristics mimic an individual's (re)scheduling behaviour that is sometimes continuous (duration operator) but most of the time highly discontinuous due to the non-incremental adjustments due to insertion, deletion and substitution operators and the constraints that introduces discontinuous choice sets to the model.

The simulation consisted of two subsequent parts, i.e. optimisation of an initial schedule optimisation and rescheduling under time pressure. Particular values of selected parameters were varied to study their behaviour. As theorised, we found that activities with higher maximum utilities and higher inflection points are more likely to provide a higher level of total utility in the final schedule. Secondly, with everything else equal, an activity's utility function with a steeper slope results in almost the same duration across simulation settings while that with a flatter slope may result in various lengths of duration depending on other activities' duration. Consequently, an activity with a flatter slope is more likely to have a lower level of total utility in the final schedule when the resulting duration of that activity is long. Finally, the imposed time pressure situation shows that, besides the impact of the parameter settings, the state of the previously optimised initial schedule is of crucial importance to the overall process of rescheduling. If the currently given optimised schedule was much in conflict with the imposed situation of time pressure, the hypothetical individuals are forced to conduct a substantial adjustment to the schedule. Overall then, it has been shown that different parameter settings characterising different individuals resulted in distinguishable, envisioned, state dependent activity scheduling and rescheduling behaviour. This implies that once the necessary parameters are obtained via some estimation procedure, the suggested model of short-term dynamics of scheduling and rescheduling behaviour can be applied to the understanding, analysis and forecasting of the activity-travel behaviour in the real world.

ACKNOWLEDGEMENTS

The research, reported in this chapter, is conducted as part of the Amadeus research programme sponsored by the Dutch Research Foundation (NWO). The details of the project are reported in Arentze *et al.* (2001).

Urban Planning Group,
Eindhoven University of Technology,
P.O. Box 513, 5600 MB Eindhoven, The Netherlands

REFERENCES

Arentze, T.A., F. Hofman, H. Van Mourik and H.J.P. Timmermans (2000), 'Albatross: A multi-agent rule-based model of activity pattern decisions.' *Transportation Research Record*, **1706**, 136–144.

Arentze, T., M. Dijst, E. Dugundji, C.H. Joh, L. Kapoen, S. Krijgsman, K. Maat, H. Timmermans and J. Veldhuisen, J. (2001), *The Amadeus Program: Scope and Conceptual Development.* Paper presented at the 9th World Conference on Transportation Research, Seoul, July.

Bhat, C.R. and F.S. Koppelman (1993), 'A conceptual framework of individual activity program generation.' *Transportation Research A*, **27**, 433–446.

Bhat, C.R. and Koppelman, F.S. (2000), *Activity-Based Travel Demand Analysis: History, Results and Future Directions.* Paper presented at the 79th Annual Meeting of the Transportation Research Board, Washington DC, January.

Bowman, J.L. and M.E. Ben-Akiva (2000), 'Activity-based disaggregate travel demand model system with activity schedules.' *Transportation Research A*, **35**, 1–28.

Burnett, P. and S. Hanson (1982), 'The analysis of travel as an example of complex human behaviour in spatially-constrained situations: Definition and measurement issues.' *Transportation Research A*, **16**, 87–102.

Damm D. and S.R. Lerman (1981), 'A theory of activity scheduling behaviour.' *Environment and Planning A*, **13**, 703–718.

Doherty, S.T. and E.J. Miller (2000), 'A computerised household activity scheduling behaviour.' *Transportation*, **27**, 75–97.

Ettema, D., A. Borgers and H. Timmermans (1993), 'A simulation model of activity scheduling behaviour.' *Transportation Research Record*, **1413**, 1–11.

and H.J.P. Timmermans (eds.), *Activity-Based Approaches to Travel Analysis*. Pergamon, Oxford, pp. 1–36.

Fujii, S., R. Kitamura and T. Monma (1998), *A Utility-Based Micro-Simulation Model System of Individuals' Activity-Travel Patterns*. Paper presented at the 77th Annual Meeting of the Transportation Research Board, Washington D.C., January.

Gärling, T., Kwan, M.P. and Golledge, R.G. (1994), 'Computational-process modelling of household activity scheduling.' *Transportation Research B*, **28**, 355–364.

Gärling, T., T. Kalen, T. Romanus and M. Selart (1998), 'Computer simulation of household activity scheduling.' *Environment and Planning A*, **30**, 665–679.

Gärling, T., R. Gillholm and W. Montgomery (1999), 'The role of anticipated time pressure in activity scheduling.' *Transportation*, **26**, 173–191.

Joh, C.H., T.A. Arentze and H.J.P. Timmermans (2001), *Towards a Theory and Simulation Model of Activity-Travel Rescheduling Behavior*. Paper presented at the 9th World Conference on Transportation Research, Seoul, July.

Joh, C.H., T.A. Arentze and H.J.P. Timmermans (2002), *Modelling Individuals' Activity-Travel Rescheduling Heuristics: Theory and Numerical Experiments*. Paper accepted for presentation at the 81st Annual Meeting of the Transportation Research Board, Washington DC, January.

Jones, P., M. Dix, M. Clarke and I. Heggie (1983), *Understanding Travel Behaviour*. Gower, Aldershot.

Kitamura, R., T. Yamamoto and S. Fujii (1996), 'A discrete-continuous analysis of time allocation to two types of discretionary activities which accounts for unobserved heterogeneity.' In J.B. Lesort (ed.), *Transportation and Traffic Theory*. Elsevier, Oxford, pp. 431–453.

McNally, M.G. (1997), 'An activity-based micro-simulation model for travel demand forecasting.' In D. Ettema and H.J.P. Timmermans (eds.), *Activity-based Approaches to Travel Analysis*. Pergamon, Oxford, pp. 37–54.

Pred, A. (1981), 'Of paths and projects: Individual behavior and its societal context.' In K.R. Cox and R.G. Golledge (eds.), *Behavioural Problems in Geography Revisited*. Methuen, London, pp. 231–256.

Recker, W.W., M.G. McNally and G.S. Root (1985), 'Travel/activity analysis: Pattern recognition, classification and interpretation.' *Transportation Research A*, **19**, 279–296.

Recker, W.W. (1995), 'The household activity pattern problem: General formulation and solution.' *Transportation Research B*, **29**, 61–77.

Supernak, J. (1988), 'A dynamic interplay of activities and travel: Analysis of time of day utility profiles.' In P. Jones (ed.), *Developments in Dynamic and Activity-Based Approaches to Travel Analysis*. Avebury, Aldershot, 99–122.

Thrift, N.J. (1983), 'On the determination of social action in space and time.' *Environment and Planning D*, **1**, 23–57.

Timmermans, H.J.P., T.A. Arentze and C.H. Joh (2001), 'Modelling the effects of anticipated time pressure on the execution of activity programs.' To appear in *Transportation Research Record*.

APPENDIX I
THE RESULTS OF THE SIMULATION OF INITIAL SCHEDULE OPTIMISATION

Person 1 (high U^{max}, large β, large γ) TU = 1455

Activity	Duration	Begin time	End time
Sleep	448	23:22	6:50
Pcare	73	6:50	8:03
Work1	207	8:03	11:30
Lunch	72	11:30	12:42
Dshop	48	12:42	13:30
Work2	296	13:30	18:26
Dinner	132	18:26	20:38
Leis_O	164	20:38	23:22

Person 2 (high U^{max}, large β, small γ) TU = 1491

Activity	Duration	Begin time	End time
Pcare	68	23:43	0:51
Sleep	432	0:51	8:03
Dshop	48	8:03	8:51
Work1	190	8:51	12:01
Lunch	71	12:01	13:12
Work2	279	13:12	17:51
Dinnr	90	17:51	19:21
Leis_O	138	19:21	21:39
Leis_I	124	21:39	23:43

Note: TU denotes total utility.

Note: The activities in grey are assumed out-of-home activities.

Person 3 (high U^{max}, small β, large γ) TU = 1338

Activity	Duration	Begin time	End time
Sleep	448	23:19	6:47
Pcare	73	6:47	8:00
Dshop	53	8:00	8:53
Work1	198	8:53	12:11
Lunch	79	12:11	13:30
Work2	296	13:30	18:26
Leis_O	162	18:26	21:08
Dinner	131	21:08	23:19

Person 4 (high U^{max}, small β, small γ) TU = 1383

Activity	Duration	Begin time	End time
Pcare	70	0:09	1:19
Sleep	437	1:19	8:36
Work1	195	8:36	11:51
Lunch	76	11:51	13:07
Work2	285	13:07	17:52
Dinnr	98	17:52	19:30
Leis_O	142	19:30	21:56
Leis_I	133	21:56	0:09

Person 5 (low U^{max}, large β, large γ) TU = 1305

Activity	Duration	Begin time	End time
Sleep	450	23:17	6:47
Pcare	74	6:47	8:01
Work1	209	8:01	11:30
Lunch	72	11:30	12:42
Dshop	48	12:42	13:30
Work2	298	13:30	18:28
Dinnr	135	18:28	20:43
Leis_I	154	20:43	23:17

Person 6 (low U^{max}, large β, small γ) TU = 1306

Activity	Duration	Begin time	End time
Sleep	452	23:11	6:43
Pcare	75	6:43	7:58
Work1	212	7:58	11:30
Lunch	72	11:30	12:42
Dshop	48	12:42	13:30
Work2	301	13:30	18:31
Dinnr	122	18:31	20:33
Leis_I	158	20:33	23:11

Person 7 (low U^{max}, small β, large γ) TU = 1188

Activity	Duration	Begin time	End time
Sleep	450	23:16	6:46
Pcare	74	6:46	8:00
Dshop	53	8:00	8:53
Work1	198	8:53	12:11
Lunch	79	12:11	13:30
Work2	298	13:30	18:28
Dinnr	134	18:28	20:42
Leis_I	154	20:42	23:16

Person 8 (low U^{max}, small β, small γ) TU = 1235

Activity	Duration	Begin time	End time
Sleep	439	23:49	7:08
Pcare	70	7:08	8:18
Work1	197	8:18	11:35
Lunch	78	11:35	12:53
Work2	286	12:53	17:39
Leis_O	135	17:39	19:54
Dinnr	100	19:54	21:34
Leis_I	135	21:34	23:49

APPENDIX II
THE RESULTS OF THE SIMULATION OF CONGESTION-INDUCED RESCHEDULING

TU = 1434 — Person 1 (high U^{max}, large β, large γ) — TU = 1443

Activity	Duration	Begin time	End time		Activity	Duration	Begin time	End time
Sleep	448	23:22	6:50		Sleep	448	23:22	6:50
Pcare	73	6:50	8:03		Pcare	73	6:50	8:03
Work1	177	8:33	11:30	→	Work1	185	8:33	11:38
Lunch	72	11:30	12:42		Lunch	65	11:38	12:43
Dshop	48	12:42	13:30		Dshop	47	12:43	13:30
Work2	296	13:30	18:26		Work2	296	13:30	18:26
Dinner	132	18:26	20:38		Dinner	132	18:26	20:38
Leis_O	164	20:38	23:22		Leis_O	164	20:38	23:22

TU = 1367 — Person 2 (high U^{max}, large β, small γ) — TU = 1473

Activity	Duration	Begin time	End time		Activity	Duration	Begin time	End time
Pcare	68	23:43	0:51		Pcare	68	23:43	0:51
Sleep	432	0:51	8:03		Sleep	432	0:51	8:03
Dshop	18	8:33	8:51	→	Work1	189	8:33	11:42
Work1	190	8:51	12:01		Lunch	70	11:42	12:52
Lunch	71	12:01	13:12		Work2	278	12:52	17:30
Work2	279	13:12	17:51		Dshop	47	17:30	18:17
Dinnr	90	17:51	19:21		Dinnr	83	18:17	19:40
Leis_O	138	19:21	21:39		Leis_O	131	19:40	21:51
Leis_I	124	21:39	23:43		Leis_I	116	21:51	23:47

TU = 1334 — Person 3 (high U^{max}, small β, large γ) — TU = 1335

Activity	Duration	Begin time	End time		Activity	Duration	Begin time	End time
Sleep	448	23:19	6:47		Sleep	448	23:19	6:47
Pcare	73	6:47	8:00		Pcare	73	6:47	8:00
Dshop	23	8:30	8:53	→	Dshop	11	8:30	8:41
Work1	198	8:53	12:11		Work1	204	8:41	12:05
Lunch	79	12:11	13:30		Lunch	85	12:05	13:30
Work2	296	13:30	18:26		Work2	296	13:30	18:26
Leis_O	162	18:26	21:08		Leis_O	162	18:26	21:08
Dinner	131	21:08	23:19		Dinner	131	21:08	23:19

TU = 1284 — Person 4 (high U^{max}, small β, small γ) — TU = 1374

Activity	Duration	Begin time	End time		Activity	Duration	Begin time	End time
Pcare	70	0:09	1:19		Pcare	70	0:09	1:19
Sleep	437	1:19	8:36		Sleep	437	1:19	8:36
Work1	165	9:06	11:51	→	Work1	192	9:06	12:18
Lunch	76	11:51	13:07		Lunch	72	12:18	13:30
Work2	285	13:07	17:52		Work2	281	13:30	18:11
Dinnr	98	17:52	19:30		Dinnr	93	18:11	19:44
Leis_O	146	19:30	21:56		Leis_O	141	19:44	22:05
Leis_I	133	21:56	0:09		Leis_I	127	22:05	0:12

TU = 1290 — Person 5 (low U^{max}, large β, large γ) — TU = 1296

Activity	Duration	Begin time	End time		Activity	Duration	Begin time	End time
Sleep	450	23:17	6:47		Sleep	450	23:17	6:47
Pcare	74	6:47	8:01		Pcare	74	6:47	8:01
Work1	179	8:31	11:30	→	Work1	186	8:31	11:37
Lunch	72	11:30	12:42		Lunch	66	11:37	12:43
Dshop	48	12:42	13:30		Dshop	47	12:43	13:30
Work2	298	13:30	18:28		Work2	298	13:30	18:28
Dinnr	135	18:28	20:43		Dinnr	135	18:28	20:43
Leis_I	154	20:43	23:17		Leis_I	154	20:43	23:17

TU = 1296 — Person 6 (low U^{max}, large β, small γ) — TU = 1298

Activity	Duration	Begin time	End time		Activity	Duration	Begin time	End time
Sleep	452	23:11	6:43		Sleep	452	23:11	6:43
Pcare	75	6:43	7:58		Pcare	75	6:43	7:58
Work1	182	8:28	11:30	→	Work1	187	8:28	11:35
Lunch	72	11:30	12:42		Lunch	68	11:35	12:43
Dshop	48	12:42	13:30		Dshop	47	12:43	13:30
Work2	301	13:30	18:31		Work2	301	13:30	18:31
Dinnr	122	18:31	20:33		Dinnr	122	18:31	20:33
Leis_I	158	20:33	23:11		Leis_I	158	20:33	23:11

TU = 1185 — Person 7 (low U^{max}, small β, large γ) — TU = 1186

Activity	Duration	Begin time	End time		Activity	Duration	Begin time	End time
Sleep	450	23:16	6:46		Sleep	450	23:16	6:46
Pcare	74	6:46	8:00		Pcare	74	6:46	8:00
Dshop	23	8:30	8:53	→	Dshop	11	8:30	8:41
Work1	198	8:53	12:11		Work1	204	8:41	12:05
Lunch	79	12:11	13:30		Lunch	85	12:05	13:30
Work2	298	13:30	18:28		Work2	298	13:30	18:28
Dinnr	134	18:28	20:42		Dinnr	134	18:28	20:42
Leis_I	154	20:42	23:16		Leis_I	154	20:42	23:16

TU = 1156 — Person 8 (low U^{max}, small β, small γ) — TU = 1228

Activity	Duration	Begin time	End time		Activity	Duration	Begin time	End time
Sleep	439	23:49	7:08		Sleep	439	23:49	7:08
Pcare	70	7:08	8:18		Pcare	70	7:08	8:18
Work1	167	8:48	11:35	→	Work1	193	8:48	12:01
Lunch	78	11:35	12:53		Lunch	74	12:01	13:15
Work2	286	12:53	17:39		Work2	283	13:15	17:58
Leis_O	135	17:39	19:54		Leis_O	130	17:58	20:08
Dinnr	100	19:54	21:34		Dinnr	95	20:08	21:43
Leis_I	135	21:34	23:49		Leis_I	130	21:43	23:53

Isao Mizuno

Geographical Model of a Self-Organising Megalopolis with Time-Space Convergence

1. Introduction

In industrialised countries, urban population and industrial production have been concentrated in some metropolitan areas, which have created a megalopolis from the expansion of their suburbs, where socio-economic functions of large cities are strongly linked with each other by high-speed transportation networks. For example, the United States megalopolis is well-known as a belt of large cities from Boston through New York to Washington D.C. on the Atlantic coast, and the Japanese megalopolis connects three metropolitan areas, Tokyo, Nagoya, and Osaka on the Pacific coast. Newly industrialised Asian countries or regions, such as South Korea, Taiwan, and Malaysia, are also developing megalopolis embryos or development belts. It can be generally stated that, as a national economy develops, one specific region becomes a megalopolis at the expense of the other regions. This phenomenon raises two questions. What economic processes differentiate the megalopolis from under-developed areas? Which geographical areas can selectively become a megalopolis? Two types of studies, the "more economic" and "more geographic" economic geographies (Lukermann, 1958), have tackled these questions in the location-theory literature.

Urban and regional economists have constructed "more economic" models of urban and industrial agglomerations by analysing market structures, transportation costs, and general equilibrium. Masahisa Fujita and his collaborators have introduced an endogenous urban agglomeration into a traditional urban economics framework (Fujita and Ogawa, 1982; Fujita, 1989). Fujita notably considered the relationship between decreasing transportation cost and the multiple equilibriums of urban agglomeration. Furthermore, Paul Krugman has recently vitalised the traditional study field of location theory from an international economic perspective in his "new economic geography" (Krugman, 1991, 1996). Krugman criticised classical location theorists for absorbing themselves in geometric solutions of industrial location without treating the market structure. He reconsidered the concept of Alfred Marshall's external economy in modelling industrial agglomeration. Krugman's proposal of a new economic geography has recently drawn much attention from neo-classical economists to the agglomeration economies of industry (Fujita and Krugman, 1995; Mori, 1997; Fujita *et al.*, 1999; Plummer, 1999; Ricci, 1999; Junis, 2000). These urban and regional economics studies enable us to integrate "space" into microeco-

B. Boots et al. (eds.), Modelling Geographical Systems, 97–117.

nomic theories of competition more analytically than the German traditional location theorists; however, it seems that these studies have little geographical perspective that is directed to uneven development, process-dependent space, and locational specificity.

On the other hand, many "more geographic" self-organisation models of urban and industrial agglomerations have appeared in urban geography and evolutionary economics. Peter Allen and Michelle Sanglier, scholars from the Brussels School, have developed the pioneering studies of the self-organising central place model (Allen and Sanglier, 1979, 1981). They simulated possible uneven development patterns of central place systems, considering an interaction between distributions of central place functions and urban population. Their studies have inspired the creation of other types of geographical self-organisation modelling (Wong and Fotheringham, 1990; White and Engelen, 1993; Bura *et al.*, 1996; Mizuno, 1998; Portugali, 2000). In addition, Allan Pred's (1966) imaginative essay precisely conceptualised the feedback mechanism of urban-industrial growth, using the concept of circular and cumulative causation (Myrdal, 1954). Furthermore, in evolutionary economics, W. Brian Arthur has presented some papers of geographical self-organisation models in the 1980's (later collected in Arthur (1994)). His papers contain many marvellous concepts, emphasising the roles of history such as "lock-in" and "path-dependence". It can be stated that all these studies have paid much attention to locational "specificity" in their modelling. As Barnes (1996) advocated, an historical geographical context matters in economic geography. An historical geographical context means not only contingent facts given in one spatial economic process, but also possible factors bifurcating the other spatial economic process.

This paper attempts to combine both economic and geographical modelling of urban-industrial agglomeration, particularly rethinking Krugman's central place model. It proposes a more geographic model that can describe the interaction between the megalopolis and "time-space convergence", based on self-organisation frameworks of "dissipative structure" (Nicolis and Prigogine, 1977) and "synergetics" (Haken, 1978). The concept of time-space convergence (Janelle, 1969) is the uneven transportation development that causes time-distance to shrink among specific places. Janelle explicates a long-term, non-linear feedback process between urban-industrial agglomeration and time-space convergence. He emphasises that one technological transport innovation continuously stimulates more urban-industrial agglomeration, and, in turn, urban-industrial agglomeration discontinuously prompts the technological innovation of transportation networks. Furthermore, many self-organisation frameworks suppose that a feedback process begins with a specific historical, geographical context in an unstable state. We can describe an uneven geographical development of the megalopolis by using self-organisation frameworks that endogenously connect the necessary processes with contingent events.

This paper is organised as follows. Section 2 presents a critical review of Krugman's central place model from a geographical perspective. Section 3 describes a short-term model, taking into account the interaction between the two distributions of urban population and the number of firms without transportation development. In Section 4, a long-term model interacting transportation development with urban-industrial agglomeration is constructed. Section 5 compares some simulation results obtained from these models with empirical details of the Japanese megalopolis and railway development since the late 19th century. Section 6 summarises the results of our model and discusses possible directions in future modelling research.

2. KRUGMAN'S CENTRAL PLACE MODEL AND A GEOGRAPHICAL PERSPECTIVE

Krugman has presented some "new economic geography" models: the core-periphery, industrial location, metropolitan, and central place models (Krugman, 1991b, 1993a, 1993b, 1996). These models treat different geographical phenomenon from each other, but share some assumptions and mechanisms for economic processes, that is, monopolistic competition with diversified production.

His basic framework assumes two kinds of production and consumption: agriculture, which is tied to the land, and manufacturing, which is mobilised to any region. On the consumption side, people's utility is formalised by a Cobb-Douglas function for the consumption of agricultural and manufactured goods, with the expenditure share parameter on manufactured goods, π. The manufacture aggregate is described by a CES (Constant-Elasticity-of-Substitution) function of a large number of manufactured goods, with the elasticity parameter of demand, σ. On the production side, it is assumed that agricultural goods are subject to constant returns, while manufactured goods are produced under increasing returns with a linear function of labour cost. Krugman considers trade costs, which are calculated by a negative exponential function of distance, with the distance-decay parameter, τ. Considering these two sides creates a general equilibrium framework.

Here, Krugman's central place model is focused on as typical (Krugman, 1996, Chapter 10), which is closely related to my model. Krugman's (1996) central place model is reduced to a simplified one concerning economic process so that I can examine the geographical implications. He assumes some given distribution of firms in a bounded, but seamless space, that is continuously laid out in a circle with a unit radius. In his simplified model, he formalises three equations composed of four basic variables: the share of the number of firms, $\lambda(x; t)$, at location x and time t, the market potential force, $P(x; t)$, the average market potential force, $P^*(t)$, and the distance between locations x and z, $D(x, z)$.

First, the firms' preference in locations is defined as the continuous market potential function, based on the Harris (1954) concept:

$$P(x; t) = \int_z [Ae^{-\gamma_1 D(x, z)} - Be^{-\gamma_2 D(x, z)}]\lambda(z; t)dz. \tag{1}$$

Parameters A and γ_1 indicate the size of the centrifugal forces and the distance-decay effect, respectively. Similarly, parameters B and γ_2 are the size of the centripetal forces and the distance-decay effect ($\gamma_1 > \gamma_2$).

Second, the average market potential force is defined as follows:

$$P^*(t) = \int_x [P(x;\ t)\lambda(x;\ t)dx. \tag{2}$$

Third, it is assumed that firms tend to gradually move toward locations with above-average market potential force and away from those with a below-average one, with the dynamics taking the form.

$$\frac{\partial\lambda(x;\ t)}{\partial t} = \gamma[P(x;\ t) - P^*(t)]\lambda(x;\ t). \tag{3}$$

Here, the parameter γ is an adjustment rate of firms. Equation (3) is a non-linear differential equation for the share of the number of firms, describing a self-reinforcing process of place advantage or disadvantage. What is more important is that Equation (3) assures the constancy of the total number of firms in an entire area.

According to Krugman's (1996) explanation, if the centrifugal force parameter A is far larger than the centripetal force parameter B, then all firms will concentrate in a single location. He also predicts that firms will come to agglomerate at multiple locations, if B is far larger than A. These results are similar with those that Clarke and Wilson (1983) previously simulated in their catastrophe model of retail spatial interaction.

It can be stated from a geographical perspective that Krugman's "new economic geography", specifically his central place model, has very strong and interrelated assumptions: global rationality, isotropic space, deterministic modelling, and continuous quantitative change. Global rationality appears to be more general than that of local rationality, but, in fact, the former is more limited and unrealistic than the latter. Global rationality, or unbounded rationality, assumes that firms and households can maximise profit and utility with perfect knowledge in a whole space (Figure 1(a)). On the contrary, local rationality, or bounded rationality, presumes that human actors consider only a limited number of alternatives in a partial space, an idea originated from Herbert Simon's (1957) "satisfiers" who are in contrast to optimisers (Figure 1(b)). Allen and Sanglier (1979, p. 256) clearly rejected the idea of "global optimisers," in their self-organisation modelling. D.M. Smith (1981) re-evaluated the "variable cost analysis" in traditional location theories, as a sub-optimal approach to industrial location concerned with spatial margins to profitability that relaxes the idea of transportation cost minimisation. Global rationality assumes a monolith.

Isotropic space is too abstract to make a relevant model of geographical phenomena. Isotropic space is an unbounded, homogeneous space, called a Cartesian coordinate system, which is *a priori* and independent of any socio-economic processes. Therefore, such a space has no concept of place and uneven

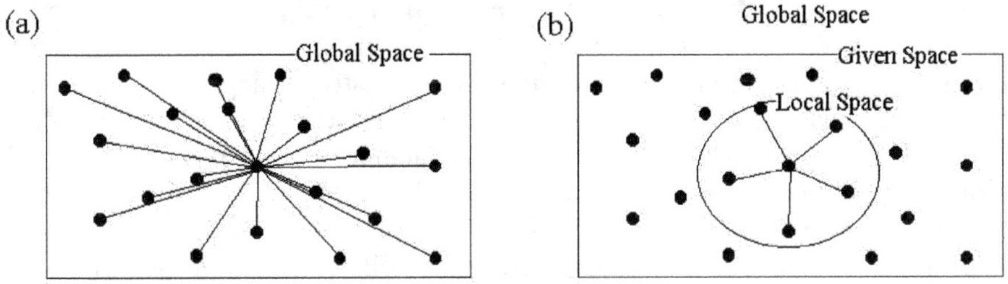

Figure 1. Relationship between global space and local space (Points and lines indicate central places and direct functional relationships between one central place and the others. (a) Global space or given space without local space, (b) Local space as a part of given space and global space as an unknown outer space).

development, which geographers pay much attention to. Krugman's new economic geography models assume an unbounded space or seamless space: two-region, unbounded one-dimensional space, and bounded but seamless space on a circle. All these spaces are chosen to avoid the complicated boundary conditions that make the initial difference of space, although Krugman treats a slightly varied, initial distribution. In contrast, geographical space has not only been given to socio-economic processes, but has also been produced and reproduced by them; thus it is a process-dependent space. Therefore, we need to internalise some spatial elements like nodes, networks, hierarchies, and surfaces in our model into a bounded area (Haggett *et al.*, 1977). In particular, the distance-decay parameter value is varied from region to region (Fotheringham, 1981), and it is an historical outcome of local interaction between urban-industrial agglomeration and transportation innovation (Janelle, 1969).

Deterministic modelling does not help us understand an historical path-dependence of spatial economic evolution. Krugman considered multiple equilibriums and varied initial distribution of industrial agglomerations. In his models, one of multiple equilibriums such as multi-polar agglomeration and one-sided concentration is determined by the initial distribution and the parameter values, exogenously. Thus, it can be said that he neglects the positive role of fluctuation played "inside" of the spatial economic process. Fluctuation means small events or slight random factors that are outside the knowledge of the observer in any time and place. According to Arthur (1994, Chapter 1), the concept of path-dependence is a combination of historical small events and necessary economic process that causes lock-in and an indeterminacy of outcome. Allen and Sanglier (1979, p. 258) previously stated, "This evolution retains the dual aspects of chance and determinism, since the form of a particular spatial instability will depend on the details of the local events that trigger its growth."

An assumption of continuous quantitative change may be useful for us to deduce a neat mathematical solution, but it differs from an assumption of areal differentiation that is characterised by a heterogeneity and boundedness. In his central place model, Krugman often solves the Differential Equation (3) mathematically. However, one non-linear "difference" equation can describe more diversified evolution than the "differential" equation (May, 1976). Urban-industrial agglomeration is a discontinuous, qualitative change. Some classical location theorists considered location problems in the setting of specific nodes and discontinuous surfaces (Weber, 1909; Palander, 1935; Hoover, 1948). A simulation of the non-linear difference model in a bounded space will give a clue to understanding areal differentiation and technological innovation.

In summary, we can see that geographical self-organisation modelling has four common features: local rationality, process-dependent space, historical path-dependence, and qualitative change. In the following section, we will construct a geographical model of urban-industrial agglomeration by rethinking Krugman's central place model in a bounded two-dimensional space.

3. Short-Term Dynamics of Distributions of Urban Population and Firms

The basic principles that determine short-term dynamics of urban-industrial agglomeration are the same as Krugman's central place model, in two points. One point is that firms tend to be located closely in a short distance, but would like to be apart from each other in a long distance. This is competitive and cooperative interaction among firms. In addition, the adjustment process holds that firms and urban population tend to move from a disadvantageous place to an advantaged one.

Our model, the "geographical megalopolis model", is different from Krugman's central place model in four respects. First, we assume a local space where urban population and firms decide whether they move or not. Second, we consider an interactive process between distributions of urban population and the number of firms. Third, we have a framework of historical path-dependence that is composed of non-linear feedback with fluctuation. Fourth, a dynamic central place system in a bounded, two-dimensional space is described.

Imagine a bounded two-dimensional lattice space of m columns and n rows. The distribution of urban population at central places does not correspond linearly to that of the number of firms. However, in order to simplify our discussion below, we assume even, initial distributions of the urban population and the number of firms in a certain number of central places, $N = 77$ (Figure 2): $m = 32$, $n = 20$, $P_i(0) = 100$ $(i = 1, 2, \ldots, N)$.

The urban population $P_i(t)$ at a place i and a time t consists of firm workers, $M_i(t)$, who are mobile in a local space, and a constant number of farmers, A, who neither move nor become firm workers.

$$P_i(t) = \mathrm{A} + M_i(t). \tag{4}$$

We also assume no natural population growth and no emigrants from other areas, which will be relaxed in the long-term dynamics later. The main variables of our model are the urban population, $P_i(t)$, and the number of firms, $E_i(t)$. We

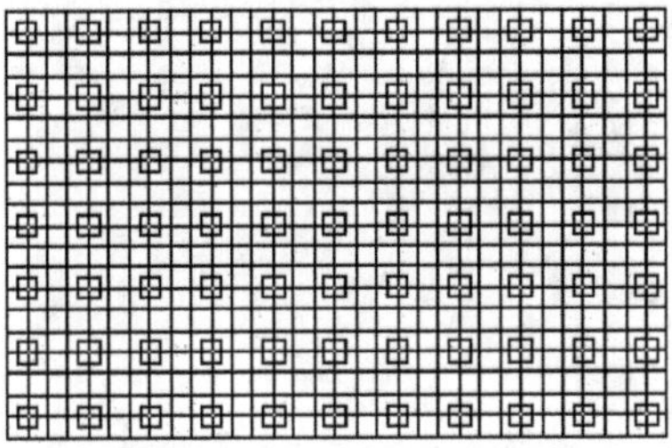

Figure 2. Initial distribution of urban population over the lattice space (small squares on the lattice indicate the location and the size of central places).

assume that potential functions are the combination of attractive forces and repulsive forces. This is almost the same idea as Krugman's Equation (2), but different from his idea concerning space. One urban population in city i affects the population potential forces on the other population in the city's own local space, C_i, though Harris's (1954) market potential was originally defined as a global one. Similarly, we define the number of firms in city i affecting the firms potential forces on the other firms in the city's own local space as D_i. These two potential forces can be formulated as the local functions from a central place with the urban population and the number of firms, $P_i(t)$ and $E_i(t)$, and distance-decay parameters for the attraction (α) and repulsion (β) ($\alpha > \beta > 0$). The local potential functions, $Q_i(t)$, for the urban population, and $R_i(t)$, for the number of firms, are calculated as the local summation of potential forces.

$$Q_i(t) = \sum_{j \in C_i} \gamma_1 e^{-\alpha d_{ij}} - \gamma_2 e^{-\beta d_{ij}})P_j(t), \tag{5}$$

$$R_i(t) = \sum_{j \in D_i} \gamma_3 e^{-\alpha d_{ij}} - \gamma_4 e^{-\beta d_{ij}})E_j(t), \tag{6}$$

where d_{ij} is the distance between central places i and j; γ_1 and γ_2 are size parameters for the attractive and repulsive forces of the urban population ($\gamma_1 > \gamma_2 > 0$); and γ_3 and γ_4 are parameters for the attractive and repulsive forces of the number of firms ($\gamma_3 > \gamma_4 > 0$). Some examples of potential functions with unit urban population are shown in Figure 3. The smaller the size parameter ratio for the attractive force of the urban population to that of the repulsive force, γ_1/γ_2, is, the stronger the repulsive force is, as shown by A and B in Figure 3. Furthermore, the smaller the distance-decay parameter ratio for the attractive force to that of the repulsive force, α/β, is, the stronger the attractive force is, as shown by A to C in Figure 3.

The boundary of the local potential space C_i for the urban population may be represented by a range of daily activities, d_{ij}^{*1}, obtained by minimising the distance-decay component in brackets on the right-hand-side of Equation (5) from city i. The local potential space D_i for firms can be defined by the distance,

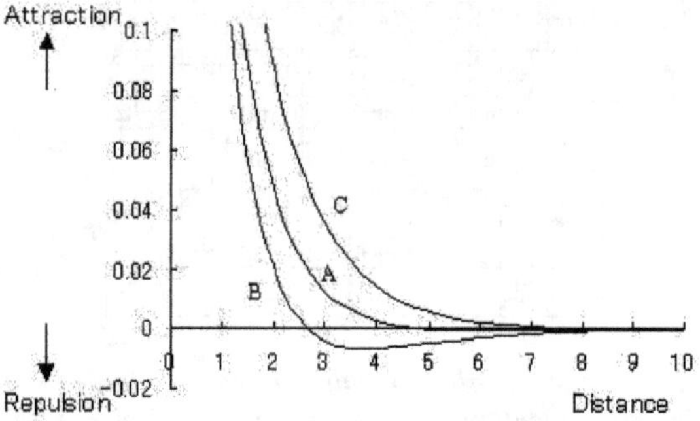

Figure 3. Potential functions of urban population or firms.
(A: $\gamma_1 = 0.8$, $\gamma_2 = 0.3$, $\alpha = 1$, $\beta = 0.8$; B: $\gamma_1 = 1.2$, $\gamma_2 = 0.7$, $\alpha = 1$, $\beta = 0.8$; C: $\gamma_1 = 0.8$, $\gamma_2 = 0.3$, $\alpha = 0.8$, $\beta = 0.7$)

d_{ij}^{*2}, obtained in the same way. These ranges of daily activities are written as:

$$d_{ij}^{*1} = \frac{1}{\alpha - \beta} \ln\left(\frac{\gamma_1\alpha}{\gamma_2\beta}\right), \tag{7}$$

$$d_{ij}^{*2} = \frac{1}{\alpha - \beta} \ln\left(\frac{\gamma_3\alpha}{\gamma_4\beta}\right), \tag{8}$$

The two critical distances depend on the distance-decay parameters and the size parameters for the urban population and the number of firms, so that a transportation innovation will make these distances larger.

In addition, the local average potentials, $Q_i^*(t)$, for the urban population and $R_i^*(t)$, for the number of firms, are calculated as weighted averages in their corresponding local spaces as:

$$Q_i^*(t) = \frac{\sum_{j \in C_i} Q_j(t)P_j(t)}{\sum_{j \in C_i} P_j(t)}. \tag{9}$$

$$R_i^*(t) = \frac{\sum_{j \in D_i} R_j(t)E_j(t)}{\sum_{j \in D_i} E_j(t)}. \tag{10}$$

Each local average potential is a criterion by which manufacturing workers or firms in place i at time t are likely to move to more preferable places or to come from less preferable places. If $Q_i(t) > Q_i^*(t)$ for the urban population, then firms look upon place i as the preferable location where firms can supply manufactured goods to more urban populations with cheaper transportation costs. On the other hand, if $R_i(t) > R_i^*(t)$ for the number of firms, then firm workers look upon place i as the preferable residential place where firm workers can gain a higher wage and cheaper manufactured goods. Farmers in place i supply agricultural goods to place i's population, and demand manufactured goods that are delivered by the firms located in their own local space D_i. We note that locational changes of workers and firms depend on their local criteria, as shown by Equations (9) and (10).

Based on Equations (4) to (10), the short-term dynamics of workers and firms are described by two non-linear difference equations with fluctuation given by

$$M_i(t+1) = M_i(t) + \lambda_1 \left\{ \frac{R_i(t)}{R_i^*(t)} - 1 \right\} M_i(t) + \delta_i M_i(t), \tag{11}$$

$$E_i(t+1) = E_i(t) + \lambda_2 \left\{ \frac{Q_i(t)}{Q_i^*(t)} - 1 \right\} E_i(t) + \delta_i E_i(t), \tag{12}$$

Here, the parameters λ_1 and λ_2 are the moving rates for workers and firms respectively. The place-specific fluctuation, δ_i, is a small percentage of random numbers $(-\delta\ \delta_I\ \delta)$ that introduces exogenous factors or uncertainty into the model, which are details of geographical endowments and historical events. Equations (11) and (12) mean that the difference between the firms' potential force and its average value in local space C_i of place i brings change in workers; in turn, the difference between the urban population potential force and its average value in local space D_i brings change in firms. The above equations imply that the total number of workers and firms are kept constant except for slight fluctuations. Our model presumes local space and endogenous fluctuation in the Difference Equations (11) and (12), whereas Krugman's central place model contains global space and a varied initial distribution in the Differential Equation (3).

The short-term model characterised by Equations (4)–(12) almost corresponds to Krugman's central place model if the assumption of space is a global one and any fluctuation is zero. This model is simulated for four cases and the results are shown time 150 in Figure 4 (a)–(d). In case (a), we assume a global rationality with even initial distribution and no fluctuation, that is a baseline for the other cases (b)–(d). The parameter values used in the simulation are as follows:

$A = 20,\ M_i(0) = 80,\ E_i(0) = 10,$
$\alpha = 1,\ \beta = 0.8,\ \gamma_1 = 0.8,\ \gamma_2 = 0.3,\ \gamma_3 = 0.8,\ \gamma_4 = 0.3,\ \lambda_1 = 0.03,\ \lambda_2 = 0.03,\ \delta = 0,$
$d_{ij}^{*1} = 999,\ d_{ij}^{*2} = 999.$

Figure 4 (a) shows that urban populations are distributed as a "planetary nebula" located at a concentric circle, although the pattern is transformed by the form and size of the rectangular area. Continuing the simulation of case (a) further, we can reach the result that urban populations are concentrated on the

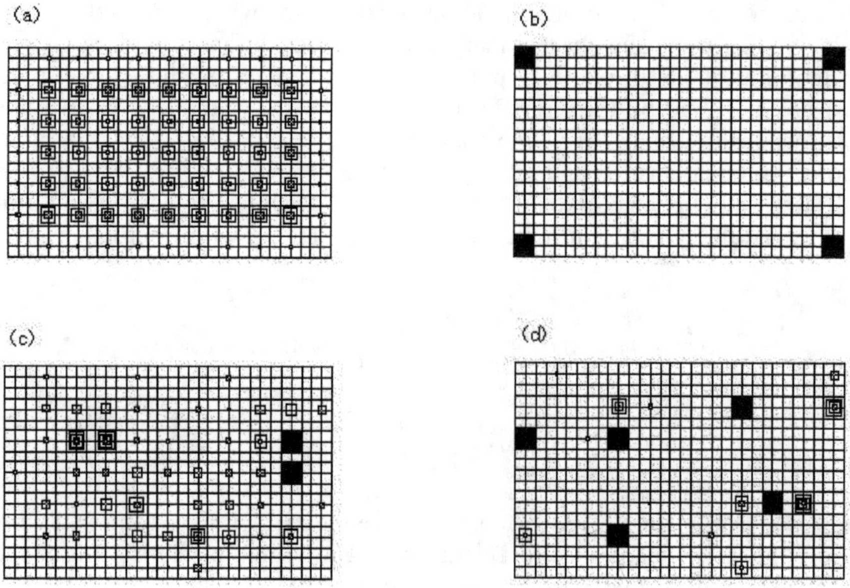

Figure 4. Several simulated distributions of short-term dynamics of urban population, t = 150 (Each iso-population line is described by 50 units of population. The shading points indicate a large number of population. (a) The case of global rationality with even initial distribution and no fluctuation as a basic case; (b) The case of strong repulsive forces for population and firms; (c) The case of global rationality with slightly varied, initial distribution and no fluctuation; (d) The case of local nationality with slightly varied, initial distribution and fluctuation).

four places close to the corners of the hypothetical rectangular area. Because of the assumptions of global rationality, even initial distribution, and no fluctuation, the distribution of urban populations symmetrically evolves around the geometrical centre of the hypothetical area. Furthermore, we simulate the model for the assumption of a stronger repulsive force than case (a) ($\gamma_2 = 0.5$, $\gamma_4 = 0.5$); all other parameter values and assumptions are the same as in case (a). According to Figure 4 (b), four agglomerated centres are distributed at the corners of the hypothetical area. On the other hand, the assumption of the stronger attractive force will lead to an evolution towards a "single-polar" urban population pattern (the result is omitted here). The results of cases (a) and (b) are largely affected by the form and the size of the bounded, hypothetical area owing to the assumption of global rationality where each range of daily activity d_{ij}^{*1} and d_{ij}^{*2} is regarded as a larger constant number, 999, than the longest distance in the bounded area.

Other cases, (c) and (d), are more complicated than the above two cases. The slightly varied initial distributions of urban population and the number of firms in case (c) differ from case (a). Case (c) is similar to Krugman's simplified central place model, except for the assumption of space and the market structure. The simulated pattern shown in Figure 4 (c) is completely different from the patterns in Figures 4 (a) and (b). Figure 4 (c) suggests that some slightly agglomerated locations that happened in the initial stage uniquely determine the later patterns of urban-industrial evolution. When we continue case (c) simulation after time 150, it is observed that the urban populations are concentrated in a single location. On the other hand, in case (d) the whole model in the short-term, assumes local rationality with a slightly varied initial distribution, and fluctuation ($\delta = 0.03$). Despite the same parameter values as cases (a) and (c), the result of case (d) describes a "multi-polar" urban system that resembles the hexagonal model of Christaller (1933) and Lösch (1940). This outcome means that some fluctuations and local rationality help to develop a multi-polar urban system. This is because some fluctuations can rapidly accumulate and amplify the polarisation process of central places in the local spaces concerned. The result of case (d) is not so much affected by the form and the size of a given area. Here, developments of some places can be seen along the boundary of the rectangular area.

In summary, we have introduced local rationality, fluctuation, and bounded space into the short-term model of the central place model from a geographical perspective. Our short-term model includes all the cases of (a)–(d), so it can produce multi-polar central place systems as well as single-polar ones. However, our model does not produce formation of the megalopolis, such as a belt of urban-industrial agglomeration. This is because we do not treat any technological innovation and process-dependent space. Thus, a long-term model with time-space convergence is constructed in the next section.

4. LONG-TERM DYNAMICS OF INTERACTION BETWEEN THE MEGALOPOLIS AND TIME-SPACE CONVERGENCE

Some recent location theorists have considered the relationship between the decline of transportation costs and the location of human activities (Ewers, 1995; Peeters *et al.*, 1997; Duranton, 1999). They share the common idea that urban-industrial agglomeration is caused by the decline of transportation costs. However, it is assumed that the transportation costs are global parameters as well as exogenous to the spatial economic process. Furthermore, Mori (1997) presented a remarkable modelling of the megalopolis by using a general equilibrium approach, but he assumed the transportation costs were given globally and exogenously to focus on scale economies of diversified goods. It is noted that urban-industrial agglomeration is caused by a continuous decline of transportation costs and, in turn, urban-industrial agglomeration causes a discontinuous decline of transportation costs among specific cities.

Let us formulate a long-term dynamic model called the "geographical megalopolis model", by introducing a time-space convergence process into the short-term one described in Section 3. According to Janelle's (1969) concept, time-space convergence implies a circular and cumulative process of shrinking time-space in a long-term model. Stated precisely, the time-space convergence means a series of processes in which, first, more demand for accessibility among specific cities appears in an historical, geographical context; second, this prompts the innovation of a faster transportation network among specific cities; third, the shrinking time-distance among specific cities prompts urban population and firms become more concentrated and specialized; fourth, these concentrated and specialised cities produce a new demand for accessibility among specific cities; and, finally, this process returns to the first stage. It is noted that this process is an endogenous, uneven development of transportation networks and urban systems.

A local space is expanded by a technological innovation of the transportation network, so the activity area of the inter-urban system gets broader. However, the cities in which this transportation network innovation takes place depends on an historical, geographical context. The historical, geographical context may be represented by uneven parameter values and place-specific fluctuation. The uneven parameter values are not given *a priori*, but are produced and reproduced by an internal process of urban population and firms. In general, the values of variables change, while those of parameters remain the same in the short-term. We also assume the values of distance-parameters, α and β, discontinuously change among specific places in the long-term. Thus, we specify the locality and level of distance-decay parameters to take account of the time-space convergence. Local potential functions and ranges of daily activities, shown by Equations (5)–(8), are reformulated to the following Equations (13)–(16):

$$Q_i(t) = \sum_{j \in C_i} \gamma_1 e^{-\alpha_{ij}^{(k)} d_{ij}} - \gamma_2 e^{-\beta_{ij}^{(k)} d_{ij}}) P_j(t), \tag{13}$$

$$R_i(t) = \sum_{j \in D_i} \gamma_3 e^{-\alpha_{ij}^{(k)} d_{ij}} - \gamma_4 e^{-\beta_{ij}^{(k)} d_{ij}}) E_j(t), \tag{14}$$

$$d_{ij}^{*1} = \frac{1}{\alpha_{ij}^{(k)} - \beta_{ij}^{(k)}} \ln \left(\frac{\gamma_1 \alpha_{ij}^{(k)}}{\gamma_2 \beta_{ij}^{(k)}} \right), \tag{15}$$

$$d_{ij}^{*2} = \frac{1}{\alpha_{ij}^{(k)} - \beta_{ij}^{(k)}} \ln \left(\frac{\gamma_3 \alpha_{ij}^{(k)}}{\gamma_4 \beta_{ij}^{(k)}} \right), \tag{16}$$

where the place-specific and technology-specific parameters $\alpha_{ij}^{(k)}$ and $\beta_{ij}^{(k)}$ for the attractive and repulsive forces are distance-decay ones among places i and j in the transportation technology level k (horse, water transportation, train and airplane).

When and where does any technological innovation of transportation networks occur? Janelle proposes the long-term model in which the transportation networks can be improved between the places and at the time when the demand for accessibility exceeds a certain threshold. Accessibility demand is revealed as the transportation flow between specific cities i and j at time t, $T_{ij}(t)$. This may be formulated as a gravity type model as in Equation (10) given by

$$T_{ij}(t) = \tau Q_i(t) Q_j(t) e^{-\alpha_{ij}^{(k)} d_{ij}} \quad . (17)$$

Here, we use local population potential instead of population size because people who reside in a metropolitan area use common transportation networks.

When the level of transportation technology improves, it is necessary to satisfy the following two conditions. First, the transportation flow, $T_{ij}(t)$, between cities i and j must exceed the threshold value of accessibility demand, $\theta^{(k)}$, in level k transportation technology. Second, the distance between cities i and j must also exceed a module distance, $\mu^{(k)}$, in level k. Here, the module distance is the minimum distance between stations that is kept to maintain a speed at each level k (k = 1, 2, and 3). Both improving and worsening conditions of transportation are described as follows:

$$\text{if} \begin{cases} T_{ij}(t) \geq \theta^{(k+1)} \\ d_{ij} \geq \mu^{(k+1)} \end{cases}, \quad \text{then the transportation is improved from level } k \text{ to } (k+1) \quad (18)$$

$$\text{if} \begin{cases} T_{ij}(t) \geq (1-\delta)\theta^{(k)} \\ d_{ij} \geq \mu^{(k)} \end{cases}, \quad \text{then the transportation is worsens from level } k \text{ to } (k-1) \quad (19)$$

The transportation networks can be developed when transportation flows and the module distance exceed each threshold value. On the other hand, the transportation technology between cities i and j declines to the lower level if these two conditions are not satisfied. Thus, distance-decay parameter values become geographically different among places through the internal spatial economic process as demonstrated by Fotheringham (1981).

In long-term dynamics, we also assume that the total population and the total number of firms increase along with a time-space convergence process. This is because national productivity increases so that agglomeration economy and spatial division of labour are strengthened by the transportation networks innovation. Equations (11) and (12) in the short-term model can be slightly modified by introducing increasing rates of total population and firms, κ_1 and κ_2, respectively, that is,

$$M_i(t+1) = M_i(t)(1+\kappa) + \lambda_1 \left\{ \frac{R_i(t)}{R_i^*(t)} - 1 \right\} M_i(t) + \delta_i M_i(t), \quad (20)$$

$$E_i(t+1) = E_i(t)(1+\kappa) + \lambda_2 \left\{ \frac{Q_i(t)}{Q_i^*(t)} - 1 \right\} E_i(t) + \delta_i E_i(t), \quad (21)$$

Equations (4), (9), (10) and (13)–(21) are all the equations of our "geographical megalopolis model." The change of distribution in urban-industrial agglomeration is simulated by a set of parameter values in our model. Only the results of urban population are shown in Figure 5 because both the distributions of the urban population and the number of firms evolve in almost the same pattern when the initial distributions are similar. The following parameter values for technological innovation of transportation networks are added, but other parameters are the same as those of Figure 4 (d):

$$\alpha_{ij}^{(1)} = 1,\ \alpha_{ij}^{(2)} = 0.8,\ \alpha_{ij}^{(3)} = 0.7,\ \beta_{ij}^{(1)} = 0.8,\ \beta_{ij}^{(2)} = 0.64,\ \beta_{ij}^{(3)} = 0.56,$$
$$\tau = 0.4,\ \kappa_1 = \kappa_2 = 0.003,\ \theta^{(2)} = 100,\ \theta^{(3)} = 10,\ \mu^{(2)} = 3,\ \mu^{(3)} = 6.$$

In a simulation of our geographical megalopolis model, we begin with the assumptions that the urban population and the number of firms are evenly distributed, and that all transportation networks are the most primitive (the level $k = 1$). Some cities develop regional city that can be one of the multi-polar centres. In addition, some transportation networks are innovated between the neighbouring growth cities. At time $t = 150$, as high-speed transportation networks have selectively been made, a megalopolis emerges in the middle of the hypothetical area (Figures 5 (c), (d)). The central districts stretch from the "southwest" to the "northeast" regions in the hypothetical area, whereas the peripheral districts lie in the "northwest" and the "southeast" regions. This implies that the process is a self-reinforced one where the intensified urban population becomes more agglomerated and the dispersed urban population becomes more scattered.

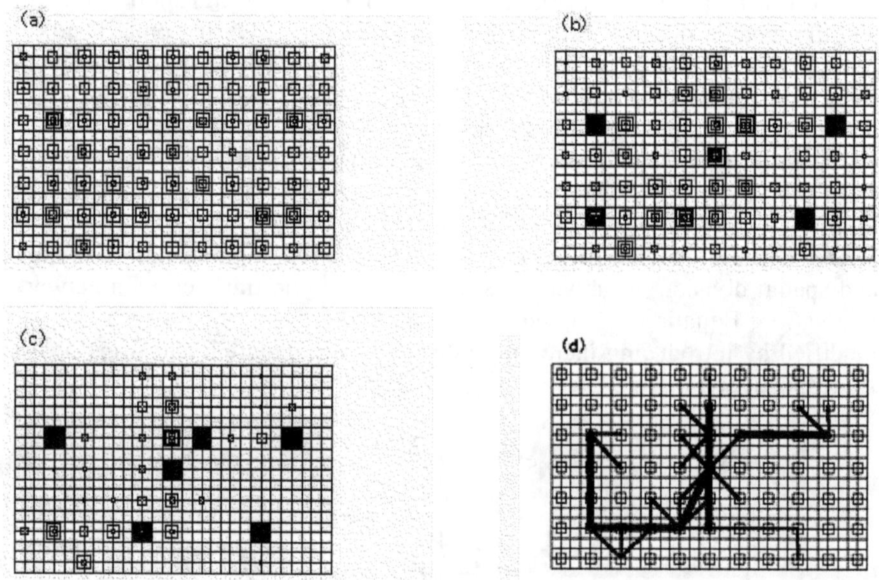

Figure 5. Long-term dynamics of urban population through time-space convergence ((a) t = 50; (b) t = 100; (c) t = 150; (d) the simulated transportation network at t = 150, where thicker lines indicate transportation level 3, and thick lines indicate level 2).

The results of Figures 5 (a)–(d) raise the following two questions. First, how is the formation of a megalopolis connected with the time-space convergence process? Second, does our model always generate the same urban systems? The first question evokes another simple one; why is the megalopolis stretching linearly towards the specific direction? Our central place model for the short-term suggests that a megalopolis cannot be formed unless the values of distance-decay parameters are changed. But it is noted again that any time-space convergence means uneven development of transportation networks. Figure 5 (c) shows a mixed pattern of multi-polar urban systems and the megalopolis. This means that the megalopolis may evolve from multi-polar urban systems. Multi-polar urban systems are mainly derived from the local rationality assumptions. When the transportation network innovation among a specific pair of cities is prompted by the excess demand in accessibility, these cities have a broader local space than ever, which enables the urban population and firms located there to gain a greater advantage compared to the other cities. The specific cities gaining more locational advantages will attract more urban population and firms for a short-term, and will save the demand in accessibility for the more improved transportation network in the long-term. These circular causality processes appear neither globally nor spontaneously. The reason why the megalopolis is distributed linearly is because the higher-speed transportation networks need straight links.

The answer to the second question stated above is "No." A different result can be simulated even for the same parameter values as those of Figure 5. Figure 6 shows an alternative simulation result of model. Another megalopolis is observed in the right-hand region that is different from the belt of Figure 6, although both results in Figures 5 and 6 show a similar formation process of a megalopolis qualitatively. When cities with some initial advantages expand their advantages slightly, they become fixed points, focal points on which the neighbouring area depends, in a self-reinforced process of development. Both Figures 5 and 6 describe the uneven development process of a megalopolis that can be called "order out of chaos" (Prigogine and Stengers, 1984). Once a specific process bifurcates, it does not reverse historically. The pattern of urban population evolved in Figures 5 and 6 is not optimal, but possible. This does not mean Figures 5 and 6 are less realistic. Rather, the author wonders whether

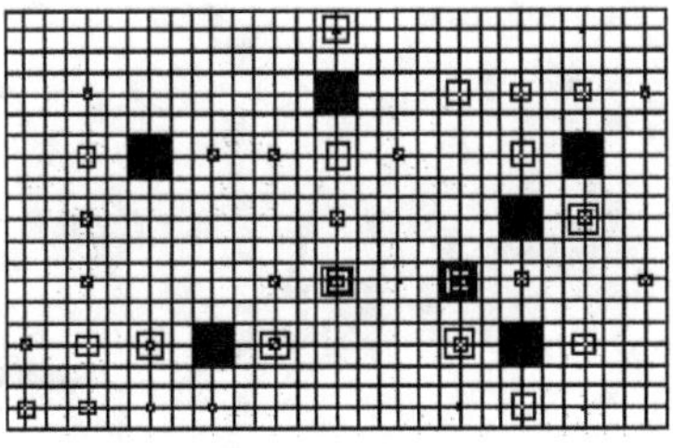

Figure 6. Alternative long-term dynamics of urban population through the time-space convergence, t = 150.

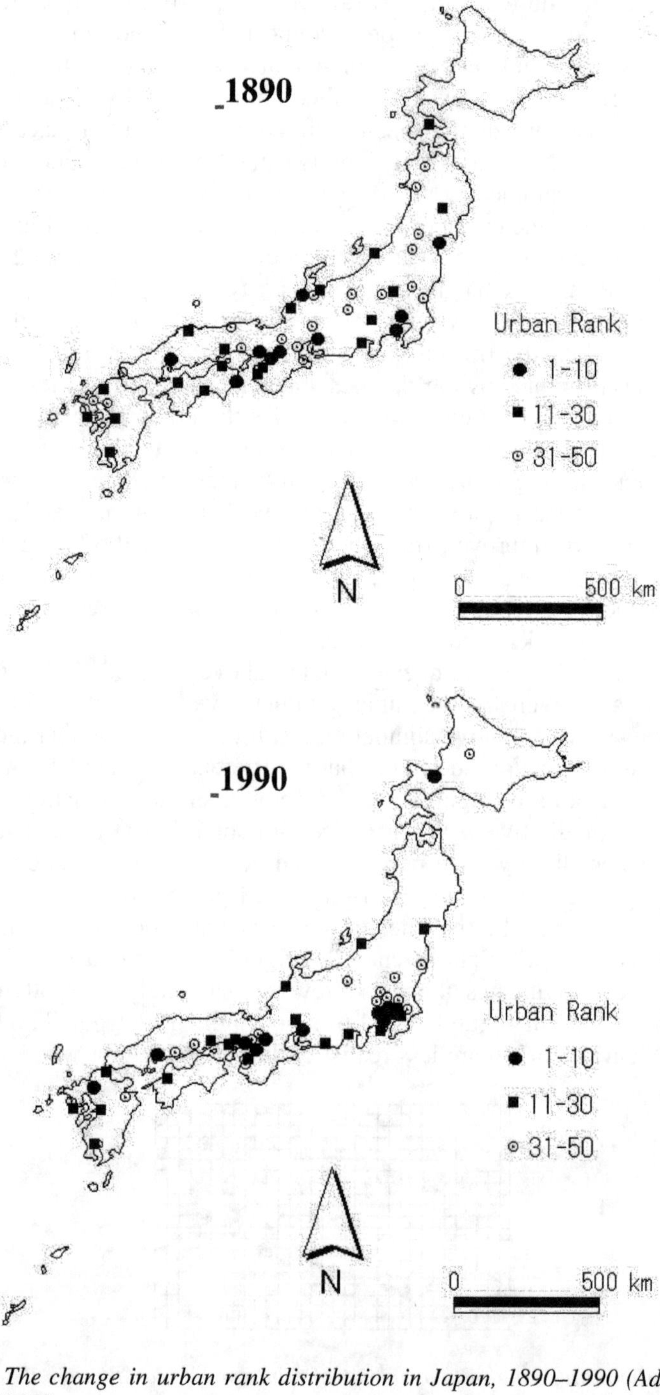

Figure 7. The change in urban rank distribution in Japan, 1890–1990 (Adapted from Yamada (1985)).

the idea of a "unique optimal" pattern is possible. What is the implication of the two possible patterns in Figures 5 and 6 obtained by the geographical megalopolis model? According to Allen and Sanglier (1979, 1981), a self-organisation model helps us not to predict but to understand the real evolution.

5. SIMULATION RESULTS AND EMPIRICAL FACTS

Let us compare the simulation results obtained in Section 4 with the observed urban agglomeration and the transport innovation during the modernisation period in Japan. Figure 7 shows the distributions of the top 50 cities for urban population in 1890 and 1990. The cities were almost evenly distributed at the end of the 19th century when Japanese society began to enter this era (Morikawa, 1962). There were a lot of harbour cities and administrative cities on the coast of the Sea of Japan and in inland areas in 1890, when the transportation media were mainly by horse and water transportation. The distance-decay parameter values in 1890 must have been far higher than those in 1990.

However, once the Tokaido route connected Tokyo with Osaka in the 1890's through the construction of the trunk railway line (Figure 8 (b)), some cities gradually grew up along the Tokaido route. The Tokaido route became an embryo that would develop into a huge megalopolis during the economic growth period. Many prefectures declined on the Sea of Japan coast, where lots of urban agglomerations had been located since the feudal period (Abe, 1997). The Sea of Japan was called the "Backside of Japan" in the1900's (Furumaya, 1997). On the other hand, the belt of huge urban-industrial agglomeration along the Pacific coast became to be called the "Pacific megalopolis belt" in the 1960's. The real history of the emerging megalopolis can be compared to the simulation results obtained from our geographical megalopolis model.

Has the centre-periphery relationship between both districts along the Pacific coast and the Sea of Japan coast been predetermined at the earliest period of

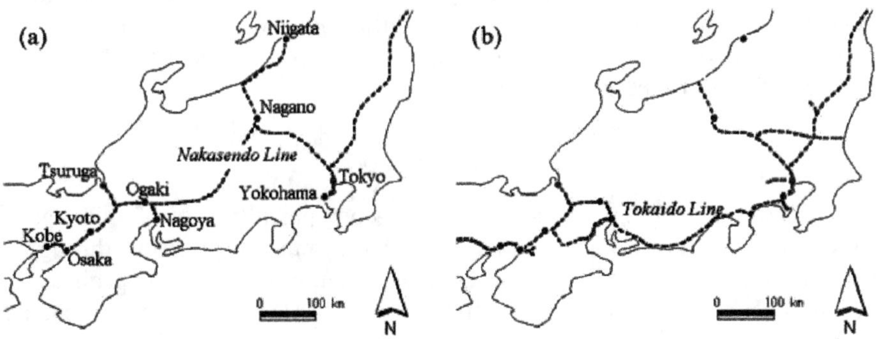

Figure 8. The planned and realised railway networks, late 19th century, Japan ((a) The planned railway networks in 1880, (b) The realised railway networks in 1892; Adapted from Abe (1997) and Furumaya (1997)).

Japanese modernisation? The author does not think so. According to the studies of modern Japanese history, some historians refer to an alternative evolution pattern of transportation networks in the late 1890's (Yamamoto, 1994; Uda, 1995).

In the first stage of modernising Japan in the late 19th century, the central government planned to make a trunk line railway not along the Tokaido route on the Pacific coast, but along the Nakasendo route in inland Japan (Figure 8 (a)). The Nakasendo was not only a traditional route that passed the inland mountainous regions from Tokyo to Nagoya through Nagano, but also areas where the textile industry, one of the most important exporting industries in the late 1890's, was far more flourishing than that in other districts. The Japanese central government in the 1880's had paid attention to the railway connecting the major harbour cities, Niigata and Tsuruga, on the Sea of Japan with the major cities, Tokyo and Nagoya, on the Pacific coast. Furthermore, the government had the idea that the trunk line railway should be protected from any military attacks out of the coasts. The government's plan in those days might have been realised. However, in reality, the Tokaido route was selected as a trunk line railway partly because the cost of railway construction was cheaper than the Nakasendo route, and partly because Japanese basic industry was rapidly shifting from light industries, the textile industry and agricultural industries, to heavy industries, the steel industry and shipping industry, in the 1890's. The 1890's Japan needed railway networks that mean the engine of rapid modernisation.

In retrospect, when you see the history of megalopolis formation, it seems as if a specific region might have been predetermined to become a megalopolis. However, bifurcations exist at any time in history, and the present formation is the cumulative result of many selected experiences. The self-organisation approach helps to understand the reality as a "possible world", as Allen and Sanglier (1979, 1981) pointed out. In the equilibrium framework, socio-economic systems "forget" any initial conditions, while in the mechanistic framework, it is considered to be "predetermined" at a certain point. Both cannot give any implications to urban and regional planning because there is no room between them to intervene in the real world. In other words, it is important to consider how socio-economic systems can "memorise" the history and how they can be "etched" in the geography.

6. Conclusion

The purpose of this paper is to propose a geographical model of self-organisation between the urban population and the number of firms through a time-space convergence in order to answer the question of how a specific area becomes a megalopolis. Our perspective has four characteristics: local rationality, process-dependent space, historical path-dependence, and qualitative change. We constructed two models of both the interaction between the urban population and firms for a short-term, and the interaction between the urban system and time-space convergence for a long-term.

The simulation results are summarised as follows. First, multi-polar urban systems are produced by a non-linear feedback mechanism with local rationality and fluctuation as well as a larger distance-decay parameter. Second, no megalopolis appears if there is no time-space convergence process. Third, where a megalopolis is formed depends on how a series of fluctuation appears in an historical geographical context. Fourth, all the same parameter values and the same initial distribution do not always generate the same urban systems. Fifth, these simulation results increase our understanding of the real megalopolis development in Japan.

Our model may be extended in several directions in economic geography, classified into more economic modelling and more geographical modelling. First, our model does not fully consider a microeconomic foundation of urban systems, while "new economic geography" models do not fully treat historical geographical diversity. There are two ways of solving these problems. One is to find out how such microeconomic variables are localised or bounded, and the other is to connect several scales of spatial economic models with each other. Second, our framework can be applied to construct additional geographical modelling of urban-industrial agglomeration. How do the parameter values, the initial condition, and the size and the form of the hypothetical area affect the evolution of urban systems? How are any hubs formed in transportation networks? How is the "inertia", the tendency for workers and firms that are unwilling to move, considered in economic geography modelling? These questions are the geographical. In short, it is important to combine "more geographic" economic geography and "more economic" new economic geography.

ACKNOWLEDGEMENTS

I am grateful to Yoshitaka Ishikawa for encouraging me to finish this paper, and to Masahisa Fujita, Tomoya Mori, Atsuyuki Okabe, Richard Thomas, the anonymous referee, and participants of the Urban Economics Workshop at Kyoto University for making valuable comments on a draft of this paper. This paper was subsidised by the Grant-in-Aid from the Japanese Ministry of Education and Science (No. 116800071) in 1999–2001.

Department of Geography,
Ochanomizu University,
Otsuka, Bunkyo-ku, Tokyo, 112-8610, Japan

REFERENCES

Abe, T. (1997), *How has the "Back side of Japan" been made?* Nihon-Keizai-Hyoronsha, Tokyo. (J)

Allen, P.M. and M. Sanglier (1979), 'A dynamic model of growth in a central place system.' *Geographical Analysis*, **11**, 256–272.

Allen, P.M. and M. Sanglier (1981), 'Urban evolution, self-organization, and decision-making.' *Environment and Planning A*, **13**, 169–183.

Arthur, W.B. (1994), *Increasing Returns and Path Dependence in the Economy*. The University of Michigan Press, Ann Arbor.

Barnes, T.J. (1996), *Logics of Dislocation: Models, Metaphors, and Meanings of Economic Space*. The Guilford Press, New York.

Bura, S., F. Guerin-Pace, H. Mathian, D. Pumain and L. Sanders (1996), 'Multiagent systems and the dynamics of a settlement system.' *Geographical Analysis*, **28**, 161–178.

Christaller, W. (1933), *Die zentralen Orte in Suddeutschland*. Fischer, Jena.

Clarke, M. and A.G. Wilson (1983), 'The dynamics of urban spatial structure: progress and problems.' *Journal of Regional Science*, **23**, 1–18.

Duranton, G. (1999), 'Distance, land, and proximity: Economic analysis and the evolution of cities.' *Environment and Planning A*, **31**, 2169–2188.

Ewers, H.J. (1995), 'The locational impact of declining costs of transportation and communication.' In H. Girsch (ed.), *Urban Agglomeration and Economic Growth*. Springer-Verlag, Berlin, pp. 183–195.

Fotheringham, A.S. (1981), 'Spatial structure and distance-decay parameters.' *Annals of the Association of American Geographers*, **71**, 425–436.

Fujita, M. (1989), *Urban Economic Theory*. Cambridge University Press, New York.

Fujita, M. and P. Krugman (1995), 'When is the economy monocentric? Von Thunen and Chamberlin unified.' *Regional Science and Urban Economics*, **25**, 505–528.

Fujita, M., P. Krugman and A.J. Venable (1999), *The spatial economy: Cities, regions and international Trade*. MIT Press, Cambridge.

Fujita, M. and H. Ogawa (1992), 'Multiple equilibria and structural transition of non-monocentric urban configurations.' *Regional Science and Urban Economics*, **12**, 161–196.

Furumaya, T. (1997), *Back Side of Japan*. Iwanami-Shoten, Tokyo. (J)

Haggett, P., A.D. Cliff and A. Frey (1977), *Locational Analysis in Human Geography*. Arnold, London.

Haken, H. (1978), *Synergetics: An Introduction, Nonequilibrium Phase Transitions and Self-Organization in Physics, Chemistry, and Biology*. Springer-Verlag, Berlin.

Harris, C.D. (1954), 'The market as a factor in the localization of industry in the US.' *Annals of the Association of American Geographers*, **44**, 311–348.

Hoover, E.M. (1948), *The Location of Economic Activity*. McGraw-Hill: New York.

Janelle, D.G. (1969), 'Spatial reorganization: A model and concept.' *Annals of the Association of American Geographers*, **59**, 348–364.

Junius, K. (1999), *The Economic Geography of Production, Trade, and Development*. Kieler Studien, Mohr Siebeck.

Krugman, P. (1991a), *Geography and Trade*. MIT Press, Cambridge.

Krugman, P. (1991b), 'Increasing returns and economic geography.' *Journal of Political Economy*, **99**, 483–499.

Krugman, P. (1993a), 'On the number and locations of cities.' *European Economic Review*, **37**, 293–298.

Krugman, P. (1993b), 'First nature, second nature, and metropolitan location.' *Journal of Regional Science*, **33**, 129–144.

Krugman, P. (1996), *The Self-Organizing Economy*. Blackwell, London.

Lösch, A.(1940), *Die räumliche Ordnung der Wirtschaft*. Gustav Fischer Verlag, Stuttgart.

Lukermann, F. (1958), 'Towards a more geographic economic geography.' *Professional Geographer*, **10**, 2–10.

May, R. (1976), 'Simple mathematical models with very complicated dynamics.' *Nature*, **261**, 459–467.

Mizuno, I. (1998), *A Regional Disequilibrium Theory of Competitive and Cooperative Central Place Systems*. Unpublished Ph.D. Dissertation, Department of Geography, Tokyo Metropolitan University.

Mori, T. (1997), 'A modelling of megalopolis formation: The maturing of city systems.' *Journal of Urban Economics*, **42**, 133–157.

Morikawa, H. (1962), 'Urban distribution in the early Meiji era, Japan.' *Jimbun-Chiri (The Human Geography)*, **14**, 377–395. (JE)

Myrdal, G. (1954), *Economic Theory and the Under-Developed Regions.* Gerald Duckworth, London.

Nicolis, G. and I. Prigogine (1977), *Self-Organization in Non-Equilibrium Systems: From Dissipative Structures to Order through Fluctuations.* Wiley Inter-Science, New York.

Palander, T. (1935), *Beitrage zur Standortstheorie.* Akademisk Avhandling, Uppsala.

Peeters, D., J.F. Thisse and I. Thomas (1998), 'Transportation networks and the location of human activity.' *Geographical Analysis*, **30**, 355–371.

Plummer, P. (1999), 'Capital accumulation, economic restructuring, and nonequilibrium regional growth dynamics.' *Geographical Analysis*, **31**, 267–287.

Portugali, J. (2000), *Self-Organization and the City*. Springer, Berlin.

Pred, A. (1966), *The Spatial Dynamics of U.S. Urban-Industrial Growth, 1800–1914: Interpretative and Theoretical Essays*. MIT Press, Cambridge.

Prigogine, I. and SI. tengers. (1984), *Order out of Chaos.* Bantam, New York.

Ricci, L.A. (1999), 'Economic geography and comparative advantage: Agglomeration versus specialization.' *European Economic Review*, **43**, 357–377.

Simon, H. (1957), *Models of Man.* John Wiley & Sons, New York.

Smith, D.M. (1981), *Industrial Location.* John Wiley, New York.

Uda, T. (ed.) (1995), *The Development of Modern Japan and the Railroad History.* Nihon-Keizai-Hyoronsha, Tokyo. (J)

Weber, A. (1909), ‹ber den Standort der Industrien 1, Tubingen.

White, R. and G. Engelen (1993), 'Cellular automata and fractal urban form: A cellular modelling approach to the evolution of urban land use patterns.' *Environment and Planning A*, **25**, 1175–1199.

Wong, D.W.S. and A.S. Fotheringham (1990), 'Urban systems as examples of bounded chaos: Exploring the relationship between fractal dimension, rank-size, and rural-to-urban migration.' *Geografiska Annaler*, **72B**, 89–99.

Yamada, M. (1985), 'Urban system in Japan.' In T. Yamaguchi (ed.), *Urban Systems in the World*. Kokon-Shoin, Tokyo, pp. 1–39. (J)

Yamamoto, H. (ed.) (1994), H*istorical Studies of the Modern Transportation Development.* Hosei University Press, Tokyo. (J)

(J): in Japanese; (JE): in Japanese with English abstract.

PHILLIP SMITH AND RICHARD THOMAS

EPIDEMIC MODELLING OF HIV/AIDS TRANSFERS BETWEEN EASTERN AND WESTERN EUROPE

1. INTRODUCTION

A recent analysis of the HIV/AIDS epidemic in Western Europe revealed a number of north-south contrasts (Thomas, 2000). The introduction of anti-retroviral drug combination therapies that suppress symptoms of AIDS in those with HIV (Feigal *et al.*, 1999; Deschamps *et al.*, 2000), for example, appears to have had the greater health impact in northern countries. Similarly, the comparatively low incidence in the north, which is predominantly among gay men, was supported by more extensive travel patterns than in the south, where intravenous drug use (IVDU) is the major risk behaviour. These findings were obtained by fitting a multiregion epidemic model to reported AIDS incidence in each country. The present paper extends this space-time analysis to all the countries of Europe with the intention of identifying further structural differences between east and west.

The key focus of interest is how the diffusion of AIDS might have been affected by the changing accessibility characteristics of the population of Eastern Europe. This concern arises from the political and economic isolation that existed in these nations prior to the major political restructuring of the late 1980's and early 1990's. Before then, travel opportunities both to and from these countries were severely restricted by the barriers imposed by the political 'Iron Curtain'. Subsequently, the East has become far less isolated and also much more attractive as a travel destination for those living in the West. It is hypothesised, therefore, that these new travel patterns are likely to have important implications for the transmission of HIV in a continent where AIDS incidence already displays considerable differentiation.

To this end, an epidemic model is fitted to recorded national AIDS incidence for two scenarios concerning population mixing. The first solution includes all European countries in the design and this large system, therefore, assumes those at risk in the east made some contact with those in the west throughout the course of the epidemic. The second is for a solution where the fit is restricted to the countries of Eastern Europe such that the epidemic in the east is regarded as a closed system with no external mixing. Possible patterns of infection transfer across the divide are discussed by comparing the feasibility of the various contact and travel parameters that support reported AIDS incidence in these two limiting scenarios.

The article is organised as follows. Section 2 outlines the multiregion mod-

B. Boots et al. (eds.), Modelling Geographical Systems, 119–138.

elling system that imitates the transmission of HIV in space and time. Section 3 gives details of the data sources and calibration procedures that are employed in the analysis. Section 4 presents the results obtained for the two limiting mixing scenarios. The discussion considers the implications of these findings for changing accessibility relations between Eastern and Western Europe.

2. A MULTIREGION MODEL FOR HIV/AIDS

This model, which employs principles of spatial interaction theory to represent regional contact rates (Thomas, 1999a; Thomas and Smith, 2000), follows a long lineage of multiregion epidemic systems (Baroyan and Rvachev, 1967; Murray and Cliff, 1977; Longini *et al.*, 1986; Flahault and Valleron, 1992; Williams and Rees, 1994; Sattenspiel and Dietz, 1995; Toubiana and Vibert, 1998). Within the interaction framework contact formation is given the following representation. Let i identify each of the regions forming the area of interest and $l = 1$ denote journeys made within a region (local travel) and $l = 2$ those made between regions (international travel). Define q_{il} as the proportion of journeys originating in region i of travel type l. Then, the local travel proportion in each region, q_{i1}, i, is linked to corresponding international travel proportion through the relationship

$$q_{i1} = 1 - q_{i2}, \quad \forall i. \tag{1}$$

These local journeys always terminate in their region of origin and, therefore, unity is the probability that the destination region (j) for a local journey started in i is $j = i$, that is,

$$p_{i=j,\, l=1} = 1 - q_{i2}, \quad \forall i = j. \tag{2}$$

Accordingly, the probability a journey of *any* travel type takes place within region i is equivalent to the local travel proportion, that is

$$p_{i=j} = q_{i1}, \quad \forall i = j. \tag{3}$$

Equivalent probabilities for travel between each pair of regions are generated according to a negative exponential function of the form $e^{-\lambda d_{ij}}$, where d_{ij} is the distance between the centres of regions i and j. Here, λ is a decay parameter calibrated in the interval $[0, \infty]$ to represent the sensitivity of travel to this spatial separation. Accordingly,

$$a_{ij} = e^{-\lambda d_{ij}}, \quad \forall ij. \tag{4}$$

is a measure of the accessibility of region i to region j. These accessibilities are subject to the condition

$$a_{ij} = 0, \quad \forall i = j. \tag{5}$$

which omits these local terms from the estimation of international travel. Moreover, define m_j as the travel opportunities present in region j such that the term $a_{ij}m_j$ is taken to represent the attractiveness of j as a destination for those travelling from region i. Consequently, the totals

$$W_i = \sum_j a_{ij}m_j , \quad \forall_i, \tag{6}$$

each measure the attractiveness of the entire system for those travelling from i. If regional contact probabilities are taken to be proportional to their accessibility levels, then the terms

$$p_{i \neq j,\, l=2} = \frac{a_{ij}m_j}{W_i}, \quad \forall i \neq j \tag{7}$$

each represent the probability that a single international journey started from region i will terminate in region j. The corresponding probability for a journey of any type started in region i is obtained by including the international travel proportion as the following weight

$$p_{i \neq j} = q_{i2}p_{i \neq j,\, l=2}, \quad \forall i \neq j. \tag{8}$$

Together, Equations (3) and (8) define the probability an i-individual will chose to journey on a particular i-j route. The next step is to represent how this traveller will select a partner upon arrival at the destination region. This task first requires identification of the sequence of states all those at risk to HIV infection in each region (n_i, $\forall i$) encounter during the course of the epidemic. Accordingly, these regional risk populations are disaggregated into the following state variables: x_i, $\forall i$, those who are susceptible; y_i, $\forall i$, those who are infectious; w_i, $\forall i$, those who are HIV+ but with antigen levels too low to be communicable to others (Nowak *et al.*, 1991); and, z_i, $\forall i$, all those ever with AIDS. Amongst these, the susceptibles and infectives are taken to be freely circulating individuals ($x_i + y_i$, $\forall i$) who, at any point in time, are available to form partnerships with each other.[1] Then, a partnership between a pair of these i-j individuals arises as the outcome of one of two possible contacts made in each direction of travel (i travels to meet j or vice versa). This relationship, together with the assumption that partners are selected with equal likelihood $[1/(x_i + y_i)]$, is known to be satisfied by the following statistic for the rate of infection between regions (Thomas and Smith, 2000)

$$s_{ji} = \frac{r\left\{p_{ji} + \left[\dfrac{(x_i + y_i)p_{ij}}{x_j + y_j}\right]\right\}}{2}, \quad \forall_{ji}. \tag{9}$$

So defined, each term s_{ji} is the expected number of partnerships among those circulating freely in i with a specific j-individual of the same status. Here, r

is the average rate of partner acquisition for the risk population of the entire system.

These regional rates are central to the specification of the transmission process which also requires the following disease parameters to be defined: β_i, $\forall_i$, the probability a partnership between an infective in region i and a susceptible results in the transmission of HIV,[2] D, the period of communicability for HIV; and A, the period from the cessation of communicability to the onset of AIDS. The transmission of infection between regions comprises the following sequence of events. Infecteds in region j form ry_j partnerships per unit of time which are made with each freely circulating i-individual at the rate $y_j s_{ji}/(x_i + y_i)$. Then, $x_i y_j s_{ji}/(x_i + y_i)$ of the infecteds' partnerships are formed with susceptibles in region i and a fraction βj of these become infectious. Summing over all j regions, defines the incidence of the infection in each region i as

$$\frac{x_i \Sigma_j \beta_j y_j s_{ji}}{x_i + y_i}. \tag{10}$$

The state diagram representing the complete epidemic process is given by

$$[x_i] \rightarrow \frac{x_i}{x_i + y_i} \Sigma_j \beta_j y_j s_{ji} \rightarrow [x_i] \rightarrow \frac{y_i}{D} \rightarrow [w_i] \rightarrow \frac{w_i}{A} \rightarrow [z_i], \quad \forall i. \tag{11}$$

where, for each region, y_i/D is the number of infectives who cease to be communicable, and w_i/A is the incidence of AIDS. The diagram defines the following set of differential equations

$$\frac{dx_i}{dt} = -\frac{x_i}{x_i + y_i} \Sigma_j \beta_j y_j s_{ji}, \quad \forall i, \tag{12}$$

$$\frac{dy_i}{dt} = \frac{x_i}{x_i + y_i} \Sigma_j \beta_j y_j s_{ji} - \frac{y_i}{D}, \quad \forall i, \tag{13}$$

$$\frac{dw_i}{dt} = \frac{y_i}{D} - \frac{w_i}{A}, \quad \forall i, \tag{14}$$

$$\frac{dz_i}{dt} = \frac{w_i}{A}, \quad \forall i. \tag{15}$$

Within this system, it can be seen that the rates s_{ji}, $\forall_{ji}$, provide the essential spatial mechanism for the transferring HIV infection between the regions.

The more generalised space-time behaviour of this system of equations is known to be governed by two possible scenarios: first, solutions with a stable Disease-Free Equilibrium (DFE) where an epidemic is unable to start; and second, solutions with an unstable DFE where all orbits of the model tend toward a unique endemic equilibrium (Lajmanovich and Yorke, 1976; Thomas, 1999b).

The DFE is of particular interest for disease control because it distinguishes between populations whose characteristic epidemic behaviour appears to protect them from the sustained transmission of HIV and those whose behaviour does not (Thomas, 1999a; Thomas and Smith, 2000). This distinction is made with reference to the reproduction number of an epidemic which, for a single geographical population, is the number of secondary infections made by the index case during the period of communicability and given by

$$R_o = \beta r D. \tag{16}$$

Equivalent regional reproduction numbers are derived from the transfer rates (s_{ji}) as

$$R_{oj} = \beta_j \textstyle\sum_i s_{ji} D, \quad \forall j, \tag{17}$$

and are interpreted as the expected number of infections in all regions during D that are initiated by a j-infected when all partnerships are with susceptibles. Here, the individual terms in this expression given by

$$R_{oij} = \beta_j s_{ji} D, \quad \forall ij, \tag{18}$$

count the equivalent infections in a single region i initiated by a j-infected. Among this set, the within region reproduction number ($R_{oj=j}$) is the count of the local infections arising in region j.

These numbers combine to provide the following approximate conditions for the stability of the DFE (Jacquez *et al.*, 1995). If $R_o > 1$ and $R_{oj} > 1$, then this outcome is sufficient to guarantee the start of a sustained epidemic in region j and the DFE is unstable. Moreover, if $R_{ojj} > 1$, this condition alone is sufficient for a start in j. Conversely, if neither of these conditions holds for any region, then the DFE will be stable and the infection will die out naturally in the population. A related statistic is given by

$$\Delta R_{oj} = R_{oj} - R_{ojj}, \quad \forall j, \tag{19}$$

which is the difference between the region and within region reproduction numbers. This value represents the total number of secondary infections exported from j to the other regions and, as such, reflects a country's diffusion potential per infective (Jacquez *et al.*, 1995; Thomas, 1999b).

3. CALIBRATION PROCEDURES

3.1. Prior Materials

The European regional system comprises the 38 countries that report annual AIDS incidence to the WHO (see Table 1). These series record incidence from

Table 1. Selected National Data.

i	Country	Date of first recorded AIDS cases	Cu. AIDS incidence per million (n_i/Pop)	Cu. HIV incidence per million[a]	Proportion of AIDS homo-sexual[b]	Proportion of AIDS IVDU[b]	Popula-tion, millions
1	France	1982	964	2221	0.45	0.25	58.54
2	Belgium	1984	285	756	0.38	0.07	10.19
3	Netherlands	1982	348	957	0.70	0.12	15.67
4	Germany	1982	249	425	0.66	0.15	82.19
5	UK	1982	352	515	0.68	0.08	58.20
6	Ireland	1982	205	562	0.35	0.43	3.56
7	Denmark	1983	467	819	0.68	0.09	5.25
8	Norway	1983	163	367	0.55	0.17	4.36
9	Sweden	1983	209	339	0.59	0.11	8.84
10	Finland	1984	62	214	0.67	0.04	5.14
11	Iceland	1985	178	741	0.81	0.07	0.27
12	Luxembourg	1985	352	786	0.51	0.17	0.42
13	Switzerland	1983	996	2335	0.37	0.40	7.28
14	Austria	1983	257	1103	0.39	0.26	8.16
15	Italy	1983	908	1660	0.13	0.64	57.24
16	Malta	1985	130	595	0.49	0.02	0.37
17	Spain	1983	1776	3021	0.14	0.66	39.72
18	Portugal	1985	806	3673	0.23	0.44	9.80
19	Greece	1983	223	760	0.57	0.05	10.52
20	Slovakia	1985	5	74	0.65	0.00	5.38
21	Yugoslavia	1985	93	472	0.14	0.49	10.59
22	Romania	1985	299	310	Na	Na	22.56
23	Poland	1986	31	341	0.26	0.50	38.15
24	Hungary	1986	40	246	0.72	0.01	10.15
25	Slovenia	1986	52	101	0.52	0.06	1.98
26	Russia	1987	4	884	0.29	0.03	147.10
27	Croatia	1987	37	78	0.46	0.09	4.49
28	Czech Rep.	1986	15	214	0.67	0.02	10.30
29	Bulgaria	1987	10	36	0.15	0.00	8.30
30	Ukraine	1988	266	4735	0.02	0.79	50.69
31	Lithuania	1988	13	135	0.68	0.07	3.70
32	Macedonia	1989	16	46	0.10	0.16	2.19
33	Moldova	1989	8	1044	0.04	0.65	4.31
34	Bosnia	1989	9	198	0.24	0.24	3.78
35	Latvia	1990	26	506	0.63	0.27	2.47
36	Belarus	1991	3	1371	0.27	0.04	10.21
37	Estonia	1992	19	345	0.73	0.00	1.45
38	Albania	1994	4	27	0.09	0.09	3.73
W Europe (1–19)			648	1336			385.72
E Europe (20–38)			71	1247			341.53
Europe (1–38)			377	1294			727.25

[a] Estimated number of those aged 15–49 years living with HIV/AIDS at the end of 1999 (WHO, 2000b) divided by their population.

[b] Proportion of all AIDS cases reported at the end of 1999 (WHO, 2000a).

1982 to 1999 (WHO, 2000a) although for some countries the first report of AIDS is after 1982. In this respect, recorded AIDS incidence in West European countries is estimated to be between 75% and 100% of all cases, whereas, in some East European countries, ascertainment is thought to be less complete (WHO, 1998). Two common factors contributing to this inconsistency are delayed reporting, which typically affects annual incidence recorded immediately prior to the present, and occasional alterations to these data caused by official changes to the surveillance definition of AIDS (Rosenberg *et al.*, 1992). Factors more specific to the East European context have included variations in the availability of HIV testing facilities and the willingness to compile and analyse information (WHO, 2000b). Most East European countries, however, have maintained a reasonably extensive surveillance infrastructure that has continued to be improved throughout the last decade (Lamptey and Tarantola, 1998). Given all these circumstances, the official series do not measure the actual incidence of AIDS in each country but, instead, enable comparisons to be made between national variations in the diagnosed burden of AIDS.

In addition to these data, some of the epidemiological parameters in the model are also given estimates taken from published sources. The duration of the periods of communicability ($D = 2$ years) and antigen suppression ($A = 4$ years), for example, are drawn from research into viral abundance and antigen concentrations in patients after HIV infection (Anderson and May, 1991). Similarly, an average value of $r = 10$ partners per year for those at high risk is taken from surveys of sexual behaviour made at different times during the course of the epidemic (Knox *et al.*, 1993; Johnson *et al.*, 1994).

Certain spatial terms in the specification are represented by given values. The co-ordinates for each country are for the major population centre. These are capital cities with the exception of Frankfurt, which was chosen to accommodate recording inconsistencies in Germany prior to reunification (Frankenberg, 1992). Distances between these locations, d_{ij}, $\forall i \neq j$, are measured along great circle routes, and distances within countries are given by the convention, $d_{ij} = 0$, $\forall i = j$.

Alongside this operational environment, information about travel frequencies across the former divide assists the interpretation of model outputs like rates of infection transfer. Statistics describing the number of visitors to a country form abroad (UN, 1984; 2000), for example, give an impression of such changes during the study period. Aggregation of these data provides strong contrasts between rates of travel to Central European countries (Czech Republic, Hungary, Poland, and Slovakia) and those to the former Soviet Union. Between 1980 and 1995, visitor rates to Central Europe were relatively frequent and increased from 0.34 per head of population per year to 0.60. The equivalent rates for the Soviet Union are much lower but have also increased over the interval from 0.03 to 0.08. Conversely, the countries of former Yugoslavia record decreased visitor rates of 0.28 and 0.11 mainly as a consequence of the impact of the Balkans War on tourism.

3.2. Optimised Estimates

The remaining terms required for the calibration of the model are obtained by an optimisation procedure comprising of two algorithms that are described fully in Thomas (2000; 2001). The first of these routines treats the epidemic in each country as a discrete entity and, therefore, employs a single-region version of Equations (12–15) where $\forall i = 1$. Upon completion of the iterations required to minimise the difference between observed and predicted AIDS incidence, best fitting estimates are obtained for the country's transmission probability (β_i), the initial size of its risk population (n_i) and the date of its first HIV infections. Here, the derivation of the optimised value of the risk population is of particular importance. These estimates are obtained by first fitting predicted cumulative AIDS incidence to the equivalent recorded incidence at the last available recording date. Then n_i is derived as the projected value of this incidence at the zero equilibrium that arises at the completion of the epidemic cycle.[3] At this time, the equality

$$z_i = n_i, \quad \forall i, \tag{20}$$

will hold to indicate that this national risk population for each country is an estimate for all those ever with AIDS. This value, therefore, will mirror any reporting inconsistencies present in the recorded data and enters the model as the size of the susceptible population ($z_{i\,t=0} = n_{i\,t=0}$) prior to the first HIV infection at $t = 0$. In addition, the values of n_i, $\forall i$, are selected as the set of travel opportunities (m_i, $\forall i$) which presumes those at risk are attracted to a country by the size of their presence there.

In the second algorithm, these country estimates are all entered into a multiregion setting designed to find optimum values for the decay parameter (λ) and travel proportions (q_{il}, $\forall il$). This routine employs a spatial search procedure intended to minimise the value of the statistic

$$T = \frac{\sum_i |P(O)_i - P(E)_i|}{I}. \tag{21}$$

Here, $P(O)_i$ is the observed (O) time of peak (P) HIV prevalence in country i and $P(E)_i$ is the corresponding predicted (E) peak time from the multiregion model.[4] I denotes the total number of countries forming the system and, therefore, T is seen to be the average time difference between observed and predicted HIV prevalence. Moreover, the minimisation of T is subject to the constraint $q_{i1} \geq q_{i2}$, $\forall_i$, to ensure that the local travel proportion in each country is always greater than or equal to the international proportion.

4. RESULTS

4.1. Geographical Distribution

The burden of both HIV and AIDS is variable across the countries of Europe (Table 1). Here, the cumulative incidence of AIDS in Western Europe is seen to exceed the continental rate (377 cases per million people) by a factor of 1.72, while the corresponding quotient for the east is just 0.19. Indeed, only Romania and the Ukraine exhibit incidence that is comparable to western rates. In contrast, cumulative HIV incidence in both the west and east is similar to the continental average of 1294 infections per million people.[5] Accordingly, each AIDS diagnosis in Western Europe is supported 2.06 HIV infections while in the east this rate is 17.56. Inspection of Table 1 will reveal that this inconsistency is largely attributable to reporting in the former republics of the Soviet Union (especially Belarus, Moldavia and Russia) where high HIV estimates occur in conjunction with very few AIDS cases. Conversely, in Romania the two totals are virtually identical. With the exception of these countries, however, the remaining HIV/AIDS quotients fall within a tolerably similar range.

The distribution of AIDS by risk behaviour is distinctly patterned in Western Europe (Table 1). In the north, the epidemic is predominantly among gay men who usually account for more than 60% of cumulative incidence, whereas, in the south, the burden of AIDS is much greater and IVDUs account for the majority of reported diagnoses. In contrast, the epidemic in Eastern Europe shows little spatial differentiation. Gay men are the dominant caseload in most countries while the Ukraine, Poland and Yugoslavia report significant incidence among IVDUs. More idiosyncratically, a dramatic epidemic of nosocomial HIV infection was discovered in Romania among orphans and hospitalised children in 1989 (Hersh, 1991) and by 1999 73% of HIV infections there were among children. The incidence of AIDS among the major transmission risks, however, is more difficult to ascertain (see Table 1) because the Romanian records contain many cases classified with unknown origins.

The graphical outcomes of fitting the single-region epidemic model to AIDS series for six selected countries are shown in Figure 1. More generally, countries with risk populations (n_i, $\forall_i$) in excess of 500 (France, Italy, Romania) tend to be characterised by a close correspondence between observed and predicted incidence. In contrast, many countries with populations below this threshold (Malta, Belarus) have a more erratic incidence that is replicated more approximately by the model. In addition, most of the series exhibit declining annual incidence by 1999 which facilitates a consistent estimate to be obtained for the size of the risk population at the projected zero equilibrium. The caseloads reported for the Ukraine and Latvia, however, were still increasing in 1999 and their risk populations are fitted without knowledge of the date of peak AIDS incidence.

The single-region fits also yield estimates for each country of the transmission probability (β_i) and date of first HIV infections (Table 2). For countries

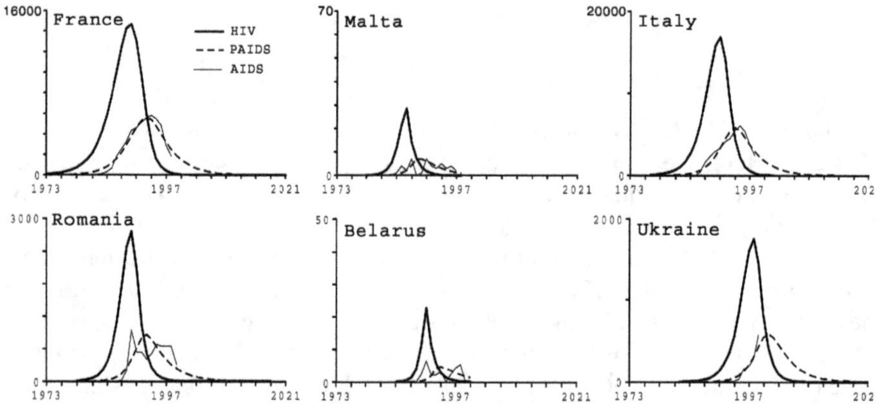

Figure 1. Single country model series for selected nations (HIV = predicted prevalence, PAIDS = predicted AIDS incidence, AIDS = recorded AIDS incidence).

with the larger risk populations, the fitted transmission probabilities are similar to a reliable epidemiological estimate of 0.1 obtained from samples of sero-positive individuals (May *et al.*, 1989). The more deviant values, however, are associated with short, low incidence series (like Albania's) where a higher rate of population turnover is required to achieve the optimum fit. Given this variation, the estimated dates of first HIV infections indicate an epidemic originating in France (1973) that had diffused to all West European countries by 1979 (Thomas, 2000). First infections in Eastern Europe begin simultaneously in Slovakia, Yugoslavia and Romania (1979) and this sequence of events is completed in Albania (1998).

4.2. All Europe Solution

Following from these results, the multiregion solution for all Europe mimics an epidemic started in France in 1973.[6] The fit for this simulation is characterised by an average time difference between observed an predicted peak HIV prevalence per country of just $T = 1.053$ years (Table 3). The estimated decay parameter of $\lambda = 0.33$ suggests travel behaviour is relatively insensitive to the effects of spatial separation. The fitted values of the international travel proportions (q_{i2}, $\forall_i$), indicate these journey types are much more prevalent in the west than in the east.

The stability of this solution is defined by the sets of reproduction numbers listed in Table 2. Here, the regional numbers (R_{oj}) are all in excess of unity to denote the presence of an unstable DFE where a start would have been possible in all the countries of Europe. The entire system has an average reproduction number of $R_o = 2.13$ for a randomly selected infection drawn from all susceptibles. Countries with numbers well above this average tend to be those with the smaller risk populations (like Malta, Estonia and Albania) where their

Table 2. Estimated National Statistics.

i Country	Date of first HIV infections	Transmission probability (β_i)	Size of risk population (n_i)	Europe solution Reproduction Numbers		E Europe solution Reproduction Numbers	
				R_{0i}	ΔR_{0i}	R_{0i}	ΔR_{0i}
1 France	1973	0.09509	57000	2.08	0.75	–	–
2 Belgium	1974	0.10099	2900	2.64	1.23	–	–
3 Netherlands	1974	0.10233	5450	2.39	1.06	–	–
4 Germany	1974	0.10116	20500	2.23	0.81	–	–
5 UK	1975	0.10053	19000	2.32	0.91	–	–
6 Ireland	1976	0.11131	730	1.98	0.76	–	–
7 Denmark	1976	0.11281	2450	2.16	0.92	–	–
8 Norway	1977	0.11965	712	2.16	0.84	–	–
9 Sweden	1976	0.11035	1850	1.87	0.66	–	–
10 Finland	1978	0.12939	320	2.18	0.89	–	–
11 Iceland	1978	0.15364	48	2.38	0.69	–	–
12 Luxembourg	1979	0.14394	148	2.91	0.06	–	–
13 Switzerland	1977	0.12007	7250	2.95	1.03	–	–
14 Austria	1977	0.12253	2100	2.93	1.46	–	–
15 Italy	1977	0.11096	52000	2.00	0.45	–	–
16 Malta	1979	0.18980	48	4.61	1.95	–	–
17 Spain	1977	0.10826	67000	1.90	0.38	–	–
18 Portugal	1978	0.10805	7900	2.24	0.12	–	–
19 Greece	1978	0.11481	2351	2.19	1.04	–	–
20 Slovakia	1979	0.12081	27	2.31	0.38	2.76	0.88
21 Yugoslavia	1979	0.12048	988	2.55	0.38	2.59	0.49
22 Romania	1979	0.13374	6751	2.55	0.46	2.65	0.11
23 Poland	1980	0.11410	1166	2.27	0.10	2.21	0.22
24 Hungary	1980	0.13259	408	2.70	0.18	2.97	0.85
25 Slovenia	1980	0.12531	102	2.55	0.07	2.54	0.54
26 Russia	1981	0.12788	516	2.11	0.45	2.25	0.31
27 Croatia	1981	0.15678	168	3.18	0.08	3.23	0.85
28 Czech Rep.	1981	0.14473	156	2.91	0.04	2.75	0.55
29 Bulgaria	1981	0.13525	81	2.79	0.17	2.77	0.09
30 Ukraine	1982	0.11802	13500	2.36	0.02	2.36	0.02
31 Lithuania	1982	0.11974	47	2.39	0.02	2.44	0.16
32 Macedonia	1983	0.17299	36	3.54	0.18	3.68	0.50
33 Moldova	1983	0.14593	36	2.98	0.15	3.01	0.12
34 Bosnia	1983	0.14657	33	2.96	0.06	2.97	0.13
35 Latvia	1984	0.16302	65	3.26	0.03	3.25	0.32
36 Belarus	1985	0.26398	30	5.27	0.04	5.32	0.09
37 Estonia	1986	0.26422	27	5.25	0.14	5.21	0.19
38 Albania	1988	0.49394	14	9.96	0.18	9.89	0.11

relatively high transmission probabilities contribute to the reproduction effect. The numbers for the diffusion potential of each country (ΔR_{oj}) reveal strong spatial variations. In the west, these exported secondary infections tend to account for between a third and half the total number and centrally located Austria has the greatest absolute potential ($\Delta R_{o14} = 1.46$). The corresponding

Table 3. Multiregion Fit Statistics.

i	Country	Europe solution $P(O)_i$	$P(E)_i$	P_i	q_{i2}	E Europe solution $P(E)_i$	P_i	q_{i2}
1	France	1990	1990	0	0.30	–	–	–
2	Belgium	1989	1989	0	0.30	–	–	–
3	Netherlands	1989	1989	0	0.35	–	–	–
4	Germany	1990	1990	0	0.30	–	–	–
5	UK	1990	1990	0	0.30	–	–	–
6	Ireland	1990	1991	–1	0.45	–	–	–
7	Denmark	1989	1991	–2	0.45	–	–	–
8	Norway	1989	1991	–2	0.45	–	–	–
9	Sweden	1989	1991	–2	0.45	–	–	–
10	Finland	1990	1991	–1	0.50	–	–	–
11	Iceland	1987	1990	–3	0.45	–	–	–
12	Luxembourg	1989	1989	0	0.01	–	–	–
13	Switzerland	1989	1989	0	0.20	–	–	–
14	Austria	1989	1992	0	0.40	–	–	–
15	Italy	1991	1991	0	0.30	–	–	–
16	Malta	1987	1988	–1	0.30	–	–	–
17	Spain	1991	1991	0	0.30	–	–	–
18	Portugal	1993	1992	1	0.02	–	–	–
19	Greece	1991	1991	0	0.50	–	–	–
20	Slovakia	1991	1991	0	0.05	1992	–1	0.22
21	Yugoslavia	1991	1991	0	0.10	1991	0	0.13
22	Romania	1990	1991	–1	0.22	1990	0	0.05
23	Poland	1993	1993	0	0.05	1993	0	0.13
24	Hungary	1991	1991	0	0.05	1992	–1	0.20
25	Slovenia	1992	1992	0	0.01	1993	–1	0.20
26	Russia	1993	1993	0	0.35	1995	–2	0.24
27	Croatia	1990	1990	0	0.01	1992	–2	0.24
28	Czech Rep.	1991	1991	0	0.01	1993	–2	0.24
29	Bulgaria	1992	1992	0	0.03	1992	0	0.01
30	Ukraine	2000	1996	4	0.01	1996	4	0.01
31	Lithuania	1995	1996	–1	0.01	1995	0	0.05
32	Macedonia	1990	1989	1	0.03	1990	0	0.08
33	Moldova	1993	1992	1	0.03	1992	1	0.01
34	Bosnia	1993	1991	2	0.01	1993	0	0.03
35	Latvia	1993	1992	1	0.01	1993	0	0.10
36	Belarus	1991	1988	3	0.01	1991	0	0.01
37	Estonia	1992	1987	5	0.01	1992	0	0.05
38	Albania	1992	1983	9	0.01	1987	5	0.01
λ				0.3300			0.5700	
Min.T, years				1.1053			1.0000	

$P_i = P(O)_i - P(E)_i$ – the difference in years between the time of the single region peak HIV prevalence in i and the predicted multiregion peak time.
q_{i2} = the international travel proportions estimated for the listed solution.

eastern values are much smaller and only Slovakia, Yugoslavia, Romania and Russia display noticeable export potentials.

The simulated pathways of HIV infection from France are illustrated in Figure 2a. Here, each link ($j \rightarrow i$) connects a country i, where a first infection has been predicted at t, with a country j, that was the major source of this incidence then. Also displayed are the dates of predicted date of first HIV infection in each country in the multiregion model and the lags (–) and leads (+) between this date and the single-country estimate. The pathway for this solution displays a characteristic hub and spoke pattern (Bryan and O'Kelly, 1999), where the infection diffuses quickly between those countries with relatively large risk populations, and then spins out from these hubs to neighbours where fewer are at risk. This hierarchical diffusion is typical of spatial systems where decay parameter is small ($\lambda = 0.33$) and, consequently, contact formation is more responsive to risk population size than distance.

Three hubs dominate the pathway. France serves as the source of initial infections in the majority of West European countries. Then, the transmission of HIV to the large risk population in Italy establishes a node for the transfer of infection to the former Yugoslav Republics and their immediate neighbours. Last, the passage of infection from Italy to Romania connects many of the Republics of the former Soviet Union to this epidemic system. Two essentially linear routes are also evident in the network. First, the path formed between the large risk populations of France, Spain and Portugal. Second, the route linking Germany to some Nordic countries and Baltic States. Among these countries, the predicted times of first infections occur later than those observed in Sweden and Finland, but some years earlier than the observed times in Latvia and Estonia.

The network structure characterising this pathway is influenced by the location of the source region. Accordingly, the posited start in France simulates any early lead in HIV incidence that enhances the likelihood of French infection transfers to other countries. An alternative representation of the epidemic structure is illustrated in Figure 2b. Here, the equivalent linkage pattern is mapped at the time of peak HIV prevalence when the influence of the source region is has been largely absorbed into the more general system structure. In this peak network each vector ($j \rightarrow i$) connects a country i, where peak prevalence has been predicted at t, with the country j that is the major source of infection then. In such a network, the vectors are not necessarily unidirectional because any pair of countries may serve as the dominant source of infection for each other (for example, France $\leftrightarrow$ Spain, Germany $\leftrightarrow$ Italy and Greece $\leftrightarrow$ Romania).

This mapping results in a more emphasised delineation of the network structure. The eastern reach of the French hub, for example, is curtailed and the previous links to Ireland and Iceland are redirected through the UK. The Italian graph retains much of its original structure except for the addition of Switzerland and the loss of Greece, Bulgaria and Romania. The major timing error on this graph is associated with Albania where predicted peak HIV preva-

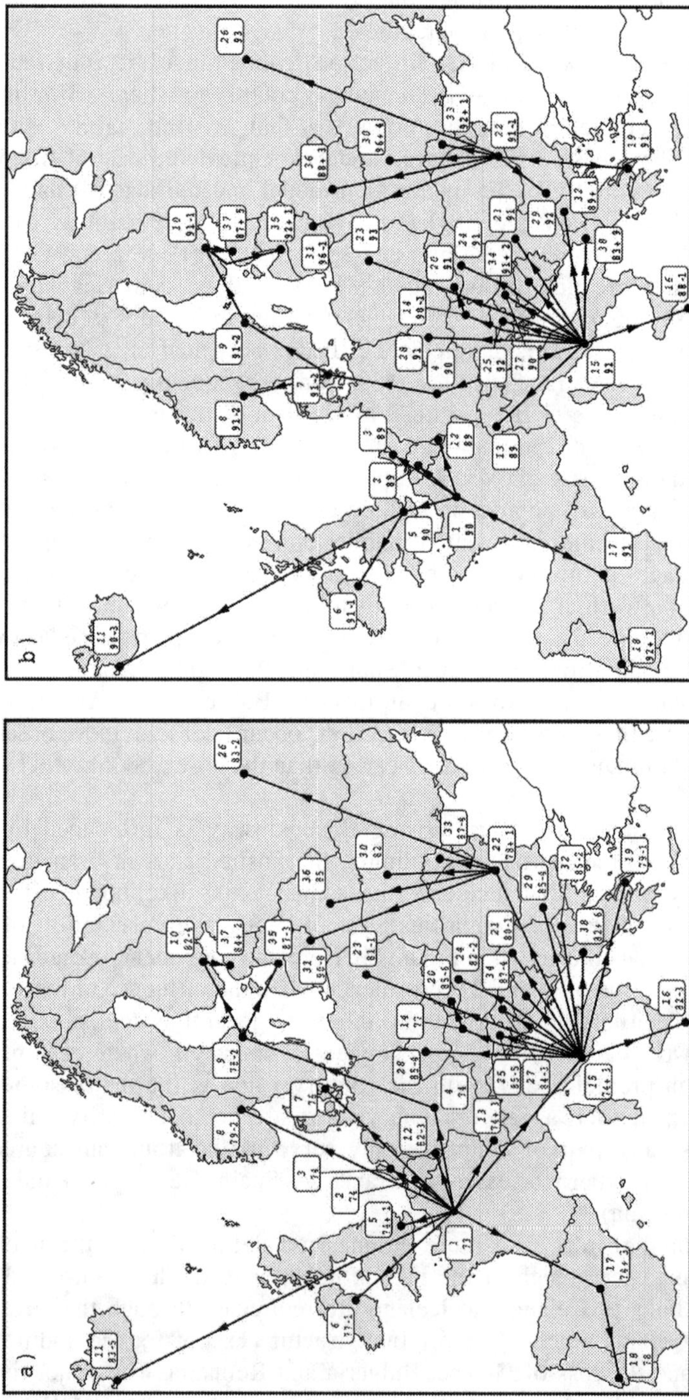

Figure 2. *Europe (a) Predicted pathway of first HIV infections (dates in Roman numbers). (b) Dominant infection vectors at predicted peak HIV prevalence (dates in bold numbers).*

lence occurs nine years before the estimated date. The Romanian network has the poorest fit and, in particular, the predicted peaks in the Ukraine and Belarus occur a number of years before their estimates. Lastly, the route from Germany is re-orientated to include all the Nordic countries where peak prevalence always occurs after the estimated time. Here, Finland now serves as a node for infection transfer to Estonia and Latvia, where predicted peak times arrive ahead of their observed estimates.

4.3. Eastern Europe Solution

This solution is simulated for an epidemic started in 1979 in Romania, which is the first country in this set to report sustained AIDS incidence. The fit is slightly improved at $T = 1.000$ years per country and the decay parameter of $\lambda = 0.57$ indicates a greater sensitivity to spatial separation than in the all Europe solution. This more restricted travel scenario, however, is obtained in conjunction with some alterations to the fitted international travel proportions (Table 3). In particular, countries closer to Western Europe (Slovakia, Poland, Hungary, Slovenia, Croatia and the Czech Republic) have their proportions increased, while some more distant countries (Romania, Russia) have reduced proportions. Many of these changes are mirrored by the East European diffusion potentials that take account of the size of each country's international travel proportion.

The simulated pathway from Romania reveals this country is the major hub in the revised system (Figure 3a). Here, Poland serves as a node for infection transfer to the Baltic States, as do Yugoslavia and Hungary for transmission towards the west. These secondary networks are all subject to negative time differences that arise because the predicted first infection dates occur later than their estimated times.

The network structure obtained at the time of peak HIV prevalence is more complex (Figure 3b). Here, the influence of Romania is much reduced, while the reaches of the hubs in Poland, Yugoslavia and Hungary are extended. In addition, the Ukraine now acts as a secondary node for transmission to Belarus and Russia. The timing errors at peak prevalence are generally much smaller. In some countries close to Western Europe the predicted peak occurs a little later than estimated. The exception here is Albania where this peak occurs five years earlier than the estimate. Similarly, predicted peak times on routes east from Romania are also ahead of their estimates. These outcomes, therefore, provide a more erratic set of vector connections for the East Europe solution than those obtained for all Europe.

5. DISCUSSION

This article has presented both open and closed modelling scenarios for evaluating the effect of the iron curtain on HIV transmission within Europe. It is evident from these solutions that the magnitude of infection transfers along this barrier has been uneven. In this respect, two simulated model outputs

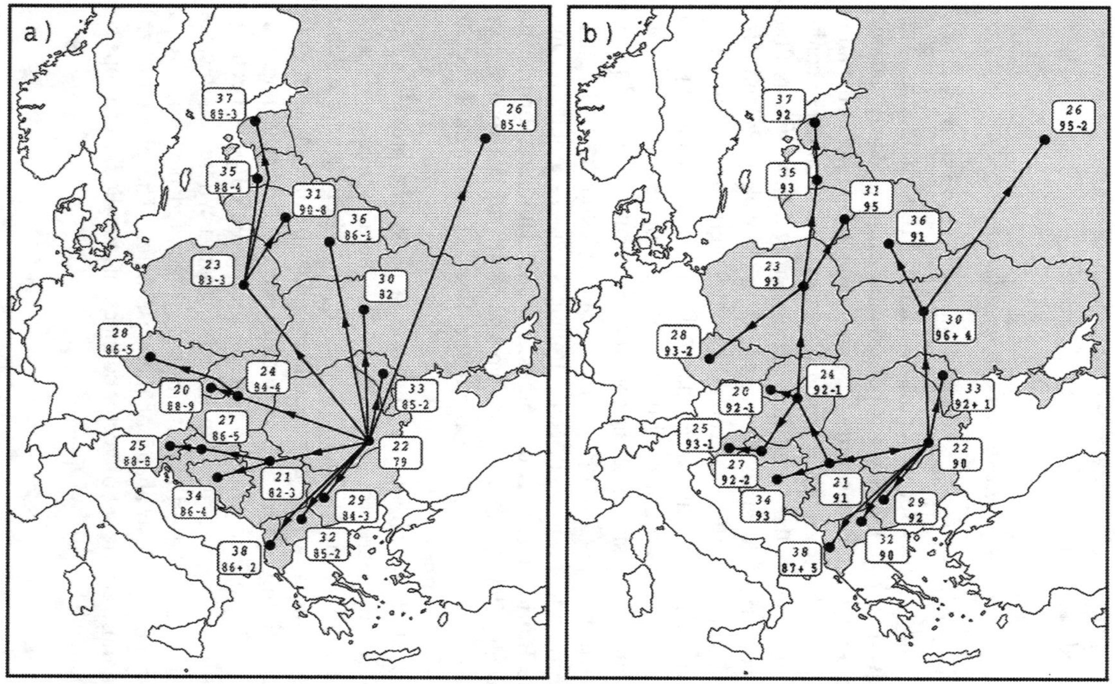

Figure 3. *Eastern Europe (a) Predicted pathway of first HIV infections (dates in Roman numbers). (b) Dominant infection vectors at predicted peak HIV prevalence (dates in bold numbers).*

have proved to be especially responsive to detecting these discontinuities in the transmission process. First, the difference between the observed and predicted times of peak HIV prevalence have indicated the presence of unsynchronised epidemic activity both across the divide and within the rest of the system. Second, the estimated sizes of the national risk populations (n_i, $\forall_i$) have been shown to exert a strong influence on the rate of contact formation between countries. Here, the implications of these statistics are examined within the context of both scenarios.

Different segments of the divide have been shown to exhibit distinctive responses to the circulation of HIV infection. The timing of peak prevalence in the Republics of former Yugoslavia, for example, is accurately explained in the all Europe solution where Italy serves as the principal infection vector for these countries. Similarly, near neighbours Hungary, the Czech Republic and Slovakia all connect more readily to West European infection sources. In the East Europe only solution, however, all these countries are distant from the Romanian origin and their predicted peak times occur later than those inferred from recorded data. These outcomes imply infection transfer to these countries were influenced more by their proximity to West European sources than to other Eastern countries. Moreover, in the Republics of former Yugoslavia this effect is probably reinforced by the comparatively high rates of tourism to this region prior to the Balkans War.

Other countries bordering the east-west divide, however, fit better within the Eastern Europe only scenario. The peak timing of the epidemic in the Baltic States, for example, is best replicated when Romania is the start location. Within the all Europe solution, however, the predicted transfer of HIV transmission to these States from their Nordic neighbours occurs some years later than that observed. In contrast to the 'Yugoslavian' segment, therefore, these outcomes reflect the low visitor rates to the Republics of the former Soviet Union and imply the barriers to travel there effectively curtailed infection transfers from Scandinavia to the Baltic States. This type of effect is even more pronounced in Albania where the population has been isolated from the rest of Europe for the majority of the time HIV has been in circulation. Here, the recorded epidemic lags many years behind predicted incidence in both scenarios, although a Romanian start was found to provide the better fit.

The simulated epidemics for both scenarios are characterised by a hierarchical diffusion process where infection transfers are influenced more by the size characteristics of countries than by their spatial separation. Consequently, risk population size (estimated from recorded incidence as all those ever with AIDS) plays a key role in the evolution of the European epidemic. In this respect, an important influence on the direction of the infection vectors linking countries across the divide is the Germans risk population. Here, this population occurs at a low rate relative to other large West European countries and its size is insufficient to form the principal vector with near neighbours like Poland and the Czech Republic. Instead, the latter all connect with the more prevalent Italian epidemic. The major German links are made with West

European countries like those of Scandinavia where risk populations with high rates of international travel assist the transfer process. The configuration of these populations along this section of the divide, therefore, has served to divert the epidemic wave away from adjacent Eastern Europe countries and buffered infection transfers to them.

Risk population size effects are also evident on the Romanian network that connects with the Republics of the former Soviet Union. Here, erratic timing between nodes occurs in conjunction with large mismatches between estimates cumulative HIV and AIDS incidence. These outcomes are difficult to relate to the epidemic process and might point to inconsistent surveillance and reporting.

Some more general features of the space-time structure of the epidemic also require comment. The nine-fold decline in recorded AIDS incidence on eastern side of the former divide would appear to be strong evidence to suggest the interruption of HIV transmission networks. This contention is reinforced by the predicted HIV prevalence series that support this recorded incidence. In many West European countries this prevalence peaks at, or soon after, 1989, to imply the force of HIV infection there was greatest prior to political restructuring. Consequently, the subsequent impact of improved travel opportunities on the magnitude of infection transfers to the east might have been mitigated by this predicted decline in HIV prevalence. Such coincidence, therefore, helps account for the persistence of relatively low rates of AIDS incidence in many East European countries. This interpretation, however, relies on the vagaries of reported AIDS incidence and might be altered when this research is extended to reconstructing the epidemic to match the more profligate HIV statistics (Smith, 2003). In this last respect, the possible effect of increasing prevalence in the former Soviet Union on the directionality of infection vectors could pose new questions for the maintenance of disease control.

ACKNOWLEDGEMENTS

Phillip Smith thanks the ESRC for postgraduate research award number R42200024393. Graham Bowden drew the figures.

School of Geography,
University of Manchester,
Oxford Road, Manchester M13 9PL, UK

NOTES

[1] The two other states are not taken to be freely circulating according to the argument that the incubation period for HIV is sufficiently long for these patients to be aware of their infectious status and, therefore, make them unlikely to engage in risky contact activity

[2] The regional subscripting is made to allow for the possibility of spatial variations in the value of the transmission probability.

[3] The model described by Equations (12–15) is a Susceptible-Infectious model with no terms for re-entry to the susceptible state. Consequently, unstable solutions will proceed through a single epidemic cycle to reach a zero equilibrium.

[4] The estimates of national HIV prevalence Europe countries are not directly comparable with recorded AIDS incidence. Therefore, the peak time of HIV prevalence in each single-country calibration is taken to be the best estimate of $P(O)_i$, $\forall_i$.

[5] Cumulative HIV prevalence is estimated from information collected at sentinel sites around Europe from individuals with known infection risks (Obadia *et al.*, 1994; Stark and Muller, 1993; Rezza *et al.*, 1993). Their reliability varies with the comprehensiveness of the country's surveillance system.

[6] The starting value given to the number of HIV infections, $y_{i,\,t=0}$, must be related to the region's risk population because their quotient affects the duration of the predicted epidemic cycle (Thomas, 2001). Here, the start population for France is $y_{i,\,t=0} = 81$. This value is scaled to a benchmark of unity for Ireland which is a country with a risk population of intermediate size. The start value for Romania in the Eastern Europe solution is $y_{22,\,t=0} = 10$.

REFERENCES

Anderson, R.M. and R.M. May (1991), *Infectious Diseases of Humans: Dynamics and Control.* Oxford University Press, Oxford.

Baroyan, O.V. and L.A. Rvachev (1967), 'Deterministic epidemic models for a territory with a transport network.' *Kibernetika*, **3**, 67–74.

Bryan, D. and M. O'Kelly (1999), 'Hub and spoke networks in air transportation: an analytical review.' *Journal of Regional Science*, **39**, 275–296.

Deschamps, D., P. Flandre and V. Calvez (2000), 'Mechanisms for virologic failure in previously untreated HIV-infected patients from a trial of induction maintenance therapy.' *Journal of the American Medical Association*, **283**, 205–211.

Flahault, A. and A.J. Valleron (1992), 'A method for assessing the global spread of HIV-1 infection based on air travel.' *Mathematical Population Studies*, **3**, 161–171.

Feigal, E.G. (1999), 'AIDS-associated malignancies: research perspectives.' *Biochimica et Biophysica Acta*, **1423**(1), c1–c9.

Frankenberg, G. (1992), 'Germany: the uneasy triumph of pragmatism.' In D.L.Kirp and R. Bayer (eds.), *AIDS in the Industrialised Democracies; Passions, Politics and Policies*. Rutgers University Press, New Brunswick, N.J., pp. 99–133.

Hersh, B.S. (1991), 'AIDS in Romania.' *Lancet*, **338**, 645–649.

Jacquez, J.A., C.P. Simon and J. Koopman (1995), 'Core groups and the R0s for subgroups in heterogeneous SIS and SI models.' In D.Mollison (ed.),. *Epidemic Models: Their Structure and Relation to Data.* Cambridge University Press, Cambridge, pp. 279–301.

Johnson, A., J. Wadsworth, K. Wellings and J. Field (1994), *Sexual Attitudes and Lifestyles.* Blackwell Scientific Publishing, Oxford.

Knox, E.G., C. MacArthur and K.J. Simons (1993), *Sexual Behaviour and AIDS in Great Britain.* HMSO, London.

Lajmanovich, A. and J.A. Yorke (1976), 'A deterministic model for gonorrhoea in a nonhomogeneous population.' *Mathematical Bioscience*, **28**, 221–236.

Lamptey, P. and D. Tarantola (1998), *HIV/AIDS in Eastern Europe*. WHO, Boston.

Longini, I.M., P.E.M. Fine and S.B. Thacker (1986), 'Predicting the global spread of new infectious agents.' *American Journal of Epidemiology*, **123**, 383–391.

May, R.M., R.M. Anderson and S. Blower (1989), 'The epidemiology and transmission dynamics of HIV/AIDS.' *Daedalus*, **118**, 163–201.

Murray, G.D. and A.D. Cliff (1977), 'A stochastic model for measles epidemics in a multi-region setting.' *Transactions of the Institute of British Geographers (New Series)*, **2**, 158–177.

Nowak, M., R.M. Anderson, A.R. McLean, T. Wolfs, J. Goudsmit and R.M. May (1991), 'Antigenic diversity thresholds and the development of AIDS.' *Science*, **254**, 963–969.

Obadia, Y., D. Rey and J.P. Moatti (1994), 'HIV prenatal screening in south-eastern France: differences in seroprevalence and screening policies by pregnancy outcome.' *AIDS Care*, **6**, 29–38.

Rezza, G., A. De Rose, M. Dorrucci, C. Arpino and I. Serafin (1993), 'Declining prevalence of HIV infection among injecting drug users entering drug treatment in Italy, 1990–1991.' *European Journal of Epidemiology*, **9**, 663–666.

Rosenberg, P.S., H.M. Gail and R.J. Carroll (1992), 'Estimating HIV prevalence and projecting AIDS incidence in the United States: a model that accounts for therapy and changes in the surveillance definition of AIDS.' *Statistics in Medicine*, **11**, 1633–1655.

Sattenspiel, L. and K. Dietz (1995), 'A structured epidemic model incorporating geographic mobility among regions.' *Mathematical Biosciences* **128**, 71–91.

Smith, P.W. (2003), *Multiregion modelling of the HIV/AIDS epidemic in Europe*. Unpublished Ph.D. thesis, School of Geography, University of Manchester (forthcoming).

Stark, K. and R. Muller (1993), 'HIV prevalence and risk behaviour in injecting drug users in Berlin.' *Forensic Science International*, **62**, 73–81.

Thomas, R.W. (1999a), 'Reproduction rates in multiregion modeling systems for HIV/AIDS.' *Journal of Regional Science*, **39**, 359–385.

Thomas, R.W. (1999b), 'Stability and mixing conditions for HIV/AIDS models with regional compartments.' *Journal of Geographical Systems*, **1**, 347–365.

Thomas, R.W. (2000), 'Reconstructing the space-time structure of the HIV/AIDS epidemic for the countries of Western Europe.' *Transactions of the Institute of British Geographers (New Series)*, **25**, 445–464.

Thomas, R.W. (2001), 'Estimated population mixing by country and risk cohort for the HIV/AIDS epidemic in Western Europe.' *Journal of Geographical Systems*, **3**, 283–301.

Thomas, R.W. and T.E. Smith (2000), 'Multiregion contact systems for modelling STD epidemics.' *Statistics in Medicine*, **19**, 2479–2491.

Toubiana, L. and J.F. Vibert (1998), 'An artificial neural network model for the spread of communicable diseases.' In L. Gierl, A.D. Cliff, A.J. Valleron, P. Farrington and M. Bull (eds.), *GEOMED '97; Proceedings of the International Workshop on Geomedical Systems*. Teubner Verlag, Leipzig, 249–259.

UN (1984), *Statistical Yearbook*, **33**, 760–763

UN (2000), *Statistical Yearbook*, **44**, 928–949

WHO (1998), 'The HIV/AIDS situation, June 1998.' *Weekly Epidemiological Record*, **73**, 195–197.

WHO (2000a), *Epidemiological Fact Sheets on HIV/AIDS and Sexually Transmitted Diseases*. UNAIDS/WHO Working Group on Global HIV/AIDS and STD Surveillance, Geneva.

WHO (2000b), *Global Surveillance of Epidemic-prone Infectious Diseases*. Geneva.

Williams, J.S. and P.H. Rees (1994), 'A simulation of the transmission of HIV and AIDS in regional populations within the United Kingdom.' *Transactions of the Institute of British Geographers (New Series)*, **19**, 311–330.

P[illegible] I

C[illegible] M[illegible]

Section A

Simulation Models

Dimitris Ballas, Graham P. Clarke and Ian Turton

A Spatial Microsimulation Model for Social Policy Evaluation

1. Introduction

Evaluation is a critical step in the analysis of social policies which, itself, can influence public thinking (Unrau, 1993; Manski and Garfinkel, 1992). Policy-relevant spatial modelling is an expanding area of research, which has a lot of potential for the evaluation of the socio-economic and spatial effects of major national social policy programmes. However, traditional modelling approaches to social policy analysis usually focus on the impact on the socio-economic structure of the population and they have tended to ignore the geographical dimensions of social policies. In particular, the focus has usually been on the redistributive effects of government policies (such as budget changes and social security benefit policies etc.) between households, but there has generally been a paucity of studies that investigate the spatial impacts of these policies.

This article highlights the importance of evaluating the spatial impact of national social policies. In particular, we demonstrate how spatial microsimulation modelling methodologies can be used to perform detailed micro-spatial social policy analysis. In Section 2, we briefly review existing geographical approaches to social policy analysis. Section 3 introduces the microsimulation method and Section 4 describes *SimLeeds*, which is a spatial microsimulation model aimed at modelling the Leeds urban economic system. Then, Section 5 shows how we used *SimLeeds* to analyse the spatial impacts of social policy change. In particular, *SimLeeds* is employed to analyse the impact of recent UK budget changes on various Leeds localities. Further, *SimLeeds* is used to estimate the impact of the proposals for social change suggested by rival political alternative parties. Finally, Section 6 offers some concluding comments.

2. Geographical Approaches to the Evaluation of Social Policies

It has been argued elsewhere (Ballas and Clarke, 2001a; Hamnett, 1997) that, although there have been a number of studies exploring the geography of the changing welfare state (Bennett 1980; Curtis 1989; Pinch, 1997) and the geography of poverty and income inequalities (Atkinson, 1996; Green, 1996 and 1998; Hills, 1996; Joseph Rowntree Foundation, 1995; Philo, 1995), there is a general paucity of information relating to the geography of household income, wealth, taxation and welfare benefits. Here, we briefly review past geograph-

B. Boots et al. (eds.), Modelling Geographical Systems, 143–168.

ical approaches to social policy and welfare analysis. Bennett's (1980) comprehensive work on the geography of public finance is a good starting point. He proposes for a geographical analysis of public finance, which comprises the analysis of the spatial patterns of revenue raising, public expenditure and the spatial balance between revenue and expenditure. Bennett adds a spatial dimension to traditional public finance analysis by including questions that are concerned with the distributions of expenditure, wealth and income between people, another set of questions concerned with the spatiality of this distribution. In other words, he puts the question of how burden and public expenditure vary as a function of geographical location:

> ". . . *who gets what benefits from public finance* as a function of *where* the individual lives and *where* the industrial enterprise is located"
>
> (Bennett, 1980, p. 1; emphasis in the original)

According to the *welfare economics* perspective to social policy analysis (Bramley and Hill, 1986; Musgrave and Musgrave, 1984), the state has three main functions: *allocation*, *distribution* and *stabilisation*. Bennett (1980) examines these state functions in a geographical context and he investigates the spatial aspects of public goods, revenue burden and benefits (see also Pinch, 1997).

It should be noted that a very important research area in the analysis of social policy is the measurement of inequalities and poverty. It has long been argued that there is a need for detailed geographical analysis of poverty at various scales. McCormick and Philo (1995) discuss the geographical dimensions of poverty and point out that much of the poverty in the UK is hidden, in the sense that *poor* people and localities are largely invisible. Further, they argue that poverty in these localities is not only the result of economic decline, reflected as shifts in demand for specific labour market skills, but also it is the cause of the decline. In particular, it seems that there is a vicious circle of poverty in these areas:

> "Reducing the purchasing power of low-income households – which are likely to spend money locally rather than saving it, spending it elsewhere or using it on expensive imported goods – damages the economies of the 'poor places'. A gradual process of uncoupling hence occurs between local and national economic trends, such that improvements in the latter may no longer feed through into any improvement in the former."
>
> (McCormick and Philo, 1995, p. 11)

In addition, Martin (1995) argues that there have been sustained inequalities in the distribution of income across the regions of Britain and stresses the increasing North-South divide. Using income inequality statistics he demonstrates the disparities between the North and South region and he discusses the political and economic causes of these disparities. Moreover, McKendrick (1995) points out that a less obvious, but equally significant division is that between the Celtic nations of the UK (Scotland, Wales and Northern Ireland) and England.

At a smaller area level, Green (1996) argues for a spatial perspective on poverty and wealth and presents selected evidence on changes and continuities in poverty and wealth between 1981 and 1991 at the spatial levels of *electoral ward, local authority district* and *local labour market area*. In addition, Goodwin (1995) examines the distribution of poverty at the intra-urban scale. He points out that one of the problems faced by those attempting to investigate urban poverty is the lack of any systematic small area level data on household income, wealth or living standards.

> ". . . an immediate problem faced by those attempting to assess the extent and the shape of urban poverty is the difficulty of obtaining reliable data on variables such as income, especially at those geographical scales which allow comparisons to be made within, as well as between, urban areas."
> (Goodwin, 1995, pp. 66–67)

Further, he addresses the problems of the most appropriate geographical scale for the analysis of urban poverty and, although he presents mapping scores of standard deprivation indices for Greater London wards, he also stresses the importance of the subjective dimensions of deprivation:

> "People experience these deprivations differently, and we should perhaps speak of their varying *experiences* of poverty. Different groups are affected in different ways, although some sections of society are more prone to poverty than are others." (Goodwin, 1995, p. 78; emphasis in the original)

Another example of poverty analysis at the ward level is the work of Dorling and Tomaney (1995). Their analysis was based on the 'five great evils' of *want*, *ignorance*, *idleness*, *squalor* and *disease*, as defined by William Beveridge in the 1940s (Dorling and Tomaney, 1995). Using data from five different sources, they generated contemporary indicators for these 'evils' at the ward level for England and Wales and they produced cartograms reflecting the geographical complexity of poverty.

Nevertheless, there is a general lack of work on analysis of poverty and the evaluation of social policies at the household or individual level. A notable exception here is the work of Noble and Smith (1996) who examined the spatial patterns of income and wealth in Oxford and Oldham at the intra-urban level, using data from the housing benefit/council tax benefit (HB/CTB) systems at the individual claimant level in an anonymised form. Further, they used these data to distinguish which claimants are in receipt of Income Support (IS) from those who are otherwise on low income and receive HB/CTB. Moreover, they used the postcode attribute of each individual record to assign to a respective enumeration district (ED). They then constructed an index of low income and performed a cluster analysis, comparing the intra-town spatial patterns of Oxford and Oldham. This analysis showed that although the overall proportions on benefit are similar, the geographical distributions at the ED level within the two towns are quite different.

Analysis at the ED and individual level, however, is rare because there is a lack of information or data at this scale. In general, there is very little published below the scale of the region. For example UK national publications such as 'Regional Trends', the 'New Earnings Survey' and the 'General Household Survey' at best disaggregate to the level of *Yorkshire and Humberside*. Green (1998) provides an excellent overview of the advantages and disadvantages of these published data sources.

Other notable exceptions of research on income and wealth at the small area level is the work of Birkin and Clarke (1989) and Vencatasawmy *et al.* (1999). In particular, Birkin and Clarke (1989) used the SYNTHESIS model to generate incomes for individuals. In other words, they added the income variable to their spatially disaggregated database. After arguing for the need for small area income data and reviewing the relevant work they followed an iterative proportional fitting (IPF) based microsimulation approach to estimate earned income at ward level for the Leeds Metropolitan District. In addition they estimated the income from transfer payments such as the *Family Income Supplement* for each household. This was probably the first successful attempt to generate income at the small area level in the UK.

Williamson (1992) has also included gross income as a variable in his OLDCARE microsimulation model. The latter was used for the spatial analysis of community care policies for the elderly. In particular, one of the main aims of OLDCARE was to estimate the prevalence and severity of disability amongst the elderly. The model aimed at synthetically reconstructing a micro-population and it applied the IPF technique on data from the 1981 Census of UK population and other sources. Moreover, Williamson and Voas (2000) report ongoing research to provide more robust and reliable estimates of income at the small area level. They argue that income estimation at the small area level may be seen as a multilevel analysis problem where variables at individual and area levels may interact. In addition, Bramley and Lancaster (1998) also developed a small area income model based largely on the number of workers in a household, economic activity (particularly working versus non-working households) and household tenure. They present their results for postal sectors in Edinburgh. Furthermore, Higgs (1994) and Longley *et al.* (1996) examined the impacts of changes to the rates of council tax in small-areas of Cardiff.

These are rare examples of income and wealth studies and impact assessments relating to tax and benefits at the small area level. It can be argued that this kind of analysis is becoming increasingly important, given that a large (may be the largest) part of the social and economic needs of residents in deprived areas is not met by area-based policies but by national social policies. As Edwards (1995) puts it:

> ". . . by far the greatest part of the social and economic needs of inner city residents will not be met by urban specific policies but by mainline housing, health, income support and education provision" (Edwards, 1995, p. 701)

This quotation suggests an agenda for urban and regional planning based on social policies rather than area-based policies. Edwards (1995) suggests that the allocation of extra resources for areas deemed to be deprived may only ever lead to cosmetic improvements. Thus he calls for an alternative research agenda based on individual households and their incomes (earned or received as transfer payments). Hence, he continues:

> ". . . inner city residents and in particular the urban deprived do receive most of their welfare by way of mainstream programmes and we know precious little about the effectiveness of such programmes either in targeting the deprived, spatially or otherwise, or in providing for the (sometime) multiplicity of needs or the different or additional needs that may be found within individual households." (Edwards, 1995, p. 711)

Spatial microsimulation provides an ideal framework to address these issues (Ballas and Clarke, 2001a). Changes in national social policies have major implications for household incomes in certain areas. Nevertheless, such changes have been rarely analysed at the small area level. In this chapter we build on previous work that has been done on this area (e.g. Ballas and Clarke, 2001a) and we use spatial microsimulation methodologies to evaluate some of the social and spatial impacts of the UK 2000 Budget changes for Leeds.

3. THE MICROSIMULATION METHOD

Microsimulation models aim at building large-scale data sets on the attributes of individuals or households (and/or on the attributes of individual firms or organisations) and at analysing policy impacts on these micro-units (Orcutt *et al.*, 1986; Birkin and Clarke, 1995; Clarke, 1996). Further, by permitting analyses at the level of the individual, family or household they provide the means of assessing variations in the distributional effects of different policies (Hancock and Sunderland, 1992; Mertz, 1991). In addition, microsimulation modelling frameworks provide the possibility of defining the goals for economic and social policy, the instruments employed and also the structural changes of those affected by socio-economic policy measures (Krupp, 1986). Micro-simulation methodologies have become accepted tools in the evaluation of economic and social policy and in the analysis of tax-benefit options and in other areas of public policy (Hancock and Sunderland, 1992).

The microsimulation procedure typically involves four major procedures:

1. The construction of a micro-data set (when this is not available);
2. Monte Carlo sampling from this data-set to 'create' a micro-level population;

3. *What-if* simulations, in which the impacts of alternative policy scenarios on the population are estimated;
4. Dynamic modelling to update a basic micro-data set.

Although the number of published microdata sets is increasing around the world, there are still many cases in which this kind of data is not available at the desired spatial scale (Clarke, 1996; Birkin and Clarke, 1995). In the case of the United Kingdom, the only official government published micro-data set is the Sample of Anonymised Records (SARs). These are samples of individual census records that are anonymised in various ways ensuring that there is no breach of the confidentiality of the census and that no individual can be identified from the data (Middleton, 1995; Marsh, 1993). However, the spatial scale at which these data sets are released is, at best, the regional or district scale. In cases when official microdata are not available at the desired spatial scale then this kind of data can be estimated with the use of existing data sets and a variety of techniques ranging from iterative proportional fitting methods to linear programming and complex combinatorial optimisation methods (Williamson, Birkin and Rees, 1998; Ballas and Clarke, 2000; Ballas, Clarke and Turton, 1999). These techniques make up stages 1 and 2 of the microsimulation procedure outlined above.

Once a spatially dissaggregated microdata set is built it is then possible to move to stage 3 and perform *what-if* policy analysis by changing the micro-unit attributes accordingly. In particular, once a spatial microsimulation database is built, the user can change any of the variables to test the impacts of various policy initiatives (Birkin, Clarke and Clarke, 1996).

This style of analysis can become even more sophisticated if dynamic procedures are incorporated into a spatial microsimulation model. In particular, the next stage in spatial microsimulation modelling is to update the micro-database and perform policy analysis in a dynamic fashion. This would involve the prediction of the micro-unit attributes and their behavioural responses over a period of time under different policy and demographic scenarios.

4. THE *SIMLEEDS* MODEL

SimLeeds is a product of on-going research and uses different approaches to conditional probability analysis for microsimulation modelling. In particular, it provides different options of microsimulation modelling ranging from Iterative Proportional Fitting (IPF) based microsimulation to Simulated Annealing reweighting approaches (for more details on the different methodologies employed by *SimLeeds* see *Ballas et al.*, 1999). The nature and number of the variables of the household or individuals modelled by *SimLeeds* depend on the selected modelling method and on the needs of the user. They also depend on time constraints, given that microsimulation modelling is highly computationally intensive.

Here, we use a version of *SimLeeds* that adopts a combinatorial optimisation simulated annealing approach to generating a spatially disaggregated population microdata set that matches given small area constraints.[1] In particular, we adopted a similar methodology to that of Williamson *et al.* (1998) who explored different solutions to finding the combination of household records from the SAR microdata, which *best fit* known small area Census tabulations. *SimLeeds* uses a similar object-oriented simulated annealing algorithm to estimate a spatially disaggregated micro-data base for Leeds (for more details on the algorithm used and the accuracy of the simulations see Ballas, Clarke and Turton, 1999).

Using the database listed in Table 1 it is possible to explore the interdependencies of any household attributes at the micro-scale. For example, Figure 1 depicts the spatial distribution of black males, aged between 16–29 and living in rented accommodation This is a typical output of spatial microsimulation models: distributions of population sub-groups for which published data do not exist.

Most of the attributes of the *SimLeeds* micro-database are census attributes. However, based on these attributes, and using various methodologies, it is possible to estimate new variables and add non-census attributes. As pointed out above, one of the most important non-census attributes is household *income*. For this study we used data from the 2000 New Earnings Survey (ONS,[2] 2000) in order to estimate the earned income of each micro-unit in our database. In particular, we used data on the income distribution by sex and occupation for the Yorkshire and Humber region and data on the income distribution by age and sex. Using a similar IPF[3] based methodology to that of Birkin and Clarke (1989), we combined different cross-tabulations from the New Earnings Survey in order to estimate the earned income for each individual on the basis of their socio-economic and demographic attributes (Figure 2). Our future work will explore more systematically different methodologies and approaches for the estimation of income and other non-census variables in a spatial microsimulation context.

Having estimated earned income, we also calculated the tax paid by each individual household based on the 1999–2000 bands of taxable income (see Table 2). Figure 3 depicts the spatial distribution of estimated tax paid in Leeds at the enumeration district level. This variable is the starting point for the analysis of many *what-if* policy changes (see next section and Ballas and Clarke, 2001). Another useful household attribute is income from benefits. The *SimLeeds* database comprises various benefit attributes such as Child Benefit and Working Families Tax Credit. The Working Families Tax Credit (WFTC) is a tax-free benefit for low-paid workers with children. It is noteworthy that the WFTC replaced the Family Credit benefit in October 1999. There are several rules for the eligibility and calculation of the amount of the WFTC (for more details see George *et al.*, 1999). These rules were incorporated in the *SimLeeds* model to assign WFTC to all the eligible households of the microsimulated database. Figure 4 represents the estimated spatial distribution of WFTC paid in Leeds.

Table 1. SimLeeds Attributes

Name	Label	Name	Label
Areahh	Household SAR area	Dhdecpos	Economic position of household head
Bath	Availability of bath/shower	Dhdage	Age of household head
Cenheat	Availability of central heating	Dhdsex	Sex of household head
Insidewc	Availability of inside WC	Dhdclass	Social class of household head
Cars	Number of cars	Qsubgrp	Subject group of highest qualification
Hhsptype	Household space type	Indusdiv	Industry (SIC Divisions)
Hhspindw	Number of household spaces in dwelling	Occmajor	Occupation: SOC Major groups
Roomsnum	Number of rooms in household space	Occsubmj	Occupation: SOC Sub-major Groups
Tenure	Tenure of household space	Occminor	Occupation: SOC Minor groups
Persinhh	Number of persons in household	Hhdcomp	Household composition type
Age	Age	Hdeptype	Household dependant type
Cobirth	Country of birth	Dfhage	Age of head of family
Econprim	Economic position (primary)	dfresid	Number of residents in family
Econsec	Economic position (secondary)	dfdepch	No. dependent children in family
Empstat	Employment status	dfolddc	Age oldest dependent child in family
Ethgroup	Ethnic group	dfyngdc	Age youngest dependent child in family
Famnum	Family number	dfadult	No. adults resident in family
Famtype	Family type	dfchild	No. under-16 year olds resident in family
Hours	Hours worked weekly	dfpensr	No. pensioners resident in family
Industry	Industry	dfltill	No. persons with ltill resident in family
Ltill	Limiting long-term illness	dfemp	No. persons in employment resident family

Mstatus	Marital status	dfecact	No. econom. active residents in family
Migorgn	Migrant – area of former residence	dfunemp	No. of unemployed residents in family
Occpatn	Occupation	dfretire	No. of retired residents in family
qualnum	No. of higher educational quals	dfpsick	No. permanently sick residents in family
qualevel	Level of highest qualification	dfinact	No. econom. inactive residents in family
residsta	Resident status	dfother	No. residents other inactive in family
sex	Sex	dfstuds	No. students enum.at tt address in family
soclass	Social Class (based on occupation)	dfdeps	No. dependants resident in family
segroup	Socio-Economic Group	dfolddep	Age oldest resident dependant in family
tranwork	Mode of transport to work	dfyngdep	Age youngest resident dependant infamily
dhresid	No. of residents in household	dfhecpos	Economic position of head of family
dhdepch	No. of dependent children in h'd	dfhsex	Sex of head of family
dholddc	Age of oldest dependent child in h'd	dfhclass	Social class of head of family
dhyngdc	Age of yngest dependent child in h'd	dhstuds	No. of students at term-time address in
dhadult	No. of adults in household	dhdeps	No. of dependants in household
dhchild	No. of under 16s in household	dholddep	Age of oldest dependant in
dhpensr	No. of pensioners in household	dhyngdep	Age of youngest dependant in hhold
dhltill	No. of LTILL persons in household	dallstud	DALLSTUD: All student household
dhemp	No. of persons in employment in h'd	dallpens	DALLPENS: All pensioner household
dhecact	No. of persons economically active in h'd	dalladlt	All-adult hhold
dhunemp	No. of unemployed persons in h'd	EarnedIncome	Earned income (annual)
dhretire	No. of retired persons in h'd	Tax	Income Tax paid
dhpsick	No. of perm. sick persons in h'd	Ret'mentPension	Retirement Pension (per week)
dhinact	No. of econ. inactive persons in h'd	WFTC	Working Families Tax Credit (per w)
dhother	No. of persons other inactive in h'd	ChildBenefit	Amount of child benefit (per week)

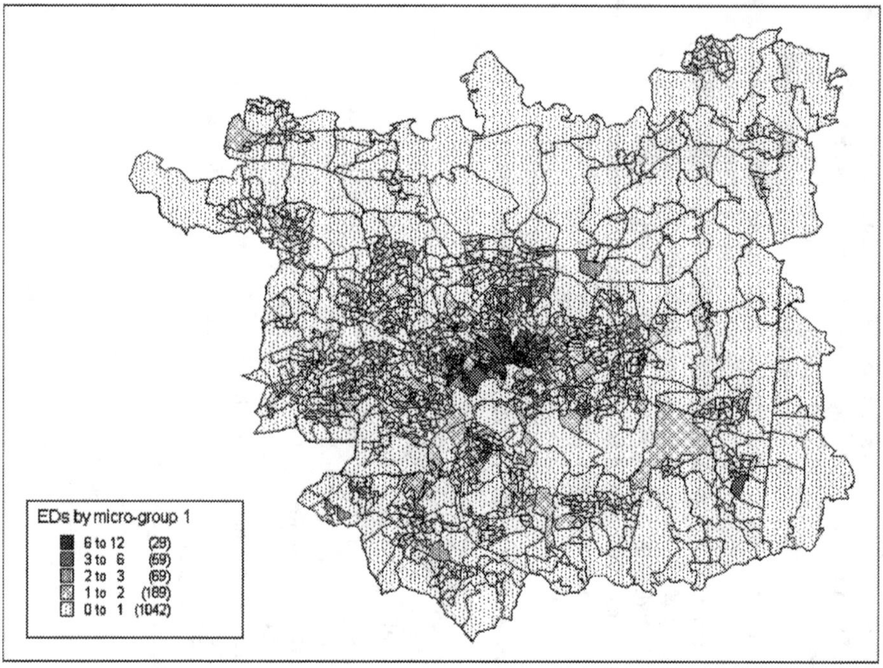

Figure 1. Estimated spatial distribution of black males, aged between 16–29 and living in rented accommodation.

Table 2. 1999–2000 Tax Bands

Bands of taxable income (£)	1999/2000
Lower rate – 10%	£0–£15,000
Basic rate – 23%	£15,001–£28,000
Upper rate – 40%	Over £28,001

(*Source:* Welfare Benefits Handbook, 1999, p. 1: xxix)

5. USING *SIMLEEDS* FOR MICRO-SPATIAL SOCIAL POLICY ANALYSIS

5.1. Evaluating the 2000 UK Budget Changes

The focus of social policy research has traditionally been on the redistributive effects of government budget changes and other policy developments, but the geographical dimension of national social policies has been frequently ignored. It can be argued that the *SimLeeds* model outputs presented in the previous section can be potentially useful for spatial policy analysis. In this section we demonstrate the policy relevance of spatial microsimulation frameworks. In par-

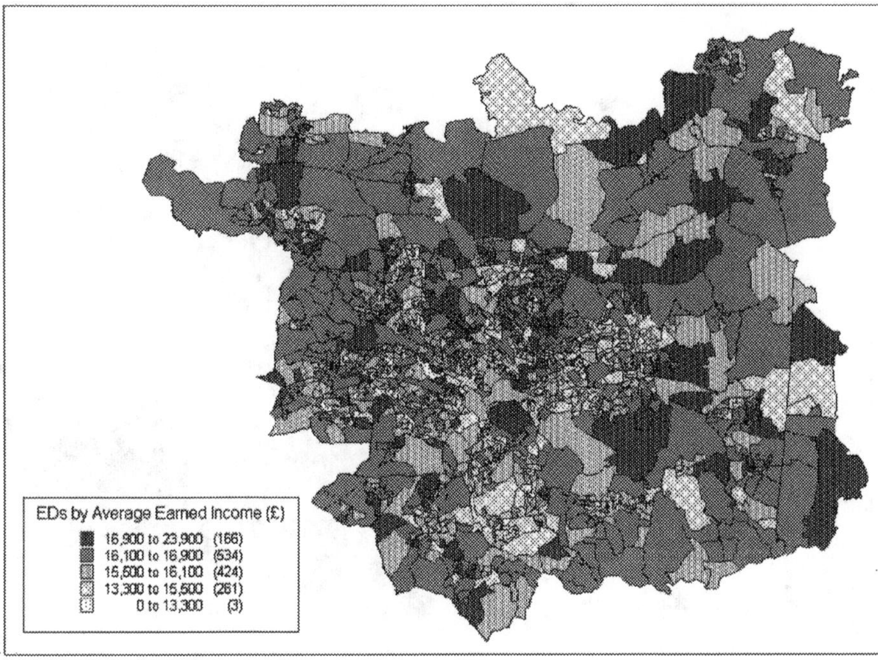

Figure 2. Estimated spatial distribution of average earned income in Leeds.

ticular, we use *SimLeeds* model outputs to evaluate the spatial and social impact of the recent UK government budget changes (HM Treasury, 2000).

The 2000 UK Budget aimed at releasing substantial new resources for education, health, transport, law and order, tackling child poverty and supporting pensioners changes (HM Treasury, 2000). There is a wide range of budget changes that can be evaluated in a spatial microsimulation framework. For this section we focused on the following changes:

- Change of the basic tax rate from 23% to 22% and withdrawal of the married couples allowance and tax allowance for children
- Increases in Working Families Tax Credit
- Increases in Child Benefit
- Concessionary TV licences for those aged 75 and over
- Increase in Winter Fuel Allowance by £50

We have presented the estimated spatial distribution of tax paid which was based on the 1999–2000 bands of taxable income and allowances. In this respect, one of the budget changes was the decrease of the basic income tax rate to 22% and the abolishment of the married couples allowance and the tax allowance for children (which were £1,970 each). Clearly, these changes will have a different impact on different types of households and individuals. They will also have different impacts in different localities. It can be reasonably expected

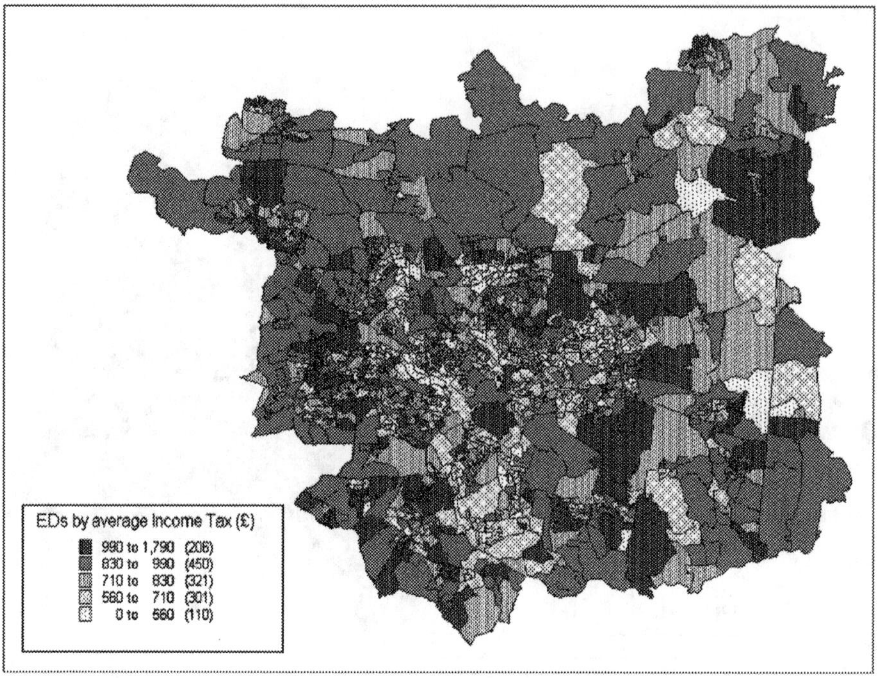

Figure 3. Estimated spatial distribution of average income tax in Leeds.

that low income as well as the comfortably secure households will be generally better off under the budget change. Nevertheless, the position of the poorest households will remain the same in absolute terms but worse relatively. This relationship presumes the relatively wealthier households will become better off in absolute terms. Also, married couples and households with dependent children will become worse off, because of the withdrawal of the tax allowances.

Given this context, we used *SimLeeds* to estimate the tax paid by each household in Leeds under the budget changes. Figure 5 depicts the spatial distribution of the difference in tax paid at the ED level in Leeds, whereas Table 3 gives these estimates at the electoral ward level. It is noteworthy that, as can be seen in Table 3, the estimated increase in tax per household in Headingley and University wards is significantly less compared to the impact on other areas. This can partly due to the fact that these areas are populated predominantly by students and younger people who are not affected by the withdrawal of the married couples allowance and tax allowance for children.

It should be noted that these total tax changes in each area translate into changes in the total disposable income of households, which will in turn have different negative multiplier effects in different localities. These effects can be estimated and analysed at the regional level (see for instance Armstrong, 1993; Armstrong and Taylor, 1993) or at the local level with the use of spatial microsimulation methods (see Ballas and Clarke, 2001b, 2000b).

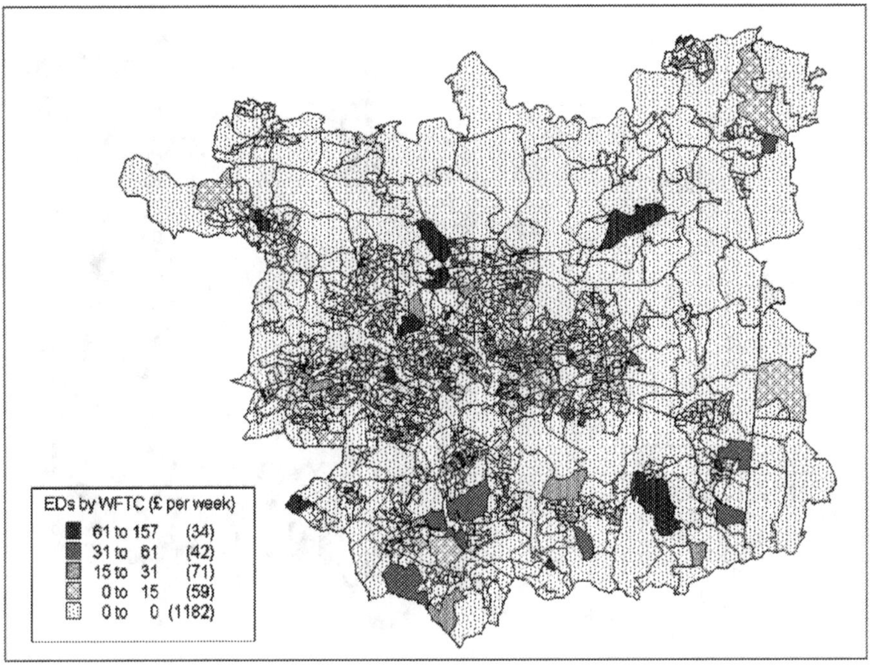

Figure 4. Estimated spatial distribution of WFTC.

Another budget change that can be evaluated in a spatial microsimulation framework is the increase in Working Families Tax Credit (which will also be integrated with the new child tax credit – for more details see HM Treasury, 2000). The spatial distribution of WFTC before the budget changes was shown in the previous section (Figure 4). Figure 6 depicts the spatial distribution of the total weekly increase of WFTC, which is proposed under the Budget. As can be seen there particular localities around Leeds with relatively high concentrations of households that will benefit from the WFTC increase. Further, although these localities are scattered all around Leeds, there are relatively more affected areas in inner Leeds and in the south of the city.

The WFTC increase will partially offset the negative impact of the withdrawal of the tax allowances for married couples with children. Table 4 shows the impact of the budget on a typical type of household that will be affected positively from the WFTC increase, whereas Figure 7 depicts the spatial distribution of this type of household in Leeds.

The new Budget included increases in the child benefit. In particular, the child benefit increased from £14.40 to £15.50 per week for the eldest child. In addition, it increased from £17.10 to £17.50 per week for lone parents. *SimLeeds* can be used to calculate the child benefit for eligible households before the budget change and to estimate the spatial impact of the increase. Figure 8 depicts the geographical distribution of the Child Benefit increase at the ED level.

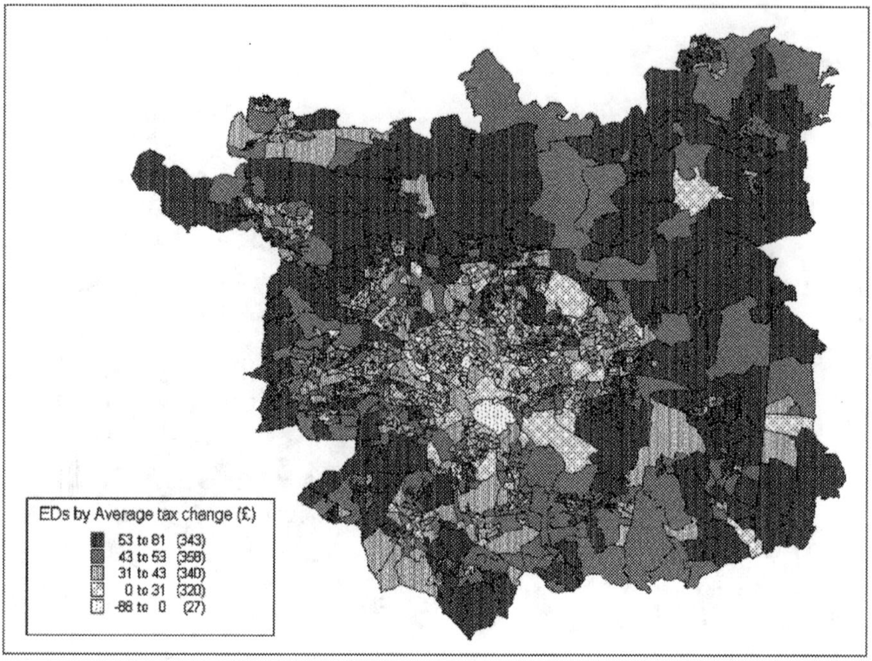

Figure 5. Spatial distribution of average change in tax paid (after Budget changes).

SimLeeds can be employed for policy analysis in the modelling of the impact that the Budget changes will have on elderly people. Recall, the 2000 Budget included increases in Winter Fuel payments for the elderly as well as free TV licences for people aged over 75 years. It also included a £3 increase in the basic state pension. However, it should be noted that the socio-economic position of pensioners is very hard to quantify, given that they are economically inactive and many of them have occupational pensions and considerable savings and capital. There is a challenge, therefore, to use spatial microsimulation techniques to estimate more variables relating to household wealth. A notable example of this kind of work is given by Caldwell *et al.* (1998) who review the geography of wealth and show how a US microsimulation model has already included many variables relating to assets and debts. Amongst our future research priorities is to translate this work from the US to British cities and regions. This would give a better picture of the socio-economic position and quality of life of population groups such as the elderly and the economically inactive. For the purposes of this paper, however, we estimated Retirement Pension crudely by using the appropriate rules:

> "You can get a retirement pension when you reach pensionable age. Pensionable age is currently 65 for a man and 60 for a woman"
>
> (Welfare Benefits Handbook, 1999, 1, p. 137)

Table 3. Estimated Average Tax Change by Ward.

Name	Total tax change	Total change per household
Aireborough	£393,097.56	£39.80
Armley	£194,546.59	£22.20
Barwick and Kippax	£293,752.42	£33.32
Beeston	£120,534.96	£16.56
Bramley	£270,621.76	£30.79
Burmantofts	£ 97,142.55	£11.81
Chapel Allerton	£153,411.06	£16.99
City and Holbeck	£125,605.34	£13.46
Cookridge	£287,179.77	£32.83
Garforth and Swillington	£393,578.35	£43.77
Halton	£383,712.57	£44.49
Harehills	£144,915.83	£17.45
Headingley	£ 27,206.98	£ 3.75
Horsforth	£286,073.17	£33.61
Hunslet	£142,821.26	£22.68
Kirkstall	£157,705.56	£18.26
Middleton	£178,351.61	£24.34
Moortown	£285,707.78	£34.17
Morley North	£291,104.79	£32.81
Morley South	£400,259.55	£38.69
North	£339,933.87	£38.63
Otley and Wharfedale	£223,941.33	£23.65
Pudsey North	£332,496.58	£36.17
Pudsey South	£247,128.18	£28.86
Richmond Hill	£162,229.38	£19.37
Rothwell	£352,524.07	£44.29
Roundhay	£241,621.71	£29.50
Seacroft	£133,564.82	£17.14
University	£ 46,612.63	£ 5.58
Weetwood	£139,457.03	£19.14
Wetherby	£399,690.42	£41.90
Whinmoor	£239,524.08	£31.36
Wortley	£323,022.22	£33.62

There are 4 different categories of retirement pensions and different rules determine the entitlement to these (see Table 5).

It can be reasonably expected that localities with large amounts of Pension paid out (Figure 9) will benefit the most from the £3 weekly increase, which was announced in the Budget speech (HM Treasury, 2000). However, the Pension estimates are crude, given they are based on the assumption that all pensioners receive pensions of the same type. In other words, private and state *Earnings Related* pension schemes were not taken into account. Nevertheless, the preliminary analysis presented here demonstrates the potential of spatial

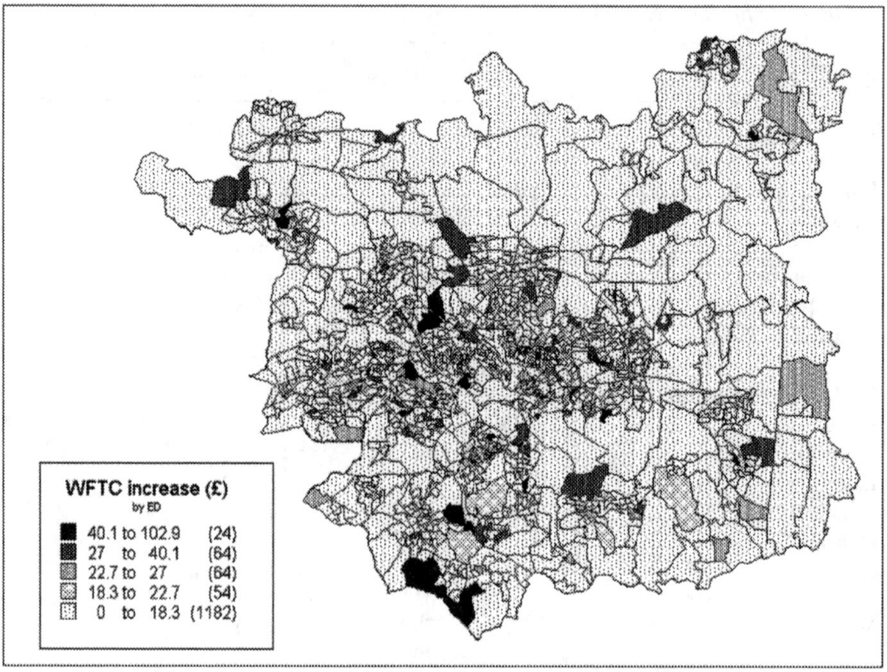

Figure 6. Estimated spatial distribution of the weekly increase in WFTC in Leeds.

Table 4. Impact On Married, Two Employed, Two Children Households.

Gross income (per person)	Pre Budget net	Win (lose)	Working families tax credit Pre Budget	Integrated child tax credit Post Budget
£5,000.00	£10,934.00	£34.00	£155.10	£159.94
£10,000.00	£17,915.00	£69.00	£81.26	£85.74
£15,000.00	£24,784.00	£169.00	£10.39	£13.81
£20,000.00	£31,584.00	£269.00	–	–
£25,000.00	£38,384.00	£369.00	–	–
£30,000.00	£45,620.00	£105.00	–	–
£40,000.00	£58,928.00	£619.00	–	–
£50,000.00	£70,928.00	£619.00	–	–
£60,000.00	£82,928.00	£619.00	–	–
£70,000.00	£94,928.00	£619.00	–	–
£100,000.00	£130,928.00	£619.00	–	–

(*Source:* After The Guardian, 22 March 2000a, p. 16.)

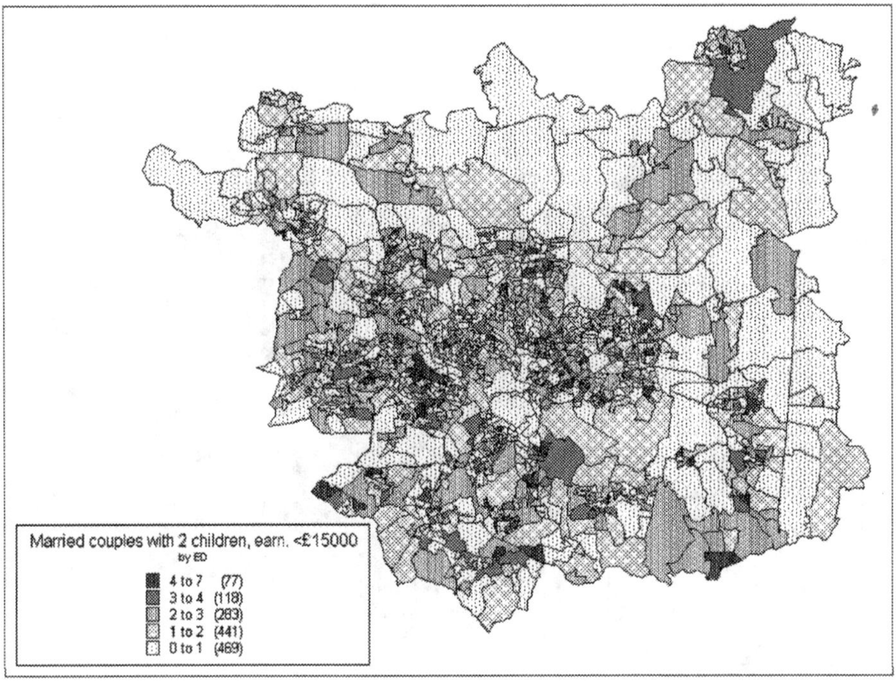

Figure 7. Spatial distribution of married couples with two children, with gross income (per person) less than £15,000.

microsimulation for the evaluation of changes to pensioner policies, especially when wealth variables are taken into account. Tables 6 and 7 show how typical pensioner households that will be affected from the Budget changes. In particular, Table 7 describes how a typical pensioner couple will be affected by the Budget.

Using *SimLeeds*, we estimated the spatial distribution of these types of households. Figure 10 depicts this distribution, whereas Figure 11 shows the spatial impact of the increase in Winter Fuel payment.

5.2. *The Conservative Party's Plans for Pensioners*

Spatial microsimulation frameworks provide a capability for evaluating the impact of social policies. Therefore, a further application of spatial microsimulation models is to compare the possible impact of competing policies. Here, we estimate the possible spatial impact of the Conservatives' propositions for pensions, as they were announced in a speech by the then leader of the Conservative Party, William Hague (The Guardian, 2000c). He dismissed the government's measures for pensioners as "gimmicks" and argued that the funds allocated for the introduction of free TV licences for the elderly and for increases in Winter Fuel payments, should be directed into the basic pension. Further,

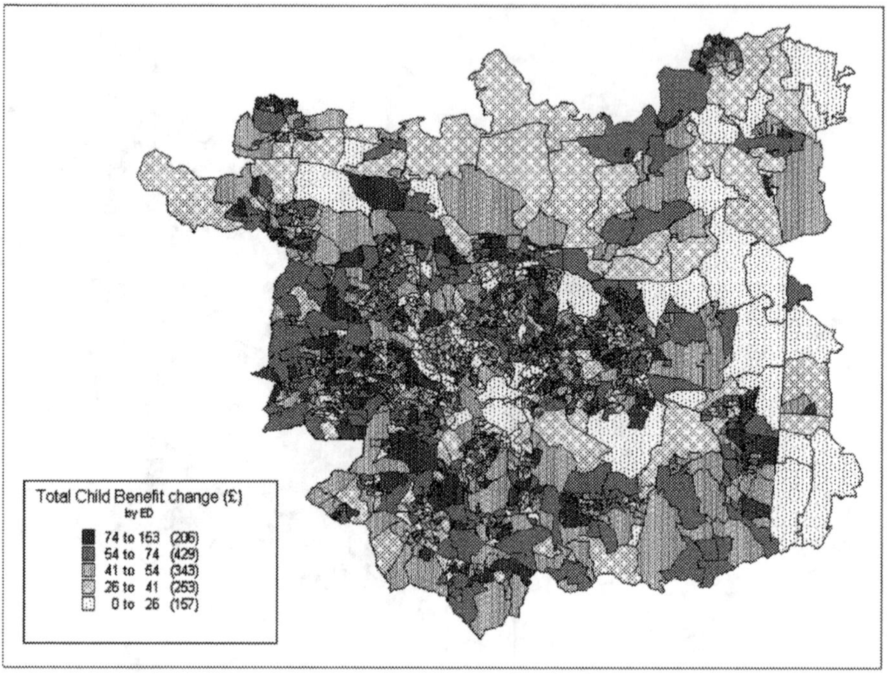

Figure 8. Estimated spatial distribution of total change in Child Benefit.

Table 5. Categories of Retirement Pension.

Retirement Pension – The amount of Benefit	
Category A retirement pension	
Claimant	£66.75
Adult dependant	£39.95
Eldest eligible child	£ 9.90
Other children(each)	£11.35
Category B retirement pension for a married woman	
Claimant	£39.95
Eldest eligible child	£ 9.90
Other children(each)	£11.35
Category B retirement Pension for a widow or widower	
Claimant	£66.75
Eldest eligible child	£ 9.90
Other children (each)	£11.35
Category D retirement pension	
Claimant	£39.95

(*Source:* Welfare Benefits Handbook, 1999, 1: pp. 143–145.)

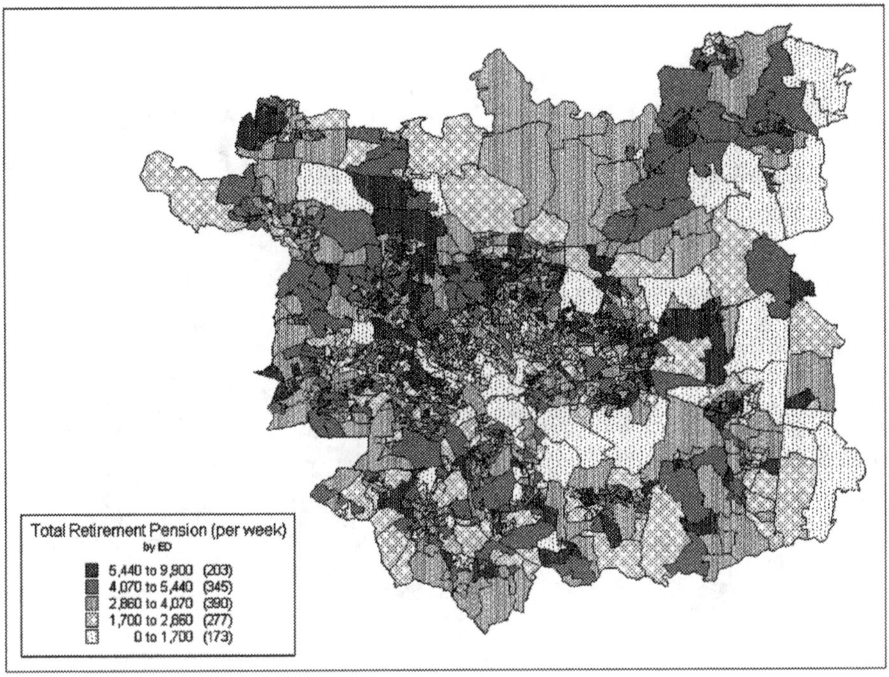

Figure 9. Estimated spatial distribution of total retirement pension.

Table 6. Budget Impact on Pensioner Couples.

Gross income (per person)	Pre Budget net	Post Budget net	Win (lose)
£4,000.00	£ 8,000.00	£ 8,000.00	£ 0.00
£8,000.00	£15,671.00	£15,681.00	£ 10.00
£12,000.00	£22,014.00	£22,120.00	£106.00
£15,000.00	£26,634.00	£26,800.00	£166.00
£20,000.00	£33,675.00	£33,863.00	£188.00
£25,000.00	£41,125.00	£41,413.00	£288.00
£30,000.00	£48,781.00	£49,094.00	£313.00

(Source: The Guardian, 22 March 2000b.)

the Conservatives criticised the government's plans for pensioners on the grounds of high administrative costs and their patronising nature. Thus, they suggested an alternative package which included an up-rating of pensions in April 2001. Table 8 summarises these proposals.

As was the case with the Budget changes, it is possible to use *SimLeeds* to estimate the spatial and social impact of the Conservatives proposals on the basis of the estimated spatial distribution of the population groups affected. Figure

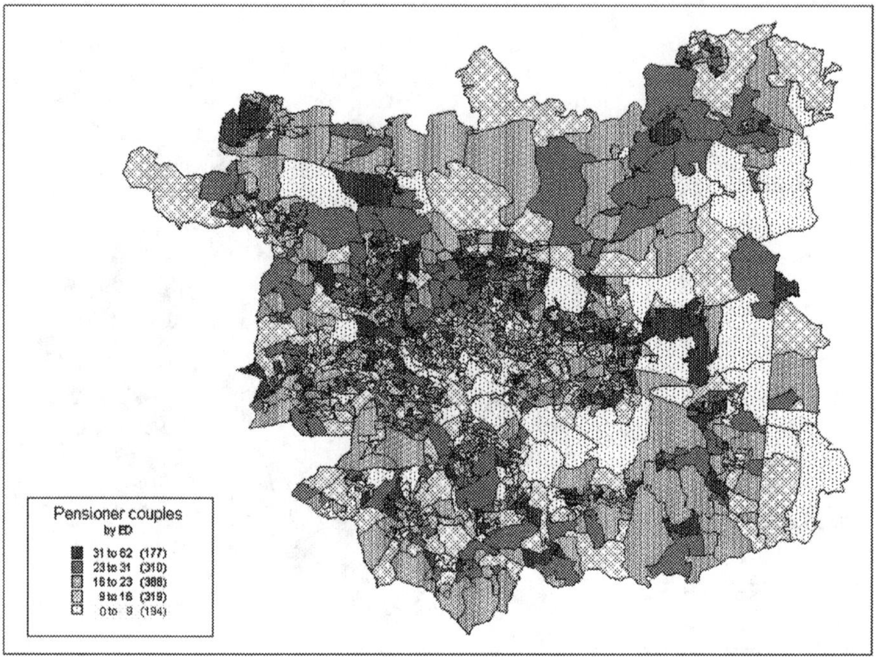

Figure 10. Estimated spatial distribution of pensioner couples.

Table 7. Single Pensioner Impacts.

Gross income (per person)	Pre Budget net	Post Budget net	Win (lose)
£4,000.00	£ 4,000.00	£ 4,000.00	£ 0.00
£8,000.00	£ 7,671.00	£ 7,681.00	£ 10.00
£12,000.00	£10,751.00	£10,801.00	£ 50.00
£15,000.00	£13,061.00	£13,141.00	£ 80.00
£20,000.00	£16,592.00	£16,747.00	£155.00
£25,000.00	£20,442.00	£20,647.00	£205.00
£30,000.00	£24,292.00	£24,547.00	£255.00

(Source: The Guardian, 22 March 2000b.)

12 depicts the estimated spatial distribution of the potential impact of these proposals in Leeds. The spatial patterns depicted in Figure 12 are similar to those of Figure 11. It is noteworthy that the Conservatives proposals were criticised on the grounds that they simply redirecting money from one source of pensioners' income to another. As a veteran pension campaigner put it:

> "Mr Jones, 87, told the BBC: 'His proposals of £10 a week increase is not an increase at all. It is pensioners' own money transferred from one pocket

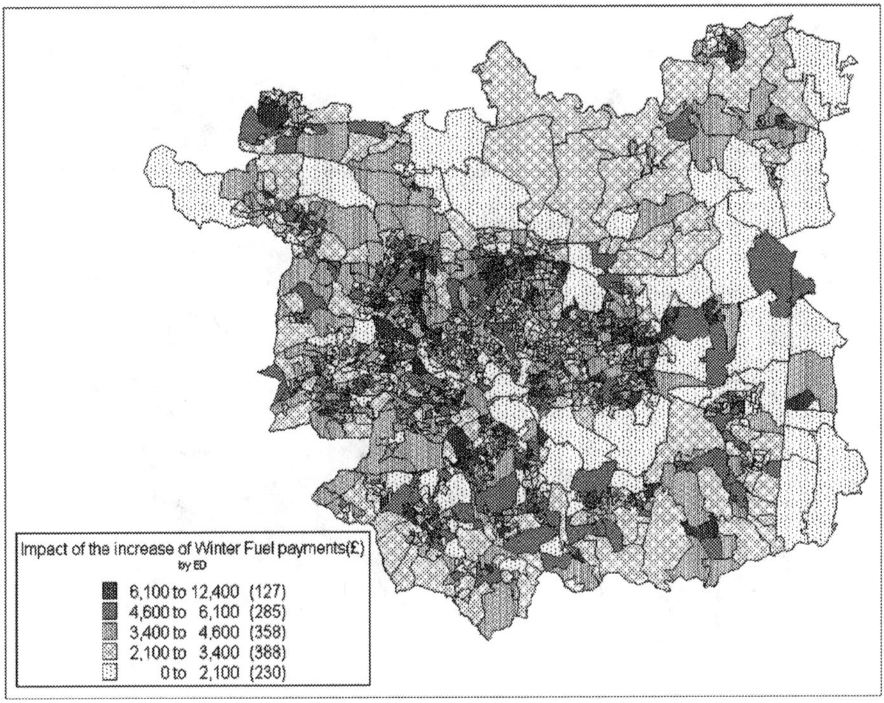

Figure 11. Estimated spatial distribution of the increase in Winter Fuel payments in Leeds.

Table 8. The Conservatives' Proposals for Pensioners

Retirement Pension – proposed increases	
Pensioner group	Amount of Increase
Single Pensioner aged under 75	£5.50
Pensioner couple aged under 75	£7
Single Pensioner aged over 75	£7.50
Pensioner couple aged over 75	£10

(Source: "William Hague's pensions speech", The Guardian, 24 May, 2000c.)

> to another. The Tories took away our link with the earnings back in 1980 – that is what we want back'." (The Guardian, 24 May 2000d)

Accordingly, spatial microsimulation frameworks can be employed to judge such criticisms and to compare the spatial as well as the social impacts of competing policy initiatives.

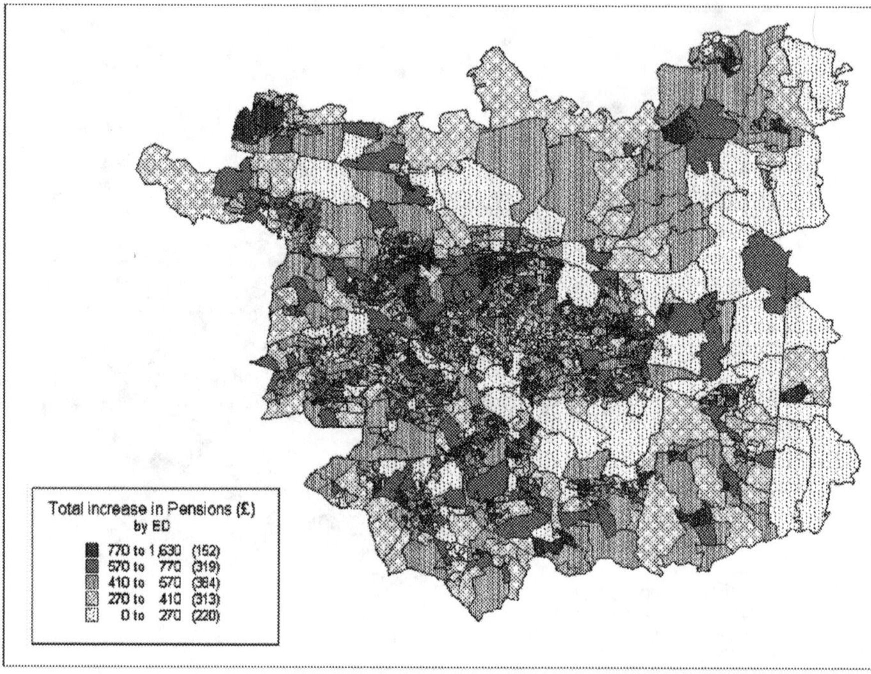

Figure 12. Estimated spatial distribution of pension increases under the Conservatives' proposals.

6. Concluding Comments

In this article, we have built on our previous work (Ballas and Clarke, 2001a) and demonstrated how spatial microsimulation frameworks can be employed for the analysis and evaluation of a wide range of social policy initiatives. It has been shown how spatial microsimulation modelling outputs can be used to compare the results of the applications of alternative and/or competing policies. In particular, we used *SimLeeds* to estimate the spatial and social impact of the recent UK government budget changes. We also briefly discussed the Opposition's proposals for pensioners and we estimated the spatial impact of these suggestions for Leeds.

It can be argued that the analysis presented in this paper is *reactive* rather than *proactive*, in the sense that it provides estimates of the spatial impact of social policies that have been designed to tackle problems at the national level. In other words, the approach to social policy analysis adopted in the context of this paper aims at answering questions such as: "What will be the spatial and social impacts of the X national social policy?"

Nevertheless, there is a challenge to use spatial microsimulation frameworks to analyse social policy in a geographically oriented *proactive* fashion. For instance, spatial microsimulation can be employed to identify deprived localities in which poor individuals and households are over-represented. Spatial microsimulation modelling can then be used to answer questions such as: "What social policy should be applied in order to improve the quality of life of residents in the inner-city localities of Burmantofts in Leeds?". In other words, new social policies can be formulated on the basis of spatial microsimulation modelling outputs. As it has been argued elsewhere (see Ballas and Clarke, 2001a) these spatially-oriented social policies can be seen as a substitute to traditional area-based policies and direct comparisons of their efficiency and effectiveness can be made.

Another interesting research area in which spatial microsimulation models can play a significant role is the measurement and analysis of income and wealth distribution. One of the ways of evaluating the impact of social policy on the welfare of society is to compute indices such as income inequality measures and study the change of these indices after a social policy programme is applied. Measures of poverty and income inequality have been used by social scientists to answer a wide range of questions. There are numerous ways to measure the degree of inequality of a personal income distribution. It can be argued that the mapping of inequality measures can highlight the degree of heterogeneity of income, household types and lifestyles within small areas. Among our future tasks is to investigate the inequalities in income and wealth distribution with the construction of inequality indices at the small area level.

Finally, it should be noted that in this article we have presented a comparative statics spatial microsimulation framework, which is ideal for comparisons of different policies at any one time. Nevertheless, spatial microsimulation models become even more powerful when they become dynamic. One of our future priorities is to develop a dynamic spatial microsimulation modelling framework for social policy analysis, which will provide the capability of predicting the short-term, medium-term and long-term spatial and social effects of different social policy scenarios.

ACKNOWLEDGEMENTS

The work reported here was part funded by the Greek State Scholarships Foundation (IKY). The authors would like to thank Paul Norman for providing his program IPFprog. The Census Small Area Statistics are provided through the Census Dissemination Unit of the University of Manchester, with the support of the ESRC / JISC / DENI 1991 Census of Population Programme. The Census Sample of Anonymised Records are provided through the Census Microdata Unit of the University of Manchester, with the support of the ESRC / JISC / DENI. All Census data reported in this paper are Crown Copyright.

School of Geography
University of Leeds
Leeds LS2 9JT, UK

NOTES

[1] Simulated annealing has been employed as an optimisation technique to solve a variety of combinatorial optimisation problems (Dowsland, 1993; Van Laarhoven and Aarts, 1987). In geography, simulated annealing has been applied in various contexts for different problems (see for instance Alvanides, 2000; Openshaw and Rao, 1995; Openshaw and Schmidt, 1996; Williamson *et al.*, 1998).

[2] See http://www.statistics.gov.uk/.

[3] For useful reviews of the IPF method and its applications see Norman (1999) and Wong (1992).

REFERENCES

Alvanides, S. (2000), *Zone Design Methods for Application in Human Geography.* Unpublished PhD thesis, School of Geography, University of Leeds.

Armstrong, H. (1993), 'The local income and employment impact of Lancaster University.' *Urban Studies*, **30**, 1653–1668.

Armstrong, H. and J. Taylor (1993), *Regional Economics and Policy* (2nd edition). Harvester Wheatsheaf, Brighton.

Atkinson, A.B. (1996), 'Seeking to explain the distribution of income.' In J. Hills (ed.), *New Inequalities: The Changing Distribution of Income and Wealth in the United Kingdom.* Cambridge University Press, Cambridge, pp. 19–48.

Ballas, D. and G.P. Clarke (2000a), 'GIS and microsimulation for local labour market analysis.' *Computers, Environment and Urban Systems*, **24**, 305–330.

Ballas, D. and G.P. Clarke (2000b), *A Spatial Microsimulation Approach to the Analysis of Local Multiplier Effects.* Paper presented at the 31st Regional Science Association British and Irish Section Conference, University of Bath.

Ballas, D. and G.P. Clarke (2001a), 'Modelling the local impacts of national social policies: a spatial microsimulation approach.' *Environment and Planning C: Government and Policy*, **19**, 587–606.

Ballas, D. and G.P. Clarke (2001b), 'The local implications of major job transformations in the city: a spatial microsimulation approach.' *Geographical Analysis*, **33**, 291–311.

Ballas, D., G.P. Clarke and I. Turton (1999), *Exploring Microsimulation Methodologies for the Estimation of Household Attributes.* Paper presented at the 4th International Conference on GeoComputation, Fredericksburg, Virginia, USA.

Bennett, R. (1980), *The Geography of Public Finance.* Methuen, London.

Birkin, M. and M. Clarke (1988), 'SYNTHESIS – a synthetic spatial information system for urban and regional analysis: methods and examples.' *Environment and Planning A*, **20**, 1645–1671.

Birkin, M. and M. Clarke (1989), 'The generation of individual and household incomes at the small area level.' *Regional Studies*, **23**, 535–548.

Birkin, M. and G.P. Clarke (1995), 'Using microsimulation methods to synthesize census data.' In S. Openshaw (ed.), *Census Users' Handbook.* GeoInformation International, London, pp. 363–387.

Clarke, G.P. (1996), 'Microsimulation: an introduction.' In G.P. Clarke (ed.), *Microsimulation for Urban and Regional Policy Analysis.* Pion, London, pp. 1–9.

Curtis, S. (1989), *The Geography of Public Welfare Provision.* Routledge, London.

Dorling, D. and J. Tomaney (1995), 'Poverty in the old industrial regions: a comparative view.' In C. Philo (ed.), *Off the Map: The Social Geography of Poverty in the UK.* Child Poverty Action Group, London, pp. 103–132.

Dowsland, K. (1993), 'Simulated annealing.' In C. Reeves (ed.), *Modern Heuristic Techniques for Combinatorial Problems.* Blackwell, Oxford.

Edwards, J. (1995), 'Social Policy and the City: A Review of Recent Policy Developments and Literature.' Urban Studies, **32**, 695–712.

Falkingham, J. and C. Lessof (1992), 'Playing God: or LIFEMOD – the construction of a dynamic microsimulation model.' In R. Hancock and H. Sutherland H (eds.), *Microsimulation Models for Public Policy Analysis: New Frontiers*. Suntory-Toyota International Centre for Economics and Related Disciplines, LSE, London.

George, C., D. Allirajah and M. Barnes (1999), *Welfare Benefits Handbook 1999/2000*, Vol. 1. CPAG, London.

George, C., D. Allirajah and M. Barnes (1999), *Welfare Benefits Handbook 1999/2000*, Vol. 2. CPAG, London.

Goodwin, M. (1995), 'Poverty in the city: "you can raise your voice, but who is listening?"' In C. Philo (ed.), *Off the Map: The Social Geography of Poverty in the UK*. Child Poverty Action Group, London, pp. 65–82.

Green, A.E. (1996), 'Aspects of the changing geography of poverty and wealth.' In J. Hills (ed.), *New Inequalities: The Changing Distribution of Income and Wealth in the United Kingdom*. Cambridge University Press, Cambridge, pp. 265–291.

Green, A.E. (1998), 'The geography of earnings and incomes in the 1990s.' *Environment and Planning C: Government and Policy*, **16**, 633–648.

Hamnett, C. (1997), 'A stroke of the Chancellor's pen: the social and regional impact of the Conservative's 1988 higher rate tax cuts.' *Environment and Planning A*, **29**, 129–147.

Hancock, R. and H. Sutherland (eds.) (1992), *Microsimulation Models for Public Policy Analysis: New Frontiers*. Suntory-Toyota International Centre for Economics and Related Disciplines, LSE, London.

Higgs, G. (1994), 'GIS in local government: analysis of the spatial implications of the council tax.' *Local Government Policy Making*, **21**, 33–37.

Hill, M. and G. Bramley (1986), *Analysing Social Policy*. Blackwell, Oxford.

Hills, J. (1996), 'Introduction: after the turning point.' In J. Hills (ed.), *New Inequalities: The Changing Distribution of Income and Wealth in the United Kingdom*. Cambridge University Press, Cambridge, pp. 1–16.

HM Treasury (2000), *Prudent for a Purpose: Working for a Stronger and Fairer Britain*. Available from http://www.hm-treasury.gov.uk/budget2000/index.html, accessed May 2000.

Joseph Rowntree Foundation (1995) *Inquiry into Income and Wealth*, Vol. 1. Joseph Rowntree Foundation, York.

Kirkpatrick, S., C.D. Gelatt Jr. and M.P. Vecchi (1983), 'Optimization by simulated annealing.' *Science*, **220**, 671–680.

Krupp, H. (1986), 'Potential and limitations of microsimulation models.' In G. H.Orcutt, J. Mertz and H. Quinke (eds.), *Microanalytic Simulation Models to Support Social and Financial Policy*. North-Holland, Amsterdam, pp. 31–41.

Longley, P., P. Higgs and D. Martin (1996), 'The rates revisited? a geographical reassignment of property valuations and local tax burdens under the council tax.' *Environment and Planning C: Government and Policy*, **14**, 101–120.

Manski, C.F. and I. Garfinkel (1992), *Evaluating Welfare and Training Programs*. Harvard University Press, Cambridge, MA.

Marsh, C. (1993), 'Privacy, confidentiality and anonymity in the 1991 Census.' In A. Dale and C. Marsh (eds.), *The 1991 Census User's Guide*. HMSO, London.

Martin, R. (1995), 'Income and poverty inequalities across regional Britain: the North-South divide lingers on.' In C. Philo (ed.), *Off the Map: The Social Geography of Poverty in the UK*. Child Poverty Action Group. London, pp. 23–44.

Mertz, J. (1991), 'Microsimulation – a survey of principles developments and applications.' *International Journal of Forecasting*, **7**, 77–104.

McCormick, J. and C. Philo (1995), 'Where is poverty? The hidden geography of poverty in the United Kingdom.' In C. Philo (ed.), *Off the Map: The Social Geography of Poverty in the UK*. Child Poverty Action Group, London, pp. 1–22.

McKendrick, J. (1995), 'Poverty in the UK: the Celtic divide.' In C. Philo (ed.), *Off the Map: the Social Geography of Poverty in the UK*. Child Poverty Action Group, London, pp. 45–64.

Metropolis, N., A.W. Rosenbluth, M.N. Rosenbluth, A.H. Teller and E. Teller (1953), 'Equation of state calculations by fast computing machines.' *Journal of Chemical Physics*, **21**, 1087–1092.

Middleton, E. (1995), Samples of anonymized Records. In S Openshaw (ed.), Census users' handbook. GeoInformation International, London, 337–362.

Musgrave, R.A. and Musgrave, P.B. (1984), Public finance in theory and practice. McGraw Hill, New York

Noble, M. and G. Smith (1996), 'Two nations? Changing patterns of income and wealth in two contrasting areas.' In J. Hills (ed.), *New Inequalities: The Changing Distribution of Income and Wealth in the United Kingdom.* Cambridge University Press, Cambridge, pp. 292–318.

Norman, P. (1999), *Putting Iterative Proportional Fitting (IPF) on the Researcher's Desk.* Working Paper 99/03, School of Geography, University of Leeds, Leeds.

Office for National Statistics (ONS) (1997), *New Earnings Survey 1997.* The Stationery Office, London.

Openshaw, S. and L. Rao (1995), 'Algorithms for reengineering 1991 Census geography.' *Environment and Planning A*, **27**, 425–446.

Orcutt, G.H., Mertz, J. and Quinke H. (eds.) (1986), *Microanalytic Simulation Models to Support Social and Financial Policy.* North-Holland, Amsterdam.

Pham, D.T. and D. Karaboga (2000), *Intelligent Optimisation Techniques: Genetic Algorithms, Tabu Search, Simulated Annealing and Neural Networks.* Springer, London.

Philo, C. (ed.) (1995), *Off the Map: The Social Geography of Poverty in the UK.* Child Poverty Action Group, London.

Pinch, S. (1997), *Worlds of Welfare: Understanding the Changing Geographies of Social Welfare Provision.* Routledge, London.

The Guardian (2000a), *The Difference in Your Pocket.* 22 March, 2000.

The Guardian (2000b), *Budget: The Main Points.* 22 March, 2000.

The Guardian (2000c), *Common Sense for Pensioners, William Hague's Full Speech to the LSE.* 24 May, 2000.

The Guardian (2000d), *Hague's Pension Plans 'Rob Peter to pay Paul'.* 24 May, 2000.

Unrau, Y.A. (1993) Expanding the role of program evaluation in social welfare policy analysis. Evaluation Review, **17**, 653–662.

Van Laarhoven, P.J.M. and E.H.L. Aarts (1987), *Simulated Annealing: Theory and Applications.* Kluwer Academic Publishers, Dordrecht.

Williamson, P. (1992), *Community Care Policies for the Elderly: A Microsimulation Spproach.* Unpublished Ph.D. Thesis, School of Geography, University of Leeds, Leeds.

Williamson, P., M. Birkin and P. Rees (1998), 'The estimation of population microdata by using data from small area statistics and samples of anonymised records.' *Environment and Planning A*, **30**, 785–816.

Williamson, P. and D. Voas (2000), Income imputation for small areas: interim progress report. Paper presented at the conference *Census of Population: 2000 and Beyond*, Manchester.

Wong, D.W.S (1992), 'The reliability of using the iterative proportional fitting procedure.' *Professional Geographer*, **44**, 340–348.

ANALYSIS OF THE EFFECT OF LAND USE PATTERS ON THE ANTHROPOGETIC ENERGY DISCHARGED FROM AIR CONDITIONING AND HOT WATER SUPPLY USING A MODIFIED CSU MESOSCALE MODEL

1. INTRODUCTION

It is observed in many large cities in the world that the temperature of the central part of a large city is warmer than that of its suburbs. This phenomenon is called a *heat island.* Figure 1 shows a heat island phenomenon in Tokyo (Yamazoe *et al.*, 1998). The temperature of the central part of Tokyo is higher than that of the suburbs by three degrees in August 1997.

The heat island phenomenon has brought many social problems. One of them is energy shortage in summer. Figure 2 shows electric energy consumption

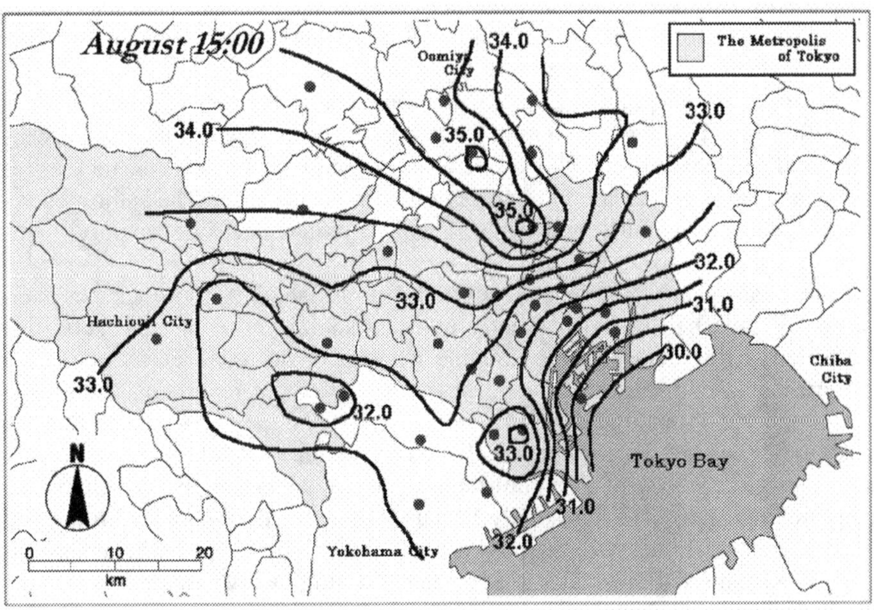

Source: Yamazoe et al., 1998.

Figure 1. The heat island phenomenon in Tokyo.

B. Boots et al. (eds.), Modelling Geographical Systems, 169–180.

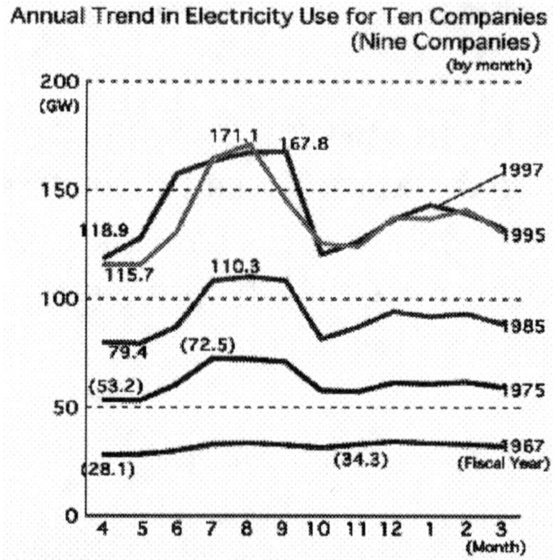

Source: http://www. fepc.or.jp/jijyou/index.html.

Figure 2. Electric energy consumption in Japan.

over twelve months for five years (The Federation of Electric Power Companies Japan, http://www. fepc.or.jp/jijyou/index.html). The consumption was almost even over a year in 1967, but in the 1990's, a high peak appeared in summer. In particular this peak is very high in large cities.

The heat island phenomenon results from many causes, and they may be classified into two types, namely, land use change and anthropogenic energy. The land use change includes a decrease in green land cover, an increase in pavements, and an increase in highrise buildings and so forth. The anthropogenic energy includes exhaust heat from air conditioning, hot water supply, factories, cars, and so forth.

The heat island phenomena could be diminished if those causes were removed, but such removal greatly restricts social and economic activities. Considering this trade-off, we examine whether or not there exists a land use pattern that diminishes heat island phenomena without restricting the present social and economic activities. The objective of this paper is to examine this question by using a 3-dimensional meteorological model, focusing on the anthropogenic energy of air conditioning and hot water supply. We choose air conditioning and hot water supply because these are the major anthropogenic energy. Related studies are numerous, such as Hjelmfelt (1982), Seaman *et al.* (1989), Kimura and Takahashi (1991), but few studies investigate a possibility of controlling heat island phenomena by land use patterns.

The paper consists of five sections. The next section, Section 2, fixes five land use patterns in a hypothetical city to be compared in this paper. Section 3 estimates temperature in the city using the Colorado State University Mesoscale model, and obtains energy consumption. Section 4 shows the results obtained from this simulation model. The paper ends in Section 5, and comments on the limitations of this study.

2. LAND USE PATTERNS IN A HYPOTHETICAL CITY

We suppose a hypothetical region (shown in Figure 3), which is square in shape with 126 km by 126 km. In this region we assume a city which faces the sea in south and is surrounded by forest north, east and west. The latitude of this region is assumed to be N35.68, which is the same as that of Tokyo.

The land uses in this city are: public land use, commercial land use, residential land use, open space (which includes parks and playgrounds), forest land use (which includes water surfaces) and industrial land use. We assume that each land use contains certain types of buildings. For example, the public land use contains public office buildings, education buildings, health care and medical buildings, and supply and processing buildings. The ratios of the six land uses are show in Figure 4.

To fix these ratios, we consider five land use patterns shown in Figure 5. These patterns have the same ratios but different spatial configurations of six land uses. Pattern 1 is the land use pattern in which all the six land uses mentioned above are mixed uniformly. This pattern is not always realistic, but it is used as a baseline land use pattern to be compared with other land use patterns. Pattern 2 is a uni-centre city pattern. The region has one centre covered with the public land use, which is surrounded by the commercial land use, which is surrounded by the residential land use, which is surrounded by the open space, which is surrounded by the forestland use. Pattern 3 is a four-centre city pattern. The region is divided into four sub-regions, each of which has a similar land

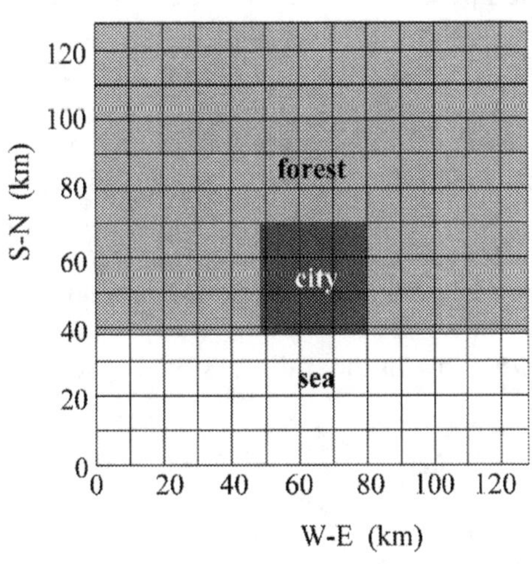

Figure 3. A hypothetical region.

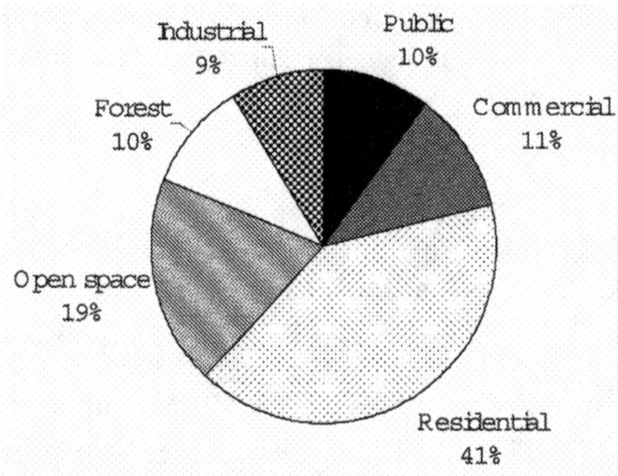

Figure 4. Ratios of the six land uses in the hypothetical city in Figure 3.

use pattern to Pattern 2. Note that the topological pattern and the proportions of the land uses are the same. Pattern 4 is a sixteen-centre city pattern. The region is divided into 16 sub-regions, each of which has a similar land use pattern to Pattern 2. Pattern 5 is a uni-centre city pattern. This pattern is the same as Pattern 2 except that open space is distributed uniformly over the region.

3. ESTIMATION OF TEMPERATURE AND ENERGY CONSUMPTION

For each land use pattern defined in the preceding section, first, we estimate change in temperature over the region during 24 hours using the Colorado State University Meso-scale Model (abbreviated to the CSU model). Second, we estimate energy consumption based upon the estimated temperature. To estimate change in temperature, we run the CSU model. The CSU model was originally developed at the Colorado State University (Pielke, 1974), and it has been modified in a few respects. Ulrickson and Mass (1990) and Kessler and Douglas (1992) take account of a turbulence factor; Uno (1995) and Ichinose *et al.* (1999) take account of high-resolution surface boundary conditions.

The model consists of three sub-models (see Figure 6). The first is the model of airflow in the 3-dimensional space. This sub-model assumes that the hydrostatic equation holds and Boussinesq approximation is applicable. The second is the model of heat conduction into the ground. The third is the model of energy balance at the surface.

The sub-model of airflow in the 3-dimensional space is formulated in terms of five equations:

(1) the equation of motion,
(2) the equation of moisture,
(3) the equation of continuity,
(4) the equation of thermodynamics, and
(5) the equation of hydrostatics.

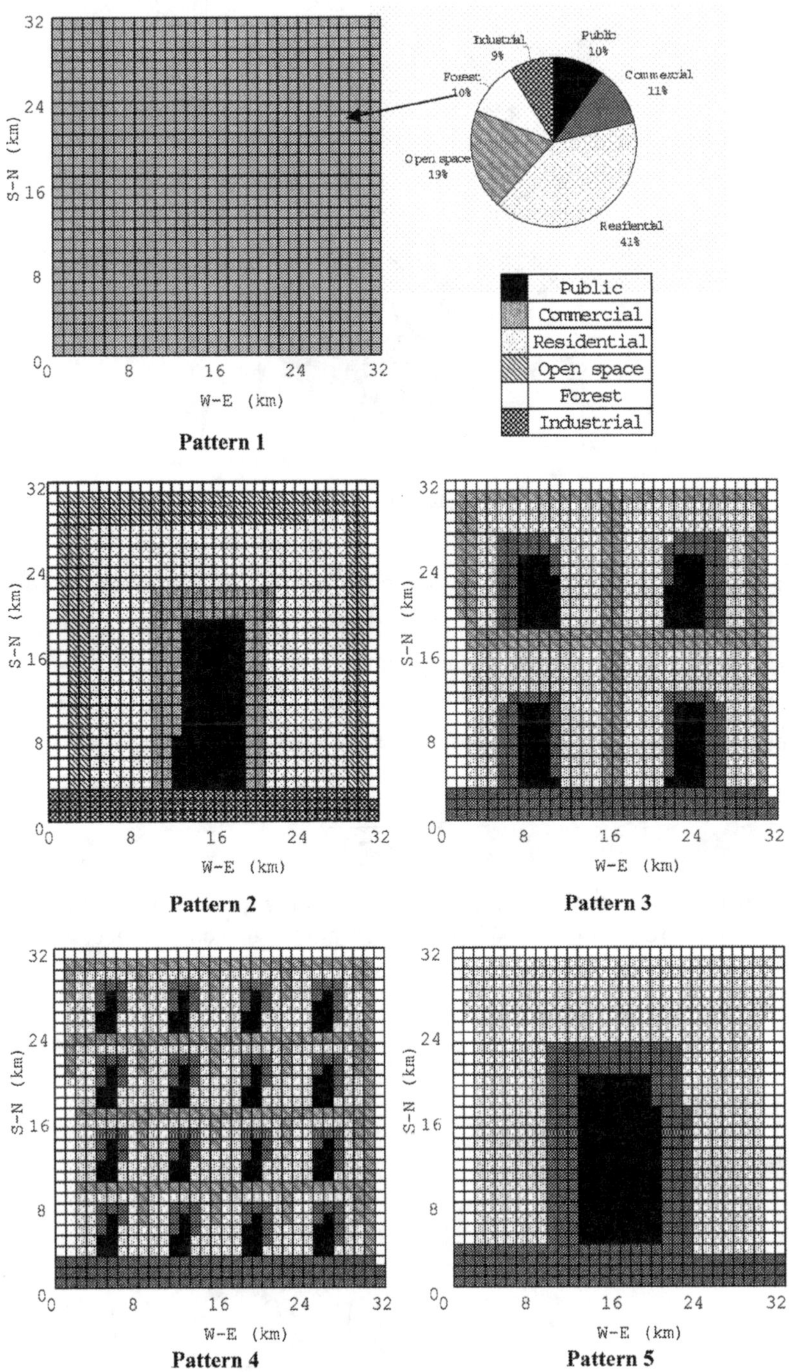

Figure 5. Five land use patterns.

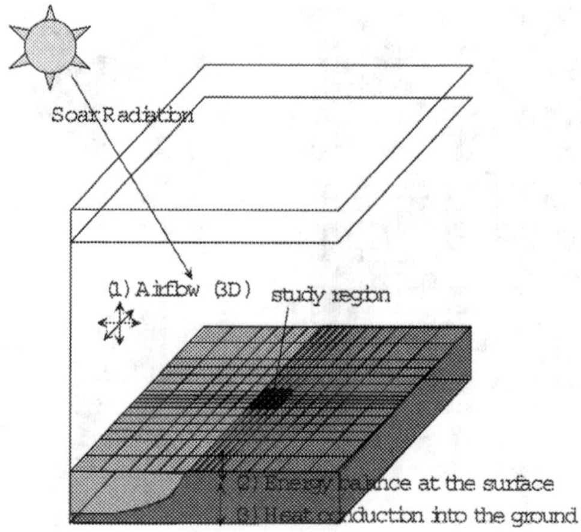

Figure 6. Three components of the CSU model.

To state these equations explicitly, we consider the 3D Cartesian coordinates (x, y, z) and let

z_g: the Cartesian height of the ground surface,
$\bar{s}$: the Cartesian height of the model top.

In these terms, we define a modified vertical coordinate, z^*, as

$$z^* = \bar{s}\,\frac{z - z_g}{\bar{s} - z_g}\,. \tag{1}$$

Note that if $z = z_g$, then $z^* = 0$ and if $z = \bar{s}$, then $z^* = \bar{s}$. This modification is a sort of standardisation to take the uneven grand surface into account, and it is commonly used in airflow models.

To show the equation of motion in a simple form, we introduce the Exner function. Let

C_p: the specific heat of air,
p: pressure,
p_o: the reference pressure, and
R_d: the gas constant for air.

Then the Exner function,, is written as

$$\pi = C_r \left(\frac{p}{p_o}\right)^{\frac{R_d}{C_p}}\,. \tag{2}$$

To define the equation of motion, let

u and *v*: the mean wind components in east-west and north-south directions, respectively,
t: time,
θ: potential temperature,
U_g and V_g: the velocity of geostorphic wind in east-west and north-south directions, respectively,
f: the Coriolis parameter; and
K_m: the vertical exchange coefficient of momentum.

In these terms with the Exner function, the equation of motion is written as

$$\frac{Du}{Dt} = -\theta\frac{\partial\pi}{\partial x} + g\left(\frac{z^* - \bar{s}}{\bar{s}}\right)\frac{\partial z_g}{\partial x} + \left(\frac{\bar{s}}{s - z_g}\right)^2\frac{\partial}{\partial z^*}\left(K_m\frac{\partial u}{\partial z^*}\right) + f(v - V_g), \quad (3)$$

and

$$\frac{Dv}{Dt} = -\theta\frac{\partial\pi}{\partial y} + g\left(\frac{z^* - \bar{s}}{\bar{s}}\right)\frac{\partial z_g}{\partial y} + \left(\frac{\bar{s}}{s - z_g}\right)^2\frac{\partial}{\partial z^*}\left(K_m\frac{\partial v}{\partial z^*}\right) + f(u - U_g), \quad (4)$$

Second, let

q: the specific humidity.

Then the equation of moisture is written as

$$\frac{Dq}{Dt} = \left(\frac{\bar{s}}{s - z_g}\right)^2\frac{\partial}{\partial z^*}\left(K_h\frac{\partial q}{\partial z^*}\right). \quad (5)$$

Third, let

w: the mean wind component in vertical direction in the z^* coordinate.

Then the equation of continuity is written as

$$\frac{\partial u}{\partial x} + \frac{\partial v}{\partial y} + \frac{\partial w^*}{\partial y} - \frac{\bar{s}}{s - z_g}\left(u\frac{\partial z_g}{\partial x} + \frac{\partial z_g}{\partial y}\right) = 0. \quad (6)$$

Fourth, let

K_h: the vertical exchange coefficient of heat, and
S_θ: the source/sink terms for potential temperature.

Then the equation of thermodynamic is written as

$$\frac{\partial\theta}{\partial t} = \left(\frac{\bar{s}}{s - z_g}\right)^2\frac{\partial}{\partial z^*}\left(K_h\frac{\partial\theta}{\partial z^*}\right) + S_\theta. \quad (7)$$

Last, to define the hydrostatic equation, let

g: acceleration of gravity.

Then the equation of hydrostatics is written as

$$\frac{\partial \pi}{\partial z^*} = -\left(\frac{\bar{s} - z_g}{s}\right)\frac{g}{\pi}. \tag{8}$$

The sub-model of energy balance at the surface is formulated in terms of seven flows:

SR_{sfc}: net short wave radiation,
IR_{sfc}: incoming infrared radiation,
$\rho C_p u^* \theta^*$: sensible heat flux,
$\rho Lvu^* q^*$: latent heat flux,
σT_{sfc}^4: outcoming infrared radiation,
$\rho_s c_s K_s \frac{\partial T_s}{\partial z}$: heat conduction into the ground, and
$Q_{m,n,0}$: anthropogenic heat.

Then heat stored in the surface layer of the ground, denoted by, Q_{sfc}, is given by

$$Q_{sfc} = SR_{sfc} + IR_{sfc} + \rho C_p u^* \theta^* + \rho Lvu^* q^* - \sigma T_{sfc}^4 - \rho_s c_s K_s \frac{\partial T_s}{\partial z} + Q_{m,n,0}. \tag{9}$$

The soil temperature in 50 cm below the ground is set constant as a lower boundary condition.

4. Simulation Results

Using the CSM model shown in section 3, we simulated temperature during 24 hours for the land use patterns 1, 2, 3, 4 and 5 defined in section 2. Part of the results is shown in Figure 7.

Using the estimated temperature, we next estimate the energy consumption. Let

i: building use i, $i = 1, \ldots, 28$,
j: be energy use $j = 1$ (heating), 2 (cooling), 3 (hot water),
t: time,
$T(t; x, y)$: the temperature at time t at (x, y), and
$N_i(x, y)$: the floor area of building use i in the cell at (x, y),

(note that is obtained from the area occupied by building use i through floor area ratios; the area occupied by building use i is obtained from the land use categories and each land use category has its own fixed ratios of the areas occupied by those building uses). Then the energy consumption, $E(x, y)$, in a

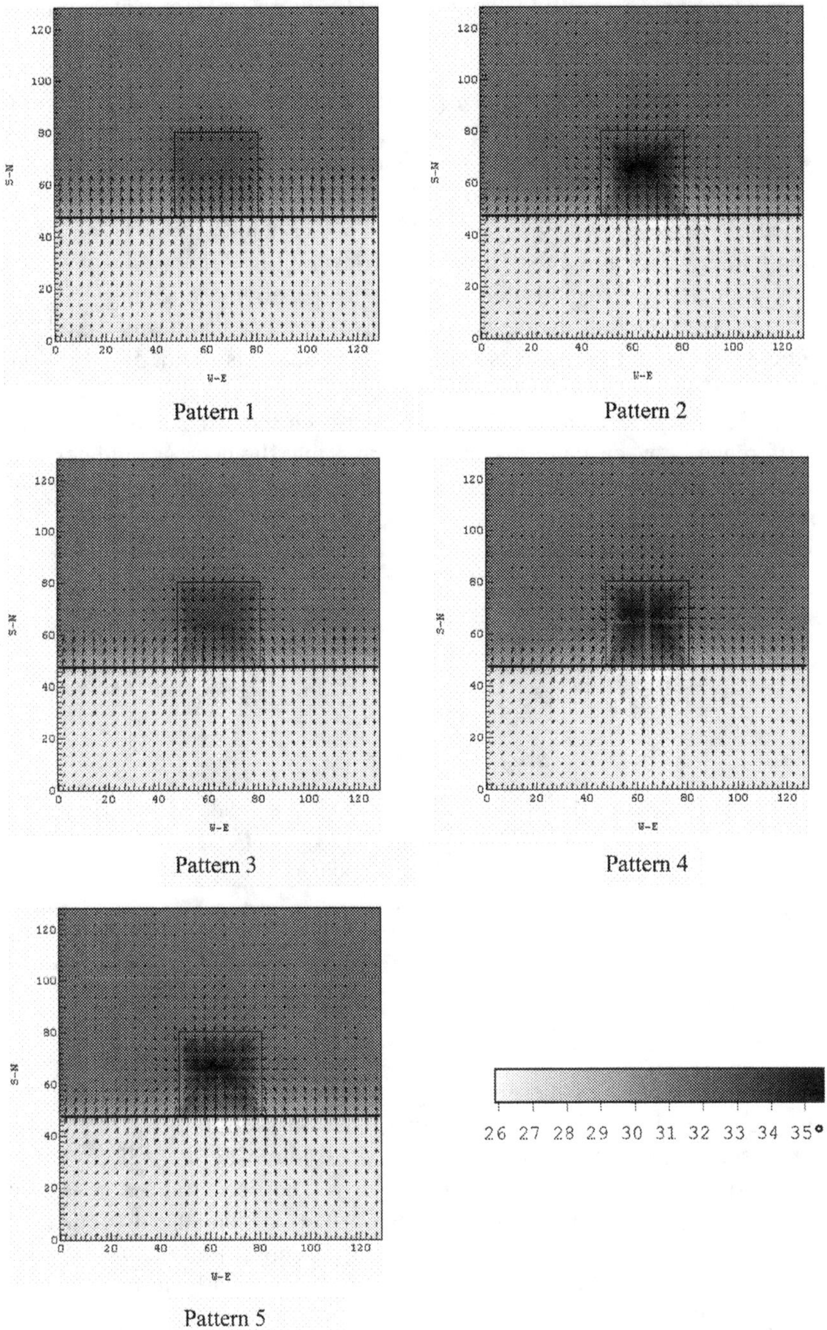

Figure 7. *Temperature at noon for the land use patterns 1, 2, 3, 4, and 5 (arrows show the direction of wind).*

square cell at (x, y) resulting from air conditioning and hot water supply is, according to Hirano (1999), given by

$$E(x, y) = \sum_i \sum_j \sum_t \{a_{ij}(t) + b_{ij}(t)T(t; x, y)\}N_i(x, y), \tag{10}$$

where $a_{ij}(t)$ and $b_{ij}(t)$ are parameter values which are the functions of time, t.

The estimated energy consumption for Patterns 2, 3, 4 and 5 in comparison with Pattern 1 are shown in Figure 8. First, we conclude from the comparison among Patterns 2, 3, and 4 that the energy consumption decreases as the number of centres increases. A multi-centre city is less energy consuming.

Second, we conclude from the comparison between Patterns 2 and 5 that the energy consumption decreases as the open space is divided and dispersed as fragments.

Last, we conclude from the difference between Pattern 2 and Pattern 4 and the difference between Pattern 2 and Pattern 5 that the decrease in energy consumption by division of open space is much larger than that by increasing the number of centres.

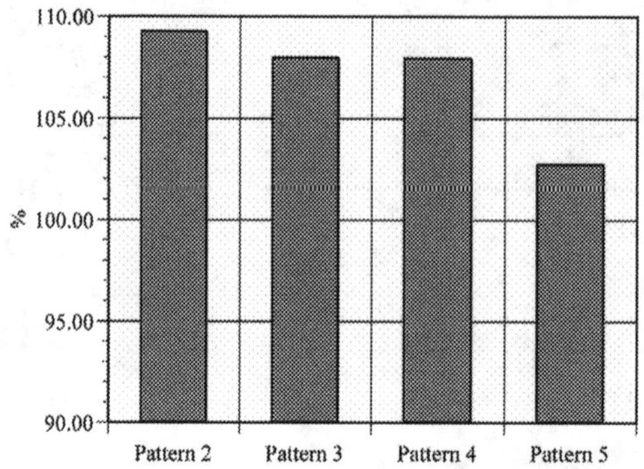

Figure 8. Energy consumption ratios of Patterns 2, 3, 4 and 5 to Pattern 1.

5. CONCLUDING REMARKS

From the examination of the preceding sections, we may conclude that: if the areas occupied by the six land uses mentioned in section 2 are fixed as they are at present, the energy consumption is decreased most effectively by dividing the open space and dispersing it over the city. The decentralised centres plan often proposed in urban planning is less effective than the dispersed open space plan. It is, however, not so easy to find actual policies to realize the dispersed open space plan. A more possible policy would be to persuade every building to make roof gardens.

As mentioned in the introduction, our model does not take all kinds of anthropogenic energy, and so the above conclusion may be arguable. It is not questionable that exhaust from air conditioning and hot water supply are the primary anthropogenic energy, but exhaust from cars may not be negligible anthropogenic energy. We will take this factor into our model and will show whether or not our conclusion is still true in a subsequent paper.

ACKNOWLEDGEMENTS

This study was partly supported by CREST (Core Research for Evolutional Science and Technology) of Japan Science and Technology Corporation (JST). We thank to Professor Dr. Keisuke Hanaki, University of Tokyo, Dr. Toshiaki Ichinose, National Institute for Environmental Studies, Dr. Yuzuru Yamazoe, Nihon University, and anonymous referees for their valuable comments on an earlier draft.

Teruhisa Watanabe
Geospatial Information Systems Division
Hitachi Software Engineering
6-81, Onoe-cho, Naka-ku, Yokohama 231-8475, Japan

Atsuyuki Okabe
Centre for Spatial Information Science
University of Tokyo
7-3-1 Hongo, Bunkyo-ku, Tokyo 113-8656, Japan

Takeki Izumi
Department of Geography
Tokyo Metropolitan University
Minami-Osawa, Hachioji, Tokyo 192-0397, Japan

Yukio Sadahiro
Department of Urban Engineering
University of Tokyo
7-3-1Hongo, Bunkyo-ku, Tokyo113-8656, Japan

REFERENCES

Hirano, Y., Y. Kaya and R. Shibasaki (1997), 'Impact assessment of heat island phenomenon on energy consumption for air conditioning, heating and hot water.' *Journal of Environmental Systems and Engineering*, **629**, VII(12), 83–96.

Hjelmfelt, M.R. (1982), 'Numerical simulation of the effects of St. Louis on mesoscale boundary-layer airflow and vertical air motion: Simulations of urban vs non-urban effects.' *Journal of Applied Meteorology*, **21**, 1239–1257.

Ichinose, T., K. Shimodozono and K. Hanaki (1999), 'Impact of anthropogenic heat on urban climate in Tokyo.' *Atmospheric Environment*, **33**, 3897–3909.

Kessler, R.C. and S.G. Douglas (1992), User's guide to the Systems Applications International Mesoscale Model (Version 2.0), Systems Applications International, SYSAPP-92-085, California.

Kimura, F. and S. Takahashi (1991), 'The effects of land-use and anthropogenic heating on the surface temperature in the Tokyo metropolitan area: A numerical experiment.' *Atmospheric Environment*, **25B**, 155–164.

Pielke, R.A. (1974), 'A three dimensional numerical model of the sea breezes over South Florida.' *Monthly Weather Review*, **102**, 115–134.

Seaman, N.L., F.L. Ludwig, E.G. Donall, T.T. Warner and C.M. Bhumralker (1989), 'Numerical studies of urban planetary boundary-layer structure under realistic synoptic condition.' *Journal of Applied Meteorology*, **28**, 760–781.

Ulrickson, B.L. and C.F. Mass (1990), 'Numerical investigation of mesoscale circulations over the Los Angels basin. Part 1, A verification study.' *Monthly Weather Review*, **118**, 2138–2161.

Uno, I. (1995), 'Quantitative evaluation of a mesoscale numerical model simulation using four-dimensional data assimilation of complex airflow over the Kanto region.' *Journal of the Japanese Society of Atmospheric Environment*, **30**(6), 351–366.

Yamazoe, Y., M. Ushiyama, T. Mikami and K. Kimura (1998), 'An analysis of diurnal variation of the horizontal distribution of temperature using high density urban climate observation network in Tokyo area: The comparison of summer with autumn.' *Proceedings of the General Meeting of the Association of Japanese Geographers*, **54**, 96–97.

KAZUYUKI KONAGAYA

GENERALISED THÜEN AND THÜEN -RICARDO MODELS FOR ASIAN LAND USE

1. GENERALISED THÜEN (GT)-MODEL

1.1. Characteristics of the GT-Model

Thüen's model explains a land use structure in an ideal world, but the model hardly explains an actual land use structure in the real world. The first difficulty in Thüen's model is its deterministic nature. When land uses are aggregated into parcels, it is impossible to determine all possible variables influencing rent. Thus, realistic models have to deal with land uses in terms of the land-use mix, which considers the uncertainty resulting from aggregation. The second difficulty results from its monocentric nature. The traditional models are strongly constrained by a single centred spatial structure, although a few studies attempt to manage a two-dimensional structure in a general and non-trivial manner.[1]

Ellickson (1981) and others have shown that Thüen's bid-rent theory can be re-interpreted as the analogous form of utility maximisation in discrete-choice theory. Their models suggest a probabilistic framework for land use modelling that reconciles the rent-maximising principle with land-use mix. As actual rent values fluctuate probabilistically (random variable), many land-use types may co-exist. This possibility is the main difference between deterministic models and probabilistic models.

There are few examples that give a general elucidation of the complex interference between non-deterministic and non-monocentric effects. To fill this gap, Konagaya (1997a, b, 1998, 1999a) extended the normative Thüen model to one that can treat real land-use ratio data. This new model with a non-deterministic and non-monocentric framework, called the *Generalised Thünen model* (abbreviated to GT-model), improves explanatory performance and provides an intuitive picture of 'movement of the land-use frontier' with a theoretical foundation. The GT-model enables us to predict future land-use changes.

A distinction between Thüen's model [2] and the GT-model is shown in Figure 1. Thüen's model has a concentric land-use structure around a monocentric city where only one land use dominates each zone (the left panel); the GT-model, which succeeds Thüen's model, explains land use mix (the right panel).

B. Boots et al. (eds.), Modelling Geographical Systems, 181–199.

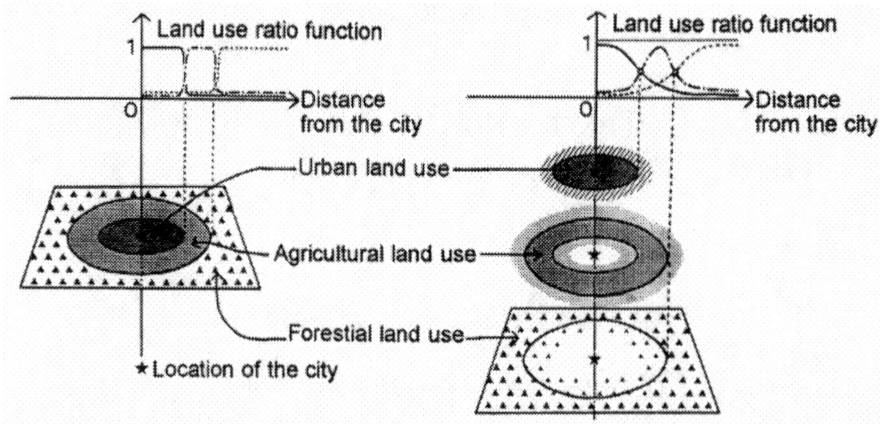

Figure 1. Classical Thünen Model (left) and Generalised Thünen Model (right).

1.2. Derivation of the GT-Model

1.2.1. Bid-rent Function

It is assumed that if a type $u(u = 1, \ldots, U)$ land user who sells his product in the city $j(j = 1, \ldots, J)$ uses the land of zone $i(i = 1, \ldots, I)$, the land user gains the 'benefit', $A^{(u)}(j)$, and pays the cost $b^{(u)} \cdot d_{ij}$ proportional to the distance between the zone i and the city j. The bid-rent function of land user u in zone i is given by

$$R_{ij}^{(u)} = A^{(u)}(j) - b^{(u)} \cdot d_{ij} + \varepsilon_{ij}^{(u)} \tag{1}$$

where $\varepsilon_{ij}^{(u)}$ is the random variable which is assumed to follow the Weibel distribution. The representative component of this probabilistic bid-rent function is written as

$$r_{ij}^{(u)} = A^{(u)}(j) - b^{(u)} \cdot d_{ij}. \tag{2}$$

1.2.2. GT-model Formula

From the assumption that ε is a random variable following the Weibel distribution, the probability, $p_{ij}^{(u)}$, that land users u exporting products to city j occupy the land in zone i is derived as

$$p_{ij}^{(u)} = \frac{\exp(r_{ij}^{(u)})}{\sum_{kl} \exp(r_{kl}^{(u)})} \cdot \tag{3}$$

Assuming that other conditions for land acquisition processes are equal, the total type u land use ratio of zone i is the sum of each user's probability over all cities. This completes the final formula of the GT-Model, i.e.

$$p_i^{(u)} = \sum_j p_{ij}^{(u)} = \frac{\sum_j (r_{ij}^{(u)})}{\sum_k \{\sum_l \exp(r_{kl}^{(u)})\}}$$

$$= \frac{K_i^{(u)}}{\sum_k K_k^{(u)}} \tag{4}$$

where

$$K_i^{(u)} \equiv \sum_j \exp(A^{(u)}(j) - b^{(u)} \cdot d_{ij}). \tag{5}$$

1.2.3. Simplification

Ideally, the value of $A^{(u)}(j)$ should be obtained for every city. This derivation, however, is intractable in practice. The benefit function $A^{(u)}(j)$ corresponds to the total revenue obtained by selling the products in a city minus the production cost, or to the profit without transport costs. If agglomeration economies are assumed to exist in each city, commodity prices will increase and $A^{(u)}(j)$ will also increase, as the city becomes larger. Under this assumption, the benefit function A is positively correlated with the population, q_j, of the city j with parameter $a^{(u)}$. Then the representative component of bid-rent is written as

$$r_{ij}^{(u)} = a^{(u)} \cdot q_j - b^{(u)} \cdot d_{ij}, \tag{6}$$

and Equations (4) and (5) become

$$p_i^{(u)} = \frac{K_i[a^{(u)}; b^{(u)}]}{\sum_{i'} K_{i'}[a^{(u)}; b^{(u)}]}, \tag{7}$$

where the weight function with two parameters is given by

$$K_i[a^{(u)}; b^{(u)}] \equiv \sum_j \exp(a^{(u)} \cdot q_j - b^{(u)} \cdot d_{ij}). \tag{8}$$

2. SOME REGULARITIES OBSERVED IN INDONESIAN LAND-USE DATA

In our studies in LUCC (Land Use/Cover Change) project (LU/GEC project),[3] we considered three land use categories, namely, forest, agriculture, and urban (3-category model). We denote the ratio of these categories by F, A, U, and the observation years 1980 and 1990 by the suffix number 8 and 9, respectively. The zonal index i is omitted. It is obvious that the equations

$$F_8 + A_8 + U_8 = 1, \quad F_9 + A_9 + U_9 = 1, \tag{9}$$

hold, and land-use change variables satisfy

$$\Delta F + \Delta A + \Delta U = 0, \tag{10}$$

where

$$\Delta F = F_9 - F_8, \quad \Delta A = A_9 - A_8, \quad \Delta U = U_9 - U_8, \tag{11}$$

In the Indonesian land use data of LUCC, we observed the following two empirical regularities.

2.1. Regularity-1: Intermediate Zone Dominance

Forest land use decreased rapidly during 1980 and 1990 in the zones where the 1990 ratio of forest land use took a middle range value between 40% and 60%, i.e., ΔF shows a large negative value in the zones where the relations $0.4 < F_8 < 0.6$ hold (Figure 2). In contrast, forest land use changes are fairly small, both in pure forest zones (i.e. $F_8 > 0.6$) and the non-forest zones (i.e.

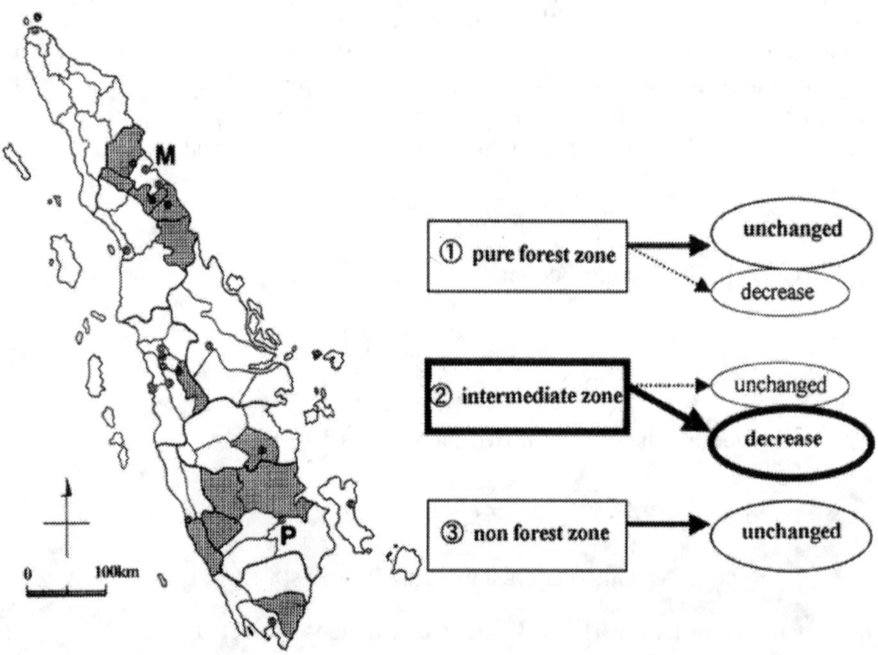

Figure 2. Intermediate zone dominance: change of forest land-use ratio. The hatched prefectures in the left diagram experienced more than 10% decrease of forest land-use ratio during 1980–1990. This radical deforestation occurs in 'intermediate' zones around Medan (M) in North Sumatra and Palembang (P) in South Sumatra. The right diagram summarises three trends : pure forest zone and non forest zone tend to remain unchanged, however intermediate zone tends to lose forest stocks.

$F_8 < 0.4$). Drastic decreases occur in intermediate zones (i.e. $0.4 < F_8 < 0.6$). For example, in Sumatra Island, the most radical deforestation occurs around Medan in the North Sumatra province and Palembang in the South Sumatra province (the largest and second largest cities in Sumatra). We call this empirical regularity the *intermediate zone dominance.*

2.2. Regularity-2: Proportional Relationship

The ratio of urban land use to agricultural land use is approximately 1:5 in most zones in Sumatra. The forest land areas converted into urban and agricultural uses have approximately the same ratio. As displayed in Figure 3, the relation

$$U{:}A = 1{:}5 \tag{12}$$

holds, and

$$\Delta U{:}\Delta A = 1{:}5, \quad -\Delta F = \Delta A + \Delta U, \tag{13}$$

We call this empirical regularity the *proportional relationship.*

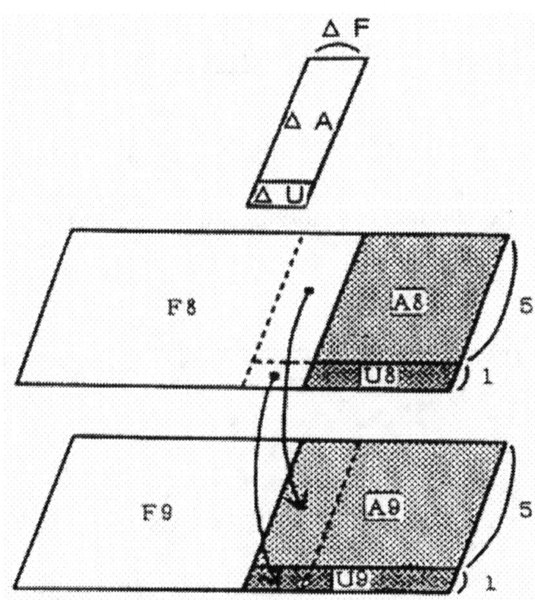

Figure 3. Proportional relationship: conversion of F to A or U.

3. APPLICATION OF THE GT-MODEL

The GT-model is a non-linear model with respect to its parameters. Thus the ordinary algorithm for linear models is not applicable. We employed a new method called the genetic algorithm or evolutional strategy (referred to it as GA) to estimate the parameters. The explanatory power (the determinant coefficient R^2) of the GT-model was between 0.75 and 0.95 for Sumatra, Jawa and Kalimantan land uses. This performance seems to be satisfactory.

4. GENERALISED THÜEN DISTANCE: INCORPORATION OF MULTI-CITY INFLUENCES

The GT-model deals with various influences from many cities. We can construct some sort of composite distance, the Generalised Thüen Distance (hereafter called GTD), which incorporates the influences from many cities into that from one virtual city (1-dimensionalization). The GTD makes it possible to interpret the GT-model as the virtual probabilistic single-city model (probabilistic classical Thüen model) through the replacement of influences from many cities by their average. The GTD of land-use categories u to $u + 1$ in zone i, '$D_i^{[u/(u+1)]}$' is defined as

$$\frac{K_i^{(u+1)}}{K_i^{(u)}} = \frac{\exp(a^{(u+1)} \cdot Q^{[u/(u+1)]} - b^{(u+1)} \cdot D_i^{[u/(u+1)]})}{\exp(a^{(u)} \cdot Q^{[u/(u+1)]} - b^{(u)} \cdot D_i^{[u/(u+1)]})}, \tag{14}$$

where

$$Q^{[u/(u+1)]} \equiv \frac{LSA_j(a^{(u)} \cdot q_j) - LSA_j(a^{(u+1)} \cdot q_j)}{(a^{(u)} - a^{(u+1)})} \tag{15}$$

$$LSA_j(X_j) \equiv \ln \textstyle\sum_j \exp(X_j) \tag{16}$$

is the averaged population. By solving equation (14) we get the explicit reprsentation of GTD as

$$D_i^{[u/(u+1)]} = \frac{\{LSA_j(a^{(u+1)} \cdot q_j) - LSA_j(a^{(u+1)} \cdot q_j - b^{(u+1)} \cdot d_{ij})\} - \{LSA_j(a^{(u)} \cdot q_j) - LSA_j(a^{(u)} \cdot q_j - b^{(u)} \cdot d_{ij})\}}{b^{(u)} - b^{(u+1)}} \tag{17}$$

5. INTERMEDIATE ZONE DOMINANCE EXPLAINED BY THE GT-MODEL

The trend of land-use change is illustrated generally in Figure 4 where three land-use categories (urban, agricultural and forest) are treated. The frontiers (transitional zones) of each land-use are shown by logistic-like curves in the

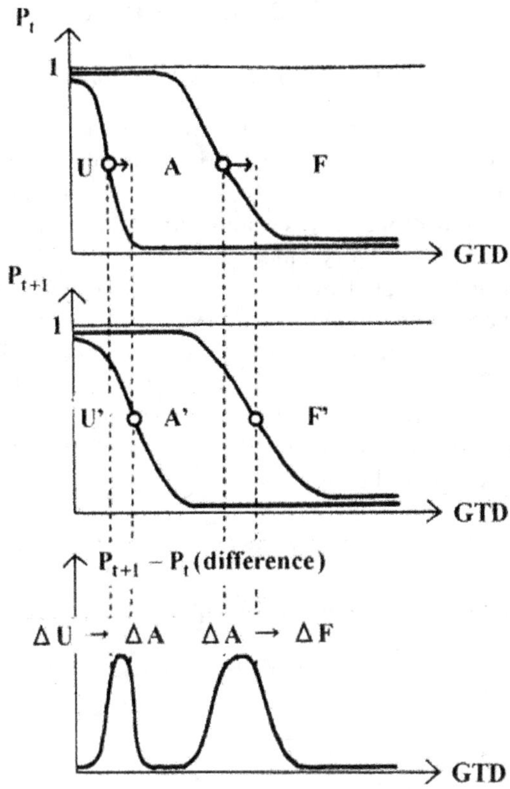

Figure 4. Explanation of intermediate zone dominance.

GT-model. The dominant land-use phases take turns where each GTD equals the critical distance. In the top graph of Figure 4, each land-use ratio P_t in the year t is depicted as the height in an accumulation style, with U, A, F expressing each value of the urban, agricultural and forest category, respectively (Konagaya 1997a, 1998).

Following the transition from year t to $t + 1$, land-use ratios in the year $t + 1$ P_{t+1} are also shown, as in the middle graph of Figure 4, with U', A', F' indicating the values of the urban, agricultural and forest categories, respectively. The logistic-like curves shift to a further location (right direction in the diagram) from P_t to P_{t+1} with the land-use borders also moving in a similar direction. As a result, two peaks may emerge in the difference graph of $P_{t+1} - P_t$ (the bottom graph of Figure 4) because they are the difference in the two logistic-like curves between different years. The major land-use changes occur in the intermediate zones, which explains the empirically observed 'intermediate zone dominance' based on the model. Thus the GT-model provides a theoretical foundation to the intuitive picture of 'the movement of land-use frontiers' and enables us to estimate future land uses.

6. EQUI-GTD APPROXIMATION

6.1. Equi-GTD Approximation Equation

Strictly speaking, the GT-model with U land use categories has $U - 1$ different GTD variables. However, they are some sort of averages with different parameters (as shown from Equation (8)) and take nearly the same values for each zone. To prove this property it is suitable to simulate the GT-model in some experimental systems (3-category model).

In this section, the regional index 'i' and the brackets enclosing the land use category number in the upper side of each variable are omitted for simplicity. After many simulation results of the 3-category model (for example, see Konagaya, 1997a for further discussion), two GTDs, D^{12} and D^{23}, have almost the same values everywhere.[4] The next question is: what results from the approximation that all GTDs is equal? The GT-model formula proves that two successive Thüen distances being equal

$$D^{12} = D^{23} \tag{18}$$

is equivalent to the equation (Equi-GTD approximation equation)

$$\ln P^2 - \{(1 - \delta) \ln P^1 + \delta \ln P^3\} = \text{const.}(\equiv \eta(D_0^{23} - D_0^{12})), \tag{19}$$

where

$$\delta = \frac{(b^1 - b^2)}{(b^1 - b^3)}, \quad \eta = \frac{1}{1/(b^1 - b^2) + 1/(b^2 - b^3)}, \tag{20}$$

and

$$D_0^{u,\,u+1} \equiv \frac{a^u - a^{u+1}}{b^u - b^{u+1}} \tag{21}$$

is the critical GTD of land-use categories to $u + 1$.

This means the addition of one more constraint. The geometrical meaning is as follows. First, the original degrees of freedom of the 3-category model are 3. Second, the trivial relationship

$$P^1 + P^2 + P^3 = 1 \tag{22}$$

reduces the dimension by one because land-use variables are in ratio form. Triangular coordinates neatly represent the remaining 2 dimensions in a symmetrical manner. The two GTDs, D^{12} and D^{23}, are interpreted as a set of coordinated transformations corresponding to these 2 degree of freedom. Third, the approximation condition $D^{12} = D^{23}$ (= D) which is equivalent to the above equation reduces the dimension one step further. If this approximation is appropriate, the final degree of freedom is 1, which means that 'the degree of freedom of the land-use structure is approximately one in GT-model'.

6.2. Land-Use Structure under the Approxmation

By considering the exponential of the approximation equation (19), we get

$$P^2 = e^{\Delta}(P^1)^{1-\delta}(P^3)^{\delta} \quad (\Delta \equiv \eta \cdot (D_0^{23} - D_0^{12})). \tag{23}$$

This equation takes the form of a Cobb-Douglas function displayed in the convex surface as in Figure 5. The cross section of this surface cut by a triangular

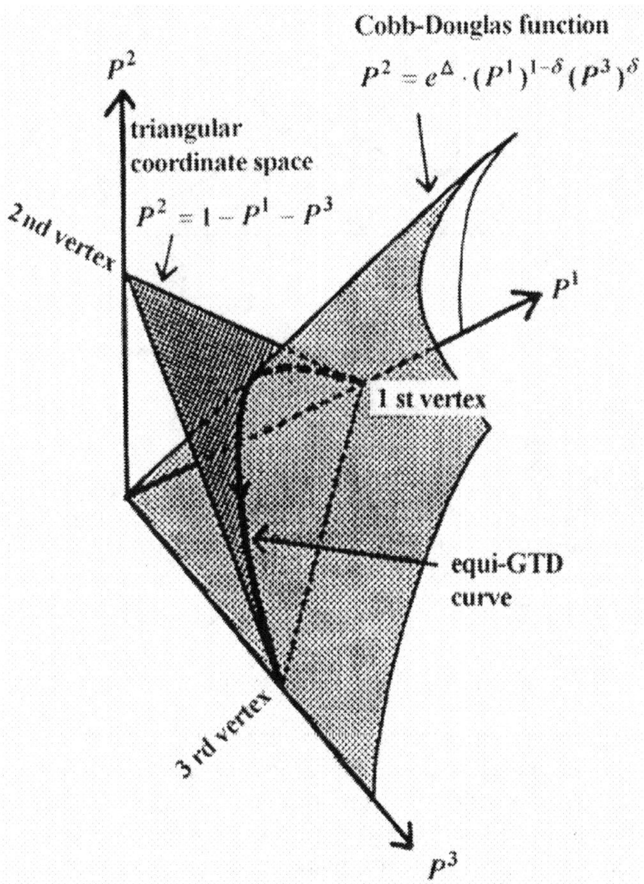

Figure 5. *Equi-GTD curve. [P_t is the land-use ratio variable in the year t, U(urban), A(agricultural), F(forest) are land-use categories in the year t. P_{t+1} is the land-use ratio variable in the year t + 1, U′(urban), A′(agricultural), F′(forest) are land-use categories in the year t + 1]*

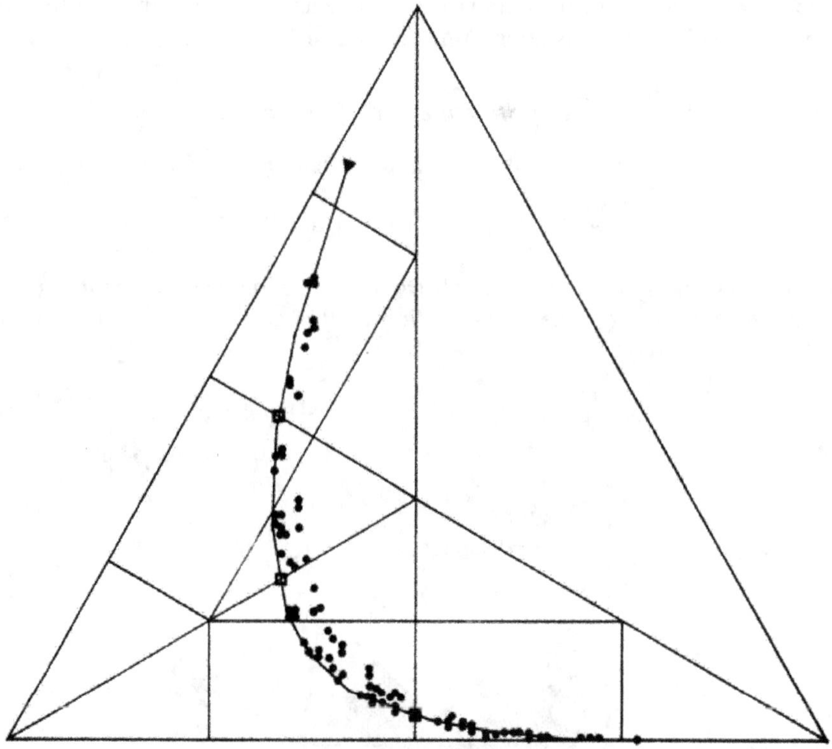

Figure 6. Simulation results (land-use mix) of 3-category model.

coordinate triangle is the equi-GTD curve in triangular coordinate space. This curve starts at the first vertex and goes along the 1–2 edge, bending upward near the second vertex, going along the 2–3 edge, and finally reaching the third vertex. Figure 6 shows one example of the simulation results of a virtual 10×10 grid space with three cities.

7. PROPORTIONAL RELATIONSHIP EXPLAINED BY THE GT-MODEL

Consider, for example, the 3-category model of forest, agricultural and urban land use, i.e., the model in the triangular coordinate space (Konagaya, 1997b; Konagaya and Otsubo, 1998). The proportional relationship stated in subsection 2.2 is equivalent to the statement that land-use composition is restricted to the line of 1:5 in the triangular coordinate space (Figure 7). If the GT-model holds exactly, the points of land-use composition are arranged along the 'equi-GTD curve' expressed by Equation (23). It starts at the 'pure forest limit' $(F, A, U) = (1, 0, 0)$ and proceeds left through the extension of agricultural land (conversion from forest to agriculture land primarily and urban land secondarily), finally approaching its 'pure urban limit' $(F, A, U) = (0, 0, 1)$.

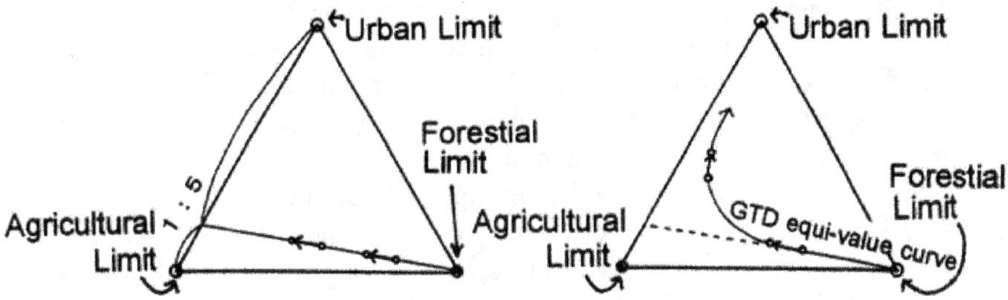

Figure 7. *Explanation of the proportional law.*

The part of the curve in the process of conversion from forest is approximated by a line originating from the 'pure forest limit' $(F, A, U) = (1, 0, 0)$. Thus the proportional relationship is explained by the GT-model.

8. PREDICTION OF 2025 INDONESIAN LAND USE

8.1. Calibration Results

By using a Genetic Algorithm (GA), the calibrated results for 3-category land use data in 43 districts and 10 cities in West Jawa and Kalimantan in 1980 were:

$$P_i^{(u)} = \frac{K_i[a^{(u)}; b^{(u)}]}{\sum_j K_j[a^{(u)}; b^{(u)}]},$$

$$K_i[a^{(u)}; b^{(u)}] \equiv \sum_j \exp(a^{(u)} \cdot q_j - b^{(u)} \cdot d_{ij}),$$

$$a^{(1)} = 0.31, \quad b^{(1)} = 1.16,$$
$$a^{(2)} = 0.56, \quad b^{(2)} = 0.45,$$
$$a^{(3)} = 0.26, \quad b^{(3)} = 0.00,$$
$$R^2 = 0.89, \quad SRMSE = 0.33,$$

and in 1990

$$a^{(1)} = 0.42, \quad b^{(1)} = 0.98,$$
$$a^{(2)} = 0.59, \quad b^{(2)} = 0.39,$$
$$a^{(3)} = 0.42, \quad b^{(3)} = 0.07,$$
$$R^2 = 0.88, \quad SRMSE = 0.33, \tag{24}$$

8.2. Some Features of the GT-Model

These results tell us about the following features.

(1) Performance: The GT-model is probably one of the best performance land-use models that use only a few parameters (such as population, distance, etc.) because the GT-model parameters are found a posteriori compared with

the parameters in the ordinary potential method that are fixed a priori. This efficiency is especially useful for studies in developing countries where available data of sufficient reliability are very limited.

(2) Non-linearity: The GT-model achieves the above performance at the cost of linearity in parameters. This non-linear character needs the GA (Genetic Algorithm) method to solve for the optimal parameters.

(3) Stability: The parameters obtained in 1980 changed little in 1990. This temporal stability in the parameters enables us to predict future land use by using the averaged parameters of past observations without resorting to more complex socio-economic models involving temporal deviations in parameters. The results of 2025 land uses predicted (by trend population growth except DKI Jakarta = 12,000,000) are depicted in Figure 8.

8.3. Trend of 2025 Prediction

The 'Botabek (Bogor, Tangerang, Bekasi: three suburban prefectures around DKI Jakarta)' region that is already highly urbanised (mature urbanised area) will have urban land use accounting for one third of total land in 2025. Additionally, 'extended suburbs' (Karawang, Pruwakarta, Serang: three prefectures around Botabek) have a drastically high speed of urbanisation until 2025. This means the emergence of a 'super megalopolis along the Jakarta-Bandung axis' and the danger of loss of excellent forest located in this region.

9. GTR(GENERALISED THÜNEN -RICARDO)-MODEL

The study of Chinese land-use change is important because there are many regions with rapid economic development and environmental change in China. Since Chinese land is much larger than a single island region in a Southeast Asian country, various physical conditions must be incorporated. The GT-model described in the preceding section is based mainly on economic conditions and influence from cities (situation variables or urban driving-force variables), however micro-physical conditions at each site (on-site variables) also need to be incorporated.

In the LUCC-project LU/GEC, we extended the GT-model to the GTR-model to explain Chinese land-use structure. It is assumed that if a type $u(u = 1, \ldots, U)$ land user who sells his product in city $j(j = 1, \ldots, J)$ whose population is X_j, uses the land of zone $i(i = 1, \ldots, I)$ whose elevation is $Z1_i$ and whose slope is $Z2_i$, the land user gains the 'benefit' $a(1, u) \cdot X_j + a(3, u) \cdot Z1_i + a(4, u) \cdot Z2_i$, and pays the cost $a(2, u) \cdot Y_{ij}$ proportional to the distance Y_{ij} between the zone i and the city j, The bid-rent function of land user u in zone i is given by

$$r_{ij}^{(u)} = a(1, u) \cdot X_j - a(2, u) \cdot Y_{ij} + a(3, u) \cdot Z1_i + a(4, u) \cdot Z2_i + \varepsilon_{ij}^{(u)}. \quad (25)$$

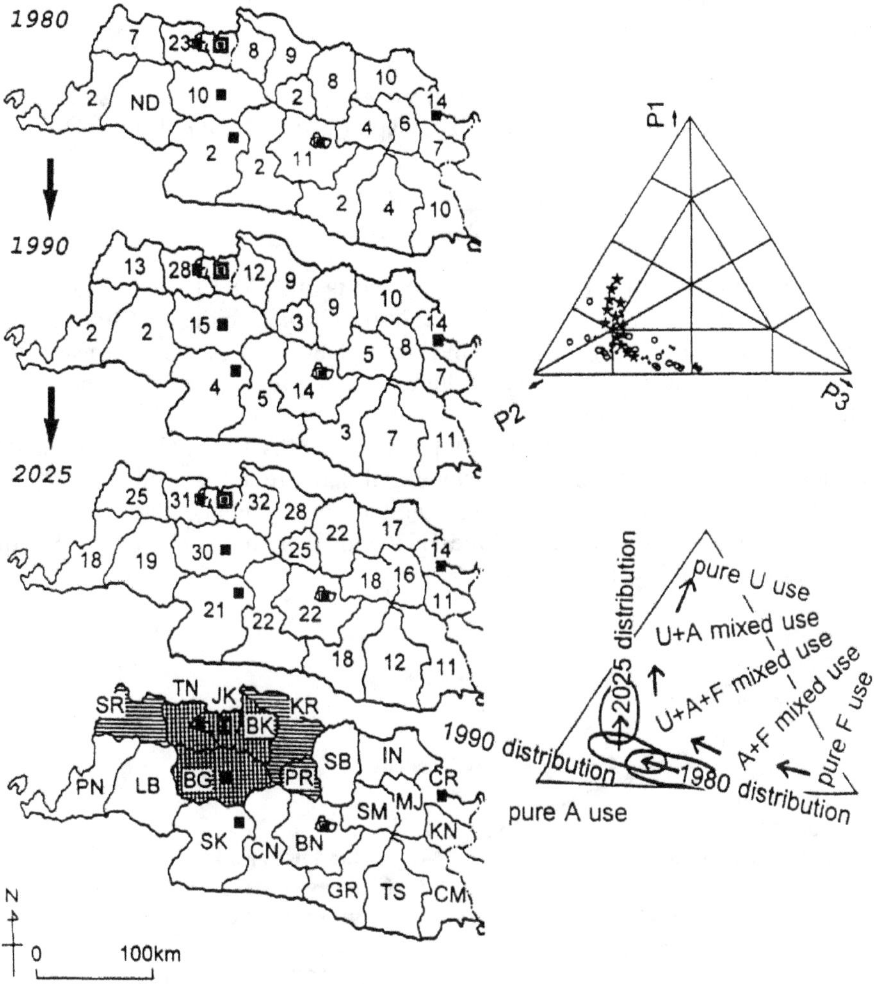

Figure 8. *Jawanese land-use prediction in West Jawa. Left four maps express the percentages of urban land-use ratio in 1980, 1990, 2025 (prediction) and each prefecture names. Capital Metropolitan Region (Jabotabek) consists of JK(Jakarta), TN(Tangerang), BG(Bogor) and BK(Bekasi). Extended Metropolitan Region consists of Jabotabek, SR(Serang), KR(Karawang) and PR(Purwakarta). Other prefectures are expressed as SB(Subang), IN(Indramayu), CR(Cirebon), TS(Takusimaraya), GR(Guruto), CN(Cianjur), SK(Sukabumi), LB(Lebak) and PN(Pandegurang).*

Upper right triangular graph indicates the values of three land-use ratio in each prefecture of West Jawa where P^1, P^2, P^3 are urban, agricultural and forest land-use ratios respectively. Each point in the triangle represents the land-use mix values in each prefecture, where black dots, white dots and stars correspond to 1980, 1990, 2025 values. Lower right triangular graph summarises the trends observed in the upper right graph.

We can integrate $\varepsilon_{ij}^{(u)}$ to produce the land-use ratio function of the GTR-model as

$$K_i[a(p, u)] \equiv \{\textstyle\sum_j \exp(a(1, u) \cdot X_j - a(2, u) \cdot Y_{ij})\}\exp(a(3, u) \cdot Z1_i + a(4, u) \cdot Z2_i)$$

$$P_i^{(u)} = \frac{K_i[a(p, u)]}{\sum_{u'} K_i[a(p, u')]}, \tag{26}$$

In the application to the Chinese case, we take 22,789 LU/GEC grids[5] as the land use zones $i(i = 1, \ldots, I = 22{,}789)$, 50 major cities in China as the set of cities $j(j = 1, \ldots, J = 22{,}789)$, three land uses as the land-use categories u: ($u = 1$: urban, $u = 2$: agricultural, $u = 3$: others (forest, steppe etc.)), and denote the four parameters corresponding to four variables as $a(1, u)$: parameter of urban population, $a(2, u)$: parameter of distance, $a(3, u)$: parameter of elevation, $a(4, u)$: parameter of slope. All data are aggregated into 22,789 LU/GEC grid[5] squares of 15 minutes longitude by 10 minutes latitude.

The most important feature of the GTR-model is that it has two kinds of explanatory variables (Figure 9).

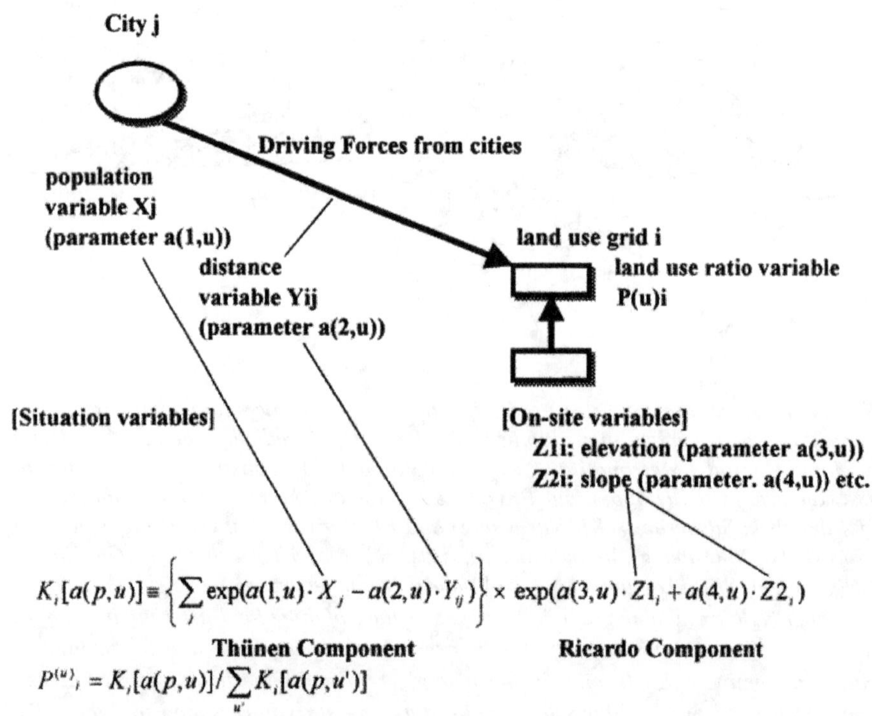

Figure 9. Structure of the GTR-Model (schematic explanation of the equations and variables).

(1) Situation variables (Urban driving force variables)
These variables represent the influence of far away cities on the land use of the grid, the driving forces. This influence consists of an indicator of the agglomeration economies of cities (such as urban population of the j-city) X_j, and an indicator of the decay effect of space (such as distance between the j-city and the i-grid) Y_{ij}. The influence of these situation variables is presented in the { } factor of the above formula (26), which is called the 'Thünen Component' because it is based on the Thünen rent theory.

(2) On-site variables
These variables represent the conditions of the same location as the land-use grid, which control the influence of driving forces. In this study we consider elevation $Z1$ and slope $Z2$ variables. The influence of these on-site variables is presented as the exponential factor of the above formula, which is called the 'Ricardo Component' because it is based on the Ricardo rent theory.

10. DATA SETS FOR THE GTR-MODEL

The first category is situation variables. The urban population of the j-city X_j is an example of a situation variable describing distant urban influence, derived in the following order:

(1) Making original data by re-aggregating county unit CIESIN[6] data into 5 minute grids.
(2) Allocating them to LU/GEC grids[5] guided by night light image as supervised data.
(3) Specifying 50 major city grids and calculating their populations.

Then, we calculate the distance matrix with 22,789 × 50 components between LU/GEC grids (i) and 50 major cities (j).

The second category is on-site variables. These are produced as follows:

(1) The elevation variable $Z1_i$ of LU/GEC grids (i) is derived from USGS[7] data.
(2) The slope variable $Z2_i$ of LU/GEC grids (i) is calculated as the standard deviation of $Z1_i$.

Last, the explanatory variable $P_i^{(u)}$ is extracted by transforming 1 km by 1km resolution NOAA data observed during April 1992 and April 1993 into LU/GEC grids.

11. CALIBRATION RESULTS FOR THE GTR-MODEL

The optimised GTR-model is as follows[8]

$$K_i[a(k, 1)] = \{\textstyle\sum_j \exp(10 \cdot X_j - 8.7 \cdot Y_{ij})\},$$

$$K_i[a(k, 2)] = \{\textstyle\sum_j \exp(9 \cdot X_j - 3 \cdot Y_{ij})\}\exp(0.3 \cdot Z1_i + 1.3 \cdot Z2_i),$$

$$K_i[a(k, 3)] = \{\textstyle\sum_j \exp(5 \cdot X_j - 3 \cdot Y_{ij})\}\exp(1.3 \cdot Z1_i + 1 \cdot Z2_i). \qquad (27)$$

This model predicts land-use profiles of the 22,789 grids in China up to almost 80% ($R^2 = 0.776$) by only 10 parameters.

Forest land use (parameter 1.3) is suitable for a high elevation more than urban land use (parameter 0) or agricultural ones (parameter 0.3). Forest (parameter 1) and agricultural (parameter 1.3) land use suits a slope more than urban land use (parameter 0).

12. PREDICTION OF 2025 CHINESE LAND USES

12.1. Prediction Method

We can predict the 2025 land uses by coupling the GTR-model with appropriate population models. One population model predicts that the 2025 population of China will be 1.464 times that of 1992. Thc lcft two maps of Figure 10 show the present agricultural land use and 2025 profiles, and the right two maps of Figure 10 show the present other land uses (forest, steppe, etc.) and 2025 profiles.

12.2. Implications

The most drastic change in the map of 2025 land use profiles is the great expansion of agricultural land in the north and north-east regions, and at the same time, the conversion and decrease of the 'others (forest, steppe, etc.)' category in these regions. This trend is most successfully captured by the Thüen-type model family (GT-model or GTR-model). In Thüen-type models, cities not only determine the urban land use (such as housing, industry, commercial, and so on), but also play a more important role as influential drivers to order all the other land uses around them, such as quasi-concentric zones of predominantly urban land use; then agricultural land use; then forest land use; and then others in diminishing order (see Figure 1).

Additionally, the Thüen-type model explains how the same phenomenon occurred in Japanese land-use structures. In Japan, the most typical production centre of rice crops ('Kome-dokoro') is in the north and northeast regions ('Hokuriku-chiho' and 'Tohoku-chiho'). This seems to contradict the tropical origin of rice, but the Thüen-type model helps explain this contradiction. In Japan after the World War II, urban land use has developed in the south region

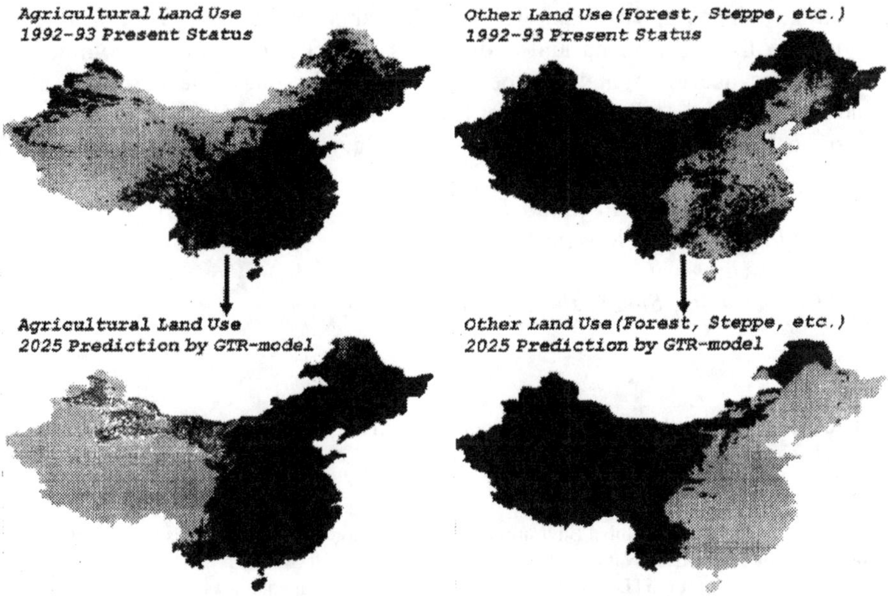

Figure 10. *2025 Chinese land-use predictions: agricultural land use (left) and other land use (right).*

('Pacific industrial belt'). It is the growth centre of economic development. Thus, the north and northeast regions far from the 'Pacific industrial belt' are suitable for rice production based on rent-bidding ability.[9]

Similarly, southern China has been the centre of agricultural production until now. However, as urbanisation based on rapid industrialisation proceeds in the south, agricultural land use is pushed north, which effects the large-scale structure of Chinese land use.

Although in this study, economic driving forces and geomorphologic conditions are taken into account, the predicted patterns could cause lack of water and degradation of soil. Thus, we are now trying to incorporate such hydrological sub models into GTR (land use model) and to evaluate the feedback effects and the interactions between these effects.

ACKNOWLEDGEMENTS

This study was funded by the Environmental Agency of Japan and the LU/GEC (Land Use/Global Environmental Conservation) Project of NIES (National Institute for Environmental Studies: Project Reader, K. Otsubo) – the main Japanese project of LUCC (Land Use/Cover Change), the joint core project of IGBP (International Geosphere-Biosphere Programme) and IHDP (International Human Dimensions Programme). The author would like to thank Commission Chair Professor Barry Boots and Professors Atsuyuki Okabe,

Yoshitaka Ishikawa, Toshitaka Tabuchi, Masahisa Fujita, Yohei Sato, Yoshio Sugiura, Yukio Himiyama and anonymous referees for valuable comments, which were useful in making the paper a fruitful one. The author is also grateful to Associate Professors Keiji Yano and Hidenri Morita, for their kind assistance in the calculations of the genetic algorithm and calibration.

Division of Urban and Regional Economies,
Institute for Economic Research,
Osaka City University,
3-3-138, Sugimoto, Sumiyoshi-ku, Osaka-City, Japan

NOTES

1 In the one-dimensional settings there are many nontrivial non-monocentric solutions found in regional science (for example see Fujita and Ogawa, 1982 and so on).

2 For the classical monocentric rent function after Thüen's original theory, see Lösch (1954) or Nerlove and Sadka (1991).

3 LU/GEC (Land Use for Global Environmental Conservation/1995.4–2001.3) is the largest LUCC project in Japan by NIES (National Institute for Environmental Studies).

4 To evaluate this equi-GTD hypothesis we use the following inequalities:
Firstly, log sum average LSA as defined in equation (16) satisfy

$$\ln J + J^{-1} \textstyle\sum_j X_j \le LSA_j(X_j) \le \ln J + \max_j \{X_j\}. \quad (28)$$

Therefore,

$$\begin{aligned} D^{23} - D^{12} &= \eta^{-1}[LSA_j(a^2q_j) - (1-\delta)\cdot LSA_j(a^1q_j) - \delta \cdot LSA_j(a^3q_j) \\ &\quad - LSA_j(a^2q_j - b^2d_{ij}) + (1-\delta)\cdot LSA_j(a^1q_j - b^1d_{ij}) + \delta\cdot LSA_j(a^3q_j - b^3d_{ij}) \\ &\le \eta^{-1}[\{\ln J + \max_j(a^2q_j)\} - (1-\delta)\cdot\{\ln J + J^{-1}\textstyle\sum_j (a^1q_j)\} \\ &\quad - \delta\cdot\{\ln J + J^{-1}\textstyle\sum_j (a^3q_j)\} - \{\ln J + J^{-1}\textstyle\sum_j (a^2q_j - b^2d_{ij})\} \\ &\quad + (1-\delta)\cdot\{\ln J + \max_j(a^1q_j - b^1d_{ij})\} + \delta\cdot\{\ln J + \max_j(a^3q_j - b^3d_{ij})\}] \\ &\le \eta^{-1}[\max_j(a^2q_j) - (1-\delta)\cdot J^{-1}\textstyle\sum_j (a^1q_j) - \delta\cdot J^{-1}\textstyle\sum_j (a^3q_j) \\ &\quad - J^{-1}\textstyle\sum_j (a^2q_j - b^2d_{ij}) + (1-\delta)\cdot\max_j(a^1q_j - b^1d_{ij}) + \delta\cdot\max_j(a^3q_j - b^3d_{ij})] \\ &\le \eta^{-1}[\{a^2 + (1-\delta)\cdot a^1 + \delta\cdot a^3\}\cdot\{\max_j (q_j) - J^{-1}\textstyle\sum_j (q_j)\} \\ &\quad + b^2\cdot\{J^{-1}\textstyle\sum_j (d_{ij}) - \min_j (d_{ij})\}]. \end{aligned} \quad (29)$$

As a land use site i to be observed goes far away from the group of cities, the quantity $(D^{23} - D^{12})/d_{ij}$ will be negligible, because the first term of the right hand side of Equation (29) is constant and the second term of the right hand side of Equation (29) is 'distance between the site i and the averaged centre of the group of cities' minus 'distance between the site i and the nearest cities' that is approximately equal to the diameter of the group of cities.

5 These data sets are produced in the LU/GEC project. LU/GEC grids are approximately equal to 20 km by 20 km square in Chinese medium latitude regions. The population data comes from Dr. Nakaya's study in the LU/GEC project.

6 CIESIN is the Centre for International Earth Science Information Network at Columbia University (URL: http://www.ciesin.org/)

7 USGS is the United States Geological Survey (URL: http://www.usgs.gov/)

8 We may safely assume restricting parameter space as

$$\begin{gathered} a(1,1) > a(1,2) > a(1,3),\;\; a(2,1) > a(2,2) > a(2,3), \\ 0 < a(3,2) < a(3,3),\;\; 0 < a(4,2) < a(4,3). \end{gathered} \quad (30)$$

And because the second factor of GTR-formula (Equation (26)) is linear, we have an arbitrariness in the third and fourth parameters because we can add an arbitrary universal constant to all elevation and slope parameters without changing land use values because they are in the ratio form. Then we may safely put the parameters of elevation and slope for arbitrary one land-use category as a criterion as zero, then

$$a(3, 1) = a(4, 1) = 0. \tag{31}$$

[9] This Thüen-like interpretation of the formation process of Japanese rice crop belt is based on Japanese agricultural geographer Dr. Toshiaki Oji's unpublished doctoral dissertation in Kyoto University.

REFERENCES

Ellickson, B. (1981), 'An Alternative Test of the Hedonic Theory of Housing Market.' *Journal of Urban Economics*, **9**, 56–79.

Fujita, M. and H. Ogawa (1982), 'Multiple Equilibria and Structural Transition of Non-monocentric Urban Configurations.' *Regional Science and Urban Economics*, **12**, 161–196.

Konagaya, K. (1996), 'An Analysis on Land-use Change in Indonesia.' In K. Otsubo (ed.), *Land Use for Global Environmental Conservation (LU/GEC) – First Report of the LU/GEC First Phase (1995–1997)*, NIES, Environmental Agency of Japan.

Konagaya, K. (1997a), 'A New Development of the LU/GEC Fundamental Model (LU/GEC-I).' In K. Otsubo (ed.), *Land Use for Global Environmental Conservation (LU/GEC) – Second Report of the LU/GEC First Phase (1995–1997)*, NIES, Environmental Agency of Japan.

Konagaya, K. (1997b), 'Land Use and the Generalised Thüen Model.' *Essays on Theoretical Geography* 10, Tokyo Metropolitan University, pp. 1–25.

Konagaya, K. (1998), 'LU/GEC fundamental model and the generalized Thüen-Alonso modelling.' In *Earth's Changing Land (Abstracts of GCTE-LUCC Open Science Conference on Global Change)*. Institute Cartogràfic de Catalunya, Barcelona, Spain.

Konagaya, K. (1999a), 'The Generalised Thüen-Alonso Model for Land-use Change in Sumatra Island.' *Geographical & Environmental Modelling*, 3-2, Carfax Publishing Limited, UK, pp. 145–162.

Konagaya, K. (1999b), 'Thüen-Type Logistic Model.' In K. Otsubo (ed.), *Land Use for Global Environmental Conservation (LU/GEC) – Final Report of the LU/GEC First Phase (1995–1997)*, NIES, Environmental Agency of Japan.

Konagaya, K. and K. Otsubo (1998), 'An Overview of LU/GEC Project and the Development of Generalised Thüen Model.' *Proceedings of the 6th Symposium on Global Environment*, Committee on Global Environment, Japan Society of Civil Engineers.

Konagaya, K., H. Morita and K. Otsubo (1999), 'Chinese land use predicted by the GTR-model.' *Abstract in the 1999 Open Meeting of the Human Dimensions of Global Environmental Change Research Community*, IGES, Shonan Village Centre.

Lösch, A. (1954), *The Economics of Location*, translated by H Woglum. Yale University Press, New Haven, CT.

Nerlove, M.L. and E. Sadka (1991), 'Von Thüen's Model of the Dual Economy.' *Journal of Economics*, **54**, 97–123.

Von Thüen, J.H. (1826), *Der Isolierte Staat in Beziehung auf Landwirtschaft und Nationalökonomie*, Hamburg: Perthes.

Section B

GIS Models

ROBERT D. FEICK AND G. BRENT HALL

BALANCING CONSENSUS AND CONFLICT WITH A GIS-BASED MULTI-PARTICIPANT, MULTI-CRITERIA DECISION SUPPORT TOOL

1. INTRODUCTION

Many land use decision problems are sufficiently complex, controversial and non-routine in nature that their resolution requires not only knowledge, expertise and data drawn from several distinct domains, but also debate of multiple and often conflicting viewpoints, objectives and solution strategies. Despite the promise of Geographic Information Systems (GIS) technology for examining a wide variety of land use issues, it has been criticised for its lack of decision support capabilities in this and other application areas (Hendriks and Vriens, 2000; Towers, 1997; Heywood *et al.*, 1994). In particular, the capacity of commercial GIS to facilitate debate and achieve some measure of balance among different viewpoints has been identified as a major weakness. This capability is constrained by several factors in commercial software, foremost of which is an intrinsic single-user perspective that disregards the multi-interest character of the decision making process and the socially constructed nature of data and analytical methods (Pickles, 1999; Flowerdew, 1998; Jones *et al.*, 1997; Goodchild, 1995).

A common, and often contentious, planning and land management problem that involves multiple interests is the task of designating sites for specific uses and activities. This problem is intrinsically difficult as it first requires identification of a set of feasible non-dominated sites for the intended use(s) and then an evaluation of non-trivial tradeoffs between the relative merits of each site. Completing these tasks in a pluralistic context is considerably more difficult than by a single decision maker as multiple participants are likely to have different values, objectives, and conceptualisation of issues which are equally likely to be non-coincident or in conflict (von Winterfeldt and Edwards, 1986).

The task of seeking a reasonable balance among conflicting viewpoints is complicated further by spatial factors such as externality effects, differentials in land capability, competition among different land uses for specific locations, and the fixed, short-term pattern of land ownership. Hence, partly in response to the criticisms of GIS noted above and partly in response to need, efforts to extend GIS technology to explore and represent the subjective elements of multi-party conflict and decision making have focused on integrating it

B. Boots et al. (eds.), Modelling Geographical Systems, 203–233.

with multiple criteria analysis (MCA) techniques (Bojórquez-Tapia *et al.*, 2001; Jankowski *et al.*, 1997). MCA methods effectively decompose choice decisions into evaluations of sets of criteria that allow the performance of alternatives to be judged. Through this process of decomposition, elements of agreement and conflict between different individuals and groups can be exposed and debated.

This paper describes how a spatial decision support system (SDSS) that incorporates MCA methods can be used to explore proactively land development conflicts among multiple participants. The focus of the paper is not on issues associated with the coupling of MCA and GIS capabilities (for a discussion of this see Feick and Hall, 2000), but rather on the problem of assessing the land development decision process in order to achieve an acceptable degree of consensus that balances conflicting viewpoints. The outline of the paper is as follows. First, a discussion of GIS and decision support is presented. Next, a sample planning problem is presented. This focuses on selecting and evaluating sites for potential tourism development in a district of the island of Grand Cayman. The specific MCA methods and consensus-building techniques applied to the sample problem are described next and the results are presented. The paper concludes with a review of the findings and a statement of future research plans.

2. GIS and Decision Support

In one of the most frequently cited models in the decision support literature, Simon (1976) identified three stages of decision making processes:

1. intelligence – defining the need for decision making or problem-solving;
2. design – preparing alternative courses of action;
3. choice – evaluating alternatives and selecting the most appropriate strategy.

When decision problems are well-structured, decision makers can state their objectives and solution space constraints in clear and computationally tractable terms and progress through Simon's decision making phases in a rationalistic manner. Semi-structured decision issues, of which complex land use issues are characteristic, are more difficult to address due to the presence of several, often noncomplementary objectives, intangibles, and multiple stakeholders who differ in terms of authority, vested interests, and access to information. Conflict arises from a host of uncertainties including:

1. the precise nature of the issue(s) to be resolved and the relative priority of each;
2. the acceptable processes for generating alternative strategies;
3. the acceptable methods for judging the relative suitability of alternatives;
4. a lack of clear standards for defining when an issue has been addressed satisfactorily;
5. and inter- and intra-group interactions and dynamics.

Given these uncertainties, the path through intelligence, design, and choice phases becomes circuitous and recursive as stakeholders engage in debate and negotiation in which information as well as their knowledge and expertise are applied selectively to the intangible aspects of the problem (Densham, 1991; O'Reilly, 1990). The element of space introduces further conflict and complexity into decision making since the cognitive models used to structure space and understand spatial processes often differ substantially between individuals and as individuals are engaged in different tasks (Couclelis and Gottsegen, 1997).

Commercial GIS can provide individual users with powerful conjunctive procedures for the intelligence, design and, to a much lesser degree, choice stages of structured decision problems. However, as the boundaries between decision phases become blurred in semi-structured decision problems, the lack of explicit mechanisms in commercial GIS to support plurality, subjectivity, and group consensus-building becomes more limiting. Interest in extending group decision support research to the spatial realm has increased in recent years and the potential of group-based SDSS (GSDSS) has been demonstrated for spatial problems such as land use planning, environmental planning, and habitat identification (Joerin *et al.*, 2001; Malczewski *et al.*, 1997; Pereira and Duckstein, 1993).

Given that land use decision making is distributed across a series of co-related decision making contexts that reflect the domain-specific interests of various individuals, groups, and organisations, several approaches to GSDSS design and use have been proposed. Some researchers have explored the feasibility of dispersed groups using web-based GIS applications to participate in land-related decision issues in an asynchronous manner, although at this point there is more focus on enabling dispersed participation than issues of conflict identification and consensus building (Kingston *et al.*, 2000). Others have focused on assisting small groups of co-located decision participants to craft land use scenarios collaboratively using GIS and subsequently apply MCA methods and group voting procedures to evaluate competing alternatives and assess the degree of group agreement (Jankowski *et al.*, 1997; Nyerges *et al.*, 1997). In these cases, each collaboratively developed scenario represents an alternative viewpoint, feasible solution strategy, or conceptualisation of the decision issue(s) that is available for subsequent comparison and analysis.

In some decision contexts, it may be preferable for participants to explore spatial issues independently from their idiosyncratic perspective and then permit them (or other individuals or groups) to examine aspects of agreement and conflict among their resultant scenarios (Hendriks and Vriens, 2000). While this dispersed approach does preclude analysis of group dynamics, it does ensure that personal biases and elements of intra-group conflict are recorded for subsequent analysis instead of being simply subsumed within a global solution. In situations of extreme conflict, during problem scoping, or the early stages of plan development, this approach can be used to maintain participant anonymity and minimise "group process losses" associated with overly dominant individuals and the inability of some participants to engage in synchronous forms of communication (Silver, 1991).

The following sections describe a demonstration problem in which this dispersed group approach to GSDSS use was employed to expose issues of conflict and agreement among the stakeholder group. Prior to discussing the specific methods used and the results that were obtained, the next section outlines the study area for the sample site selection problem.

3. SAMPLE PROBLEM

3.1. Setting

Grand Cayman is the largest of the three islands that comprise the British Crown Colony of the Cayman Islands (CI). The island is located approximately 480 miles south of Miami and is divided into five administrative districts, occupying a total land area of 76.4 square miles (Figure 1).

Over the last three decades, fundamental socio-demographic and economic changes have spurred rapid land use change and development on Grand Cayman. During this period, the population more than trebled from 10,652 in 1970 to approximately 36,000 in 2000 and annual tourist arrivals surpassed 1.3 million (CIA World Factbook, 2002).

This fast pace of growth has led to corresponding pressures on the island's limited land base and has introduced substantial and, in some cases, irreversible changes to its biophysical and socio-cultural character. Tourism development has had a particularly important impact on the morphology of land use change in Grand Cayman since it is highly concentrated spatially along the sandy extents of Seven Mile Beach (Figure 1). More than 80% of all hotel beds and rental condominiums in Grand Cayman are situated along this stretch of coast. The vast majority of this accommodation is sited on the waterfront (west) side of the road linking George Town, the capital, with West Bay District. By the end of 1996, virtually all parcels with water frontage on Seven Mile

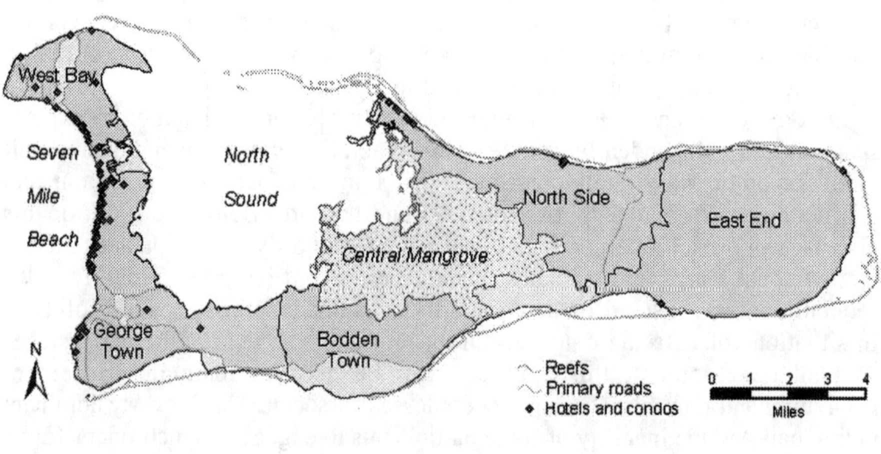

Figure 1. Grand Cayman Administrative Districts.

Beach and capable of supporting a hotel or condominium complex had been developed.

3.2. Problem Description

Demand for additional tourism development in Grand Cayman has led to increasing levels of land-related conflict as various stakeholders have expressed growing concerns over habitat loss, infrastructure limitations, opportunities for future development, and degradation of marine ecosystems, among other factors. Much of this conflict is centred on the construction of additional tourism accommodation in West Bay District (Figure 1).

To evaluate conflict and consensus balancing in land development options, a land use planning exercise was conducted using a GIS-enabled decision support tool with a group of participants representing various land and/or tourism development interests. Although the group of participants was small (n = 7), they were selected purposively based on their different perspectives on and involvement in the tourism and land development process. Table 1 shows that many of the participants were engaged in more than one facet of land use decision making and/or tourism planning. This 'multiple role' phenomenon is not surprising given the relatively small size and population of Grand Cayman and is most likely characteristic of other small island states.

Table 1. Participants and Perspectives Represented.

Participant	Perspective(s) Represented
Central Planning Agency Member, CIG	Government decision maker in planning process, private sector business owner and West Bay resident
Deputy Director of Planning, CIG	Government Planning Department – decision maker, advisor to Central Planning Agency and West Bay resident
Director of Environment, CIG	Government Environment Department – decision maker, researcher, advisor to government decision makers
Cayman Islands National Trust – Scientific Programs Manager	Researcher and program developer for preservation (ecological and cultural) and education
Hotel Manager & Land Developer	Private sector: land development and tourism sectors
Real Estate & Land Developer	Private sector: land development and real estate association
Local Business Person	Private sector: commercial sector and West Bay resident

Note: CIG = Cayman Islands Government.

A two-stage procedure was employed in the research. Participants first identified, using standard GIS functions (attribute selection, spatial selection, point and click selection), viable sites for new tourist accommodation (*site selection stage*). Subsequently, MCA techniques were used to rank the relative suitability of candidate development sites selected by all participants according to each individual's preferences and objectives (*site evaluation stage*). It should be noted that although the software developed for this purpose does permit multiple concurrent users to interact within a single site selection or evaluation scenario, separate scenarios were created by each participant in this instance using the dispersed group approach described in Section 2. For consistency, each participant's site selections and evaluations were moderated by the same researcher who, while guiding participants through the use of the tool, was careful not to bias any aspect of the process.

3.3. Data Sets

In any decision support context, it is vital to the defensibility of decision outcomes for decision makers to have access to data that encompass all relevant aspects of the decision problem. Further, since land use issues tend to be multidisciplinary and pluralistic, it is important to ensure an adequate balance between data sets that represent different facets of the issue at hand (e.g. environmental, economic, social). When quantitative MCA methods are applied to choice decisions, it is also important to ensure that the decision criteria generated from available data represent valid and measurable translations of planning objectives (Voogd, 1983). Hence, data availability and accuracy are important concerns.

Since many land-related planning decisions are based on individual land parcels, the digital cadastre for West Bay District was used as a foundation to build both the site selection and evaluation stages of the data collection and analysis. In 1997, the study area totalled some 4,268 acres and was divided into 4,025 surveyed land parcels. Most of the data sets for this area were provided by the CI Government's Lands and Survey Department in Arc/Info export format (Table 2). These were subsequently converted into the public domain shapefile format for use with the decision support tool developed for this research. Data were augmented by shapefiles supplied by the CI National Trust describing historical, cultural, and environmental features, and a tourism facilities map layer that was created by the researchers. Attribute data for the shapefiles were supplemented and updated to record items such as the extent of coastal mangroves, parcel-specific zoning designations and the development status of individual parcels.

As noted above, data availability inevitably constrains decision outcomes in group decision contexts where data requirements may differ substantially between individuals. In Grand Cayman, some data such as land prices, occupancy rates for individual hotels, and parcel ownership, existed but could not be used in order to ensure either personal privacy or fair marketplace competition. Other data that could have been relevant to the decision problem either

Table 2. Digital Data Sets.

Item	Feature Type	Map Scale	Description
Land parcels	Polygon	1:2,500	17 cadastral map sheets containing 4,025 individual parcels coded for acreage, vacancy status, zoning, length of water frontage
1977 and 1997 land use zones	Polygon	1:25,000	2 map layers extracted from 1977 and 1997 National Development Plan
Building footprints	Polygon	1:2,500	Extracted from digital aerial photographs coded with major commercial and tourism structures
Roads	Line	1:2,500	Extracted from cadastral layers
Historic structures (National Trust)	Polygon	1:2,500	Historic and culturally-significant structures managed by the Cayman National Trust
Protected Areas (National Trust)	Polygon	1:2,500	Environmentally-significant features/areas managed by the Cayman National Trust
Marine Parks	Polygon	1:25,000	Marine parks and replenishment zones
Wetlands	Polygon	1:25,000	Location and extent of main wetlands
Aquatic features	Line	1:25,000	Lakes, reefs, coastline coded by type
Accommodation	Point	1:2,500	Hotels, condominiums and guest houses extracted from cadastre and coded with name, number of rooms, structure type, estimated tourist occupants

existed only in hardcopy format (e.g. terrestrial habitats, hard services infrastructure), were restricted to internal use because of known quality limitations, lacked spatial referencing, or had not been collected. Hence, definition of substantially dissimilar objectives, use of different problem solving approaches, and different problem definitions were limited to the data sets listed in Table 2.

4. METHODS

4.1. MCA Techniques

MCA methods effectively decompose choice decisions by indicating the performance of alternatives across different aspects of the decision problem through a series of criteria scores and capturing the relative importance of each criterion to a given decision maker through assignment of a criterion weighting factor. Weighted summation (WS) and concordance analysis (CA) are two commonly used methods for evaluating alternatives. Weighted summation is a straightforward technique that is suitable for the evaluation of simple decision problems (Hobbs *et al.*, 1992). However, the method assumes that the evaluation criteria are both substitutable and non-complementary; conditions that often can be difficult to satisfy and, if violated, can generate misleading results (Pereira and Duckstein, 1993). Further, in order to permit evaluation of a utility index by aggregating across criteria that are measured in different units, attribute values must be standardised (0 to 1) by dividing criterion values by the maximum value for that criterion. On this basis, a utility score for each alternative in a set of J alternatives can be calculated as:

$$U_i = \sum_{j=1}^{J} (w_j \hat{x}_{ji}) \tag{1}$$

where: U_i is the summed score for each alternative i across all criteria J,
$\hat{x}_{ji}$ is the standardised attribute score of criterion j for alternative i,
w_j is the relative preference weight assigned to criterion j.

The family of concordance analysis (CA) methods offers a more robust set of procedures for evaluating decision problems than WS. In general, CA methods evaluate outranking relationships between alternatives. These relationships seek to establish whether an alternative is at least as good as any other in the evaluation set. With this approach, a concordance index is established to determine the relative dominance of alternative i over alternative i' through pairwise comparison. This index is calculated as the sum of the priority weights in the concordance set, which is defined to include only criteria where alternative i performs better than or equal to alternative i' (Bodini and Giavelli, 1992):

$$c_{ii'} = \sum_{j \in C_{ii'}} w_j \tag{2}$$

where: $c_{ii'}$ is the concordance score for alternative i relative to alternative i'
$C_{ii'}$ is the concordance set of criteria for alternatives i and i'
$\hat{x}_{ji'}$ is the weight assigned to criterion j.

A discordance index $d_{ii'}$ can be viewed as the reciprocal of $c_{ii'}$ as it reveals the maximum difference between pairs of criteria scores where alternative i' outperforms alternative i. Hence, whereas the concordance index in Equation 2 deals with differences in criteria weights, the equivalent discordance index centres on criteria scores. In this instance, data values must be standardised (0 to 1) in order for the discordance component to be calculated, since it depends on inter-criteria comparisons. Specifically,

$$d_{ii'} = \max_{j \in D_{ii'}} (\hat{x}_{ji'} - \hat{x}_{ji}) \tag{3}$$

where: $d_{ii'}$ is the discordance score for alternative i relative to alternative i'
$D_{ii'}$ is the discordance set of criteria for alternatives i and i'
$\hat{x}_{ji'}$ is the standardised score of alternative i' on criterion j.

In general, the most preferred alternative in a CA-based decision outcome will have a high concordance index relative to other alternatives and a low discordance value. Since it is not always possible to generate unique and optimal classifications of alternatives in this manner, user-defined threshold values, $\bar{c}$ and $\bar{d}$ are commonly applied to the first order concordance and discordance matrices in order to produce the respective dominance matrices. Hence, following Nijkamp and van Delft (1977), with the use of a threshold value, for alternative i to outrank alternative i', the following condition must hold:

$$c_{ii'} \geq \bar{c} \quad \text{and} \quad d_{ii'} \leq \bar{d} \tag{4}$$

An analyst can progressively alter these thresholds to eliminate dominated alternatives and develop partial or complete rankings of non-dominated alternatives. Based upon its consistent and full use of information concerning alternatives and its thorough method of comparing each alternative against every other candidate, CA methods have been applied successfully to many decision problems (for example, Phaneuf, 1990; Hwang and Yoon, 1981). However, the arbitrary and often ambiguous nature of threshold definition in CA has been criticised, especially in light of its influence on outcomes (Yoon and Hwang, 1995; Jankowski and Richard, 1994). Additionally, the concept of threshold values has proven to be difficult for some participants to understand, leading Hobbs *et al.* (1992) to suggest that this lack of understanding may undermine the confidence that participants have in the results.

Both WS and a form of CA approaches were implemented in the software developed for this research. The specific CA technique is based on Nijkamp and van Delft (1977), who utilise pair-wise concordance and discordance indexes to produce measures of net dominance instead of outranking. This net concordance discordance (NCD) method does not require thresholds to be used, produces complete rankings of alternatives, and is easier for non-experts to conceptualise than more complex methods, such as fuzzy outrankings (Vinke, 1992). The latter consideration (i.e. non-expert use) is particularly important for land use decision problems that are evaluated by participants who lack specialised training and expertise in the planning process as well as evaluation methods.

Concordance and discordance scores are calculated for pairs of alternatives, as described in Equations (2) and (3). The net concordance dominance value for choice candidate i, denoted here as *net* c_i indicates the difference between the degree to which alternative i dominates all other alternatives and the extent to which they in turn dominate it. This is calculated as:

$$net\ c_i = \sum_{i'=1,\, i' \neq i}^{I} c_{ii'} - \sum_{i'=1,\, i' \neq i}^{I} c_{i'i} \qquad (5)$$

where: I is the set of choice alternatives.

Net discordance dominance values are calculated in a similar manner. Candidate sites can then be ranked according to their $net\ c_i$ or $net\ d_i$ values, or by a composite ranking can be produced by averaging the two set of ranks with the highest average rank being the most preferred (Yoon and Hwang, 1995).

4.2. *Criteria Selection and Weighting*

In the procedure adopted for this research, each participant first selected land parcels based on his/her idiosyncratic tourism development objectives (*site selection*). They were then required to complete three additional tasks (*site evaluation*). First, the criteria upon which the selected sites would be evaluated were established. Second, data normalisation parameters were set for each criterion. Third, criteria weights were set using one of two possible methods described below.

Most of the criteria used to judge the suitability of candidate sites were generated by the researchers due to time constraints of the participants (that precluded field assessment of the sites for scenic quality, etc.), data confidentiality concerns (e.g. land ownership, price, hotel occupancy rates), and the lack of spatial-referencing in some data sets. Given these limitations, most of site evaluation criteria in Table 3 are proximity-oriented and represent the shortest straight-line distance in feet between any point on the perimeter of a candidate site and any feature in a second map layer of interest. A value of zero was assigned to a parcel either if it contained or was adjacent to a feature in a second map layer. Hence, a site that fronted directly onto a sandy beach and was zoned for Hotel/Tourism in the CI Development Plan would have a value of zero recorded for both the D2Sand and D2TourZone criteria.

Although limited to proximity measures, some of the criteria in Table 3, such as D2Hotel, D2Replent, and D2Wetland, may act as surrogates for the attractiveness of a property for tourist accommodation or for the potential impacts that tourism activities may have on nearby fragile ecosystems. However, several of the criteria are equally constrained in that they not only assume linearity in the effect of distance (e.g. D2Reef and D2Marine), but also possibly subsume "functional" effects within the proximity effects (e.g. reefs suffocated by sediment transported by longshore drift).

With these constraints in mind, each participant was required to select at least two criteria that would enable them to rank the relative attractiveness of selected sites, given their own objectives and priorities. Criteria that were not selected were assumed explicitly to be of no significance in the participant's decision process and were assigned a value of zero in the subsequent criteria weighting. Since it is possible for redundant criteria to skew evaluation results by "double-

Table 3. Site Evaluation Criteria.

Criterion	Description	Data source
Acres	Size of parcel in acres	Cadastral map layers
D2Hotel	Distance from parcel to nearest hotel, condominium, or guest house	Calculated from accommodation layer
D2Mangrove	Distance from parcel to nearest mangroves	Calculated from coastline map layer
D2Marine	Distance from parcel to nearest marine park	Calculated from marine parks map layer
D2PrimRd	Distance from parcel to nearest primary road	Calculated from road network layer
D2Reef	Distance from parcel to nearest reef	Calculated from coastline map layer
D2Replent	Distance from parcel to nearest replenishment zone	Calculated from marine parks map layer
D2Rock	Distance from parcel to nearest ironshore coastline	Calculated from coastline map layer
D2Sand	Distance from parcel to nearest sand coastline	Calculated from coastline map layer
D2SeaPond	Distance from parcel to Sea Pond	Calculated from coastline map layer
D2SecRd	Distance from parcel to nearest secondary road	Calculated from road network layer
D2Shop	Distance from parcel to nearest commercial zone	Calculated from 1995 zoning map layer
D2Tourzone	Distance from parcel to nearest parcel zoned for Hotel/Tourism	Calculated from 1995 zoning map layer
D2Wetland	Distance from parcel to nearest wetland	Calculated using wetlands map layer
NumSelect	Number of participants that designated the parcel as a potential accommodation site	Aggregated from the individual site selection scenarios
Vacant	Indicator if any structures were found on parcel	Based on 1994 aerial photographs
Waterfront	Length of water frontage	Cadastral map layers

counting" a single facet of the decision problem, participants were discouraged from selecting both the D2PrimRd and D2SecRd indicators of a parcel's proximity to roadways (Malczewski, 1999).

After selecting their criteria set, each participant completed tasks two and three noted above in order to standardise data values (Malczewski, 1996). To achieve this, participants had to indicate for each criterion whether high data values would contribute to the relative attractiveness of a site (i.e. a maximising or 'benefit' criterion) or detract from it (i.e. a minimising or 'cost' criterion). Standardised data values increased or decreased monotonically between 0 and 1 depending on this setting.

A simple approach was adopted in the site evaluation process where each person determined whether criterion data values should be scaled relative to calculated minima and/or maxima or whether user-specified values should be substituted into the scaling process. Participants could adopt the latter approach for a given criterion in order to recognise that there are practical limits beyond which no further benefit or cost is realised (Massam, 1988). To illustrate, although the calculated maximum D2Sand value (greatest straight line distance to a sandy beach) for any parcel in the West Bay data set was 5,185 feet, many tourists have a much lower acceptable threshold for beach access. Given this, an alternative value (e.g. 1,000 feet) could be substituted into the scaling process such that the site(s) closest to a sand beach would receive the highest standardised score of 1, all candidate sites beyond the user-specified limit of 1,000 would have a score of 0, and the D2Sand values for the remaining sites would be scaled relative to these extreme points.

Next, the participants experimented with and applied either a seven-point scale or used the eigenvector method for generating criteria weights. The former is based on psychological scaling research which suggests that individuals are generally able to distinguish reliably between seven classifications, plus or minus two (Voogd, 1983). Each evaluator could generate his/her own set of criteria weights relatively quickly and easily with the dialog implemented for this method by assigning to each criterion a value ranging from very low importance (1) through to very high importance (7) (Figure 2). The numeric values are then standardised to produce criteria weights. This method has also been used in spatial decision-making by Carver (1991) and by Jankowski and Stasik (1997), although the latter use a nine-point, rather than a seven-point, scale.

The second method for estimating criteria weights in this research is the eigenvector method, popularised by Saaty (1980) in the Analytical Hierarchy Process. In this case, evaluators compare each pair of criteria and indicate the relative importance of criterion to each other on the following nine-point scale with even numbers being considered as intermediate points between adjacent values:

1. Criteria i and i' are equally important
3. Criterion i is moderately more important than criterion i'
5. Criterion i is strongly more important than criterion i'
7. Criterion i is very strongly more important than criterion i'
9. Criterion i is extremely more important than criterion i'

For each pair of criteria, ratios of their relative importance are entered into

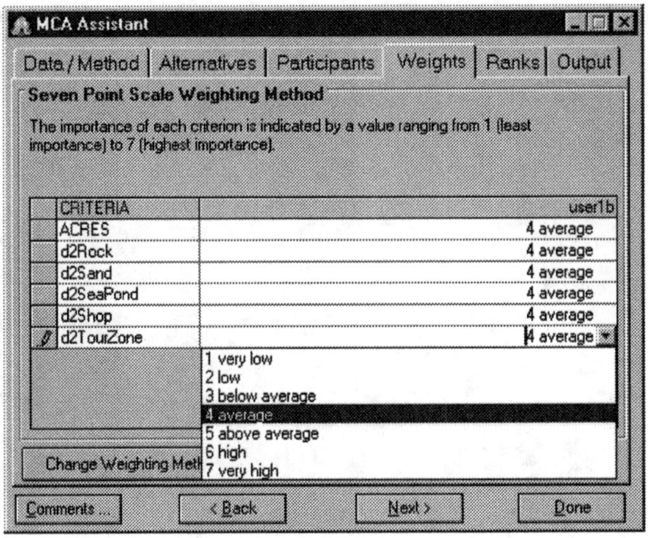

Figure 2. Seven Point Criteria Weighting Method.

a positive pair-wise reciprocal matrix such that $a_{ii'} = a_{i'i}^{-1}$ (Van Huylenbroeck and Coppens, 1995). A total of $n(n - 1)/2$ comparisons are required to complete the matrix, making this method more demanding when comparing a large number of criteria.

If decision-makers are completely consistent in their assessments, the same weight vector could be created simply by normalising each matrix element by its corresponding column total (Olson, 1996). Since this is usually not the case, especially when larger numbers of criteria are involved (i.e. A > B, B > C, and C > A), criteria weights are derived from the eigenvector of the matrix that has the largest eigenvalue, which is denoted as $^{\lambda}$max (Harker, 1989). Since $^{\lambda}$max is always greater than or equal to n for a positive reciprocal matrix and $^{\lambda}$max is only equal to n if the matrix is consistent, following Saaty's notation (1980) a consistency index (CI) can be defined as:

$$CI = \frac{^{\lambda}\text{max} - n}{n - 1} \tag{6}$$

A consistency ratio (CR) can then be determined by dividing CI by an appropriate random index associated with a sample of purely random matrices (Banai, 1993). As a general rule, CR values less than or equal to 0.10 signify that the pair-wise comparisons are reasonably consistent. It is usually recommended, although not required, that pair-wise comparisons be revised if the CR value exceeds a threshold of 0.10 (Harker, 1989).

In the software developed for this research, the relative importance of each criteria pair is reflected by the height of corresponding graphical bars supplemented by descriptive pair-wise statements. The relevant dialog, shown in

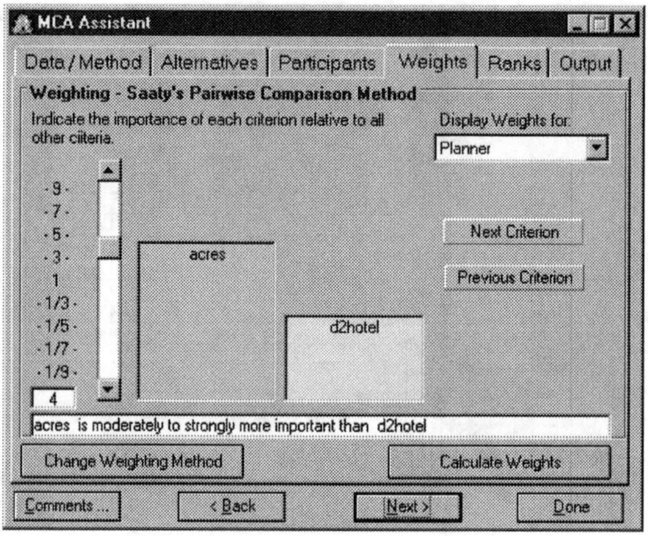

Figure 3. Pair-Wise Criteria Weighting Method.

Figure 3, illustrates that the 'Planner' user has designated the size of a site (acres) as "moderately to strongly more important" than the distance from the site to other tourist accommodation (d2hotel). The reciprocal of this assessment between the same criteria, if consistent, should reflect the inverse. Once the criteria weights were set and reviewed for all criteria, ranks were calculated for selected sites using the evaluation method established previously for display in tabular and map-based form.

4.3. Consensus Approaches

To supplement the procedures noted above, two general approaches have been developed to extend the consensus-building aspect of pluralistic MCA decision making with group-based rankings of alternatives. The first approach requires the relative suitability of choice alternatives be judged against a common set of evaluation criteria that is formed as a result of administrative or regulatory requirements, or a group's shared objectives and interests. Group-wide weights may be defined for this criteria set by negotiating definition of a common set or by aggregating the criteria weights of specific participants. In the latter case, the participant's weights may contribute equally or differentially to the group weighting vector. Differential contributions may be realised by applying an influence factor to each person's criteria weights to reflect the relative importance of their expertise to the group's decision-making process (Malczewski *et al.*, 1997).

This approach captures and subsumes conflict balancing or consensus building within the negotiation processes that result in a common criteria set and, to a lesser degree, the arithmetic methods that produce a single group weighting vector. However, in situations where stakeholder objectives and priorities are in conflict, it is difficult in practice to reach agreement on the relative importance of individual criteria or even the set of criteria to be used to assess the attractiveness of alternatives. In these cases, it is more appropriate to explore the various dimensions of conflict, as represented in criteria choices and weightings, by producing group rankings that are based on the ranks generated by individual group members. This comprises the second method noted above and is the approach that is described below in results section.

The problem of deriving a consensus ranking of alternatives from a set of individual ranks in a just and theoretically consistent manner has a long history in the study of elections and voting (Borda, 1781). Hwang and Lin (1987), for example, detail almost a dozen social choice techniques that are applicable to this problem. The approach developed for this research incorporates the Borda and Copeland counting functions to produce group-wide rankings of alternatives.

The Borda function uses a simple sum of ranks rule to calculate an aggregate ranking from a series of individual ranking vectors. For m alternatives, the Borda counting function awards $m - 1$ points to the alternative ranked highest by an individual, $m - 2$ points for the second highest ranked alternative, and so on. This is repeated across all group members and summed to produce overall scores with the highest value being associated with the most preferred alternative (Massam, 1993). It is assumed that all individual ranks have equal influence on the aggregate ranks and that the differences between adjacent alternatives (e.g. 3rd and 4th position) are considered to be equal in the absence of any contrary information (Arrow, 1963).

In contrast, the Copeland method for generating aggregate ranks is based on the Condorcet condition which asserts that the only fair way to determine which alternative is supported by the majority of all voters is to examine each pair of alternatives and select the alternative that has a simple majority over all other candidates, should such a candidate exist (Hwang and Lin, 1987; Fishburn, 1971). Whereas a Condorcet-based count can be determined for each alternative based on the number of times it is preferred to another alternative, the Copeland method produces aggregate ranks by subtracting from this sum the number of times each alternative is ranked lower than other alternatives (Massam, 1993). Alternatives are ranked from highest net score (i.e. number of 'wins' – number of 'losses') to lowest net score. Compared to the Borda count, the Copeland method is more likely to produce a higher number of tied ranks, particularly as the number of alternatives increases.

5. RESULTS

In selecting sites for tourism development in West Bay District, each participant was required to designate sufficient land to support a minimum of 100 additional rooms of tourist accommodation. From the resultant site selection scenarios, a set of 53 candidate tourism sites (land parcels) selected by at least two participants and of at least one acre in size was established (Feick and Hall, 2000). This section examines the MCA-based rankings of this set of sites and explores how the methods described above permit the issue of multi-party consensus and conflict balancing to be investigated in practice.

5.1. Criteria Choices Weightings

Through the process of disaggregating choice decisions into sets of criteria for comparing alternatives and their corresponding weight vectors, MCA methods provide insight into how different individuals approach a decision problem as well as into areas and intensity of consensus or conflict among individuals. Three fundamental elements of consensus and conflict in multi- participant decision making are shown in Table 4:

1. the extent of agreement concerning the parcel attributes that the participants used as criteria for comparing the alternative sites,
2. data normalization settings, and
3. differentials in the relative importance of individual criteria, as expressed through weight settings.

Some understanding of each participant's objectives and concerns is required to understand the origin of conflict and the rationale underlying criteria choices and weightings. For example, the three private sector business participants (Business owner, Hotel operator/Developer, Real estate/Developer) and, to a lesser degree, the CPA Member preferred new tourist accommodation to be in the form of large beachfront hotels. The primary concerns they expressed throughout the exercise were: high land costs and a lack of waterfront development opportunities on Seven Mile Beach, congestion on the main road leading to West Bay from the capital city (George Town) and the international airport, the limited economic feasibility of small hotels, and the need for more hotel management opportunities for young Caymanians. Their criteria selections reflected these concerns as they concentrated almost exclusively on factors that addressed a site's tourism development potential (D2TourZone, Acres, Waterfront) and its attractiveness to tourists (D2Shop, D2Sand).

In contrast, the National Trust Scientist, Director of Environment and the Deputy Director of Planning all favoured low density and small scale tourism development (e.g. small hotels and guest houses) due to concerns about the ecological and social effects of further tourism development in West Bay in particular and the country in general. The National Trust Scientist focused exclusively on criteria that represented the potential environmental impacts of development (e.g. D2Mangrove, D2Marine, D2SeaPond). The Director of Environment and the Deputy Director of Planning selected many of the same criteria (see Table 4), but also recognised the importance of the employment opportunities tourism development could provide to local residents through many of the same criteria selected by the pro-development participants.

Table 4. Criteria Weightings and Benefit/Cost Settings.

	Director of Environment	National Trust	Real Estate/ Developer	Hotel Oper./ Developer	Business Owner	CPA Member	Deputy Dir. of Planning
Acres	**0.075367**	0	**0.166667**	**0.185185**	**0.130435**	**0.192308**	0.051478
D2Hotel	0	0	**0.166667**	0	0	0	0.099735
D2Mangrove	**0.143221**	**0.131579**	0	0	0	0	**0.120503**
D2Marine	**0.040849**	**0.131579**	0	0	0	0	0
D2PrimRd	0	0.052632	0	0	0	0.115385	**0.123595**
D2Reef	0	**0.131579**	0	0	0	0	**0.05236**
D2Replent	**0.044319**	**0.105263**	0	0	0	0	**0.025511**
D2Rock	**0.078903**	**0.105263**	0	0	0.086957	0	0
D2Sand	0.096823	0	0.25	0.259259	0.26087	0.269231	0
D2SeaPond	0	**0.184211**	0	0	**0.130435**	0	0
D2SecRd	0.034237	0	0	0	0	0	0
D2Shop	0.019675	0	0.166667	0.111111	0.086957	0.153846	0.070991
D2Tourzone	0.148688	0	0	0.259259	0.304348	0	0.201688
D2Wetland	**0.239397**	**0.157895**	0	0	0	0	0
NumSelect	0	0	0	0	0	0	**0.068527**
Vacant	0.039633	0	0.25	0	0	0	0.035312
Waterfront	0.038864	0	0	**0.185185**	0	**0.269231**	**0.150301**

1. Criteria weights may not sum exactly to 1.0 due to rounding
2. Benefit criteria (high data values preferred) are shown in bold; Cost criteria (low data values preferred) are not in bold.

Based on the above perspectives, the private sector and CPA member participants are considered to be relatively uniform in their approach to tourism development and are henceforth referred to as comprising a *pro-development group*. The National Trust Scientist and Director of Environment in turn are referred to as a *pro-conservation group* while the Deputy Director of Planning sits between the two in his attempt to balance conservation issues with controlled land development for tourism.

The relatively few differences in cost/benefit settings shown in Table 4 can, to some degree, be accounted for by each participant's individual tourism development preferences. For example, the Deputy Director of Planning wanted to promote sites that would be reasonably close to existing development, near but not on primary roads, and suitable for small hotels and guest homes. Hence, in contrast to the other participants, he set Acres as a cost (smaller sites preferred), D2PrimRd as a benefit (lower proximity to primary roads preferred) and D2Hotel as a cost. Interestingly, the Director of Environment shared many of these goals but designated Acres as a benefit to reflect her preference for low density development.

As mentioned earlier, individuals had the option of determining how criterion data values were scaled by specifying their own practical minima and maxima (Table 5). With proximity-based criteria, these values provided some insight into the extent that the perceived spatial bounds of particular externality fields varied across the site evaluations. For example, four participants shared a similar view on how far tourists would be prepared to walk to shops by stipulating maximum D2Shop values in the range of 1,000 to 1,500 feet. The two participants who were most concerned with environmental issues (Director of Environment and the National Trust Scientist) were consistent when defining a development-free area of 200 or 300 feet around features such as mangroves, marine parks, replenishment zones, and wetlands. However, their perception of the spatial impact of development on these features differed considerably (i.e. 1,000 feet versus 5,280 feet). These differences, even among individuals with similar planning objectives and professional expertise, illustrate some of the potential sources of conflict and consensus in multi-party decision making.

Once each participant's criteria set had been established, either the seven-point scale or pair-wise comparison method was chosen to develop criteria preferences. From a practical perspective, the usefulness of different criteria weighting methods is determined to a large degree by the amount of time required to complete the procedure and the cognitive demands the method places on a user (Voogd, 1983). Only the Deputy Director of Planning and the Director of Environment opted for pair-wise comparisons because of its comprehensiveness in comparing criteria and its more rigorous mathematical structure. The National Trust Scientist favoured the pair-wise comparison method for these reasons as well, but elected to use the seven-point technique due to time constraints. No participants opted to change weighting methods during their site evaluations.

Other indications of group-wide consensus and conflict are evident in criteria weights used by the participants, although given the differences in their criteria sets it is necessary to consider both the ordinal rankings of criteria and their metric weighting values. In terms of the former, factors relating to a site's development or tourism potential were most frequently designated as the most important (D2Sand, D2TourZone) or second most importance criteria (Acres, D2Shop, Waterfront) by the pro-development group. By excluding other considerations from their criteria sets, these individuals assigned by definition the lowest priority weighting (0) to criteria that represented environmental concerns. On the other hand, the pro-conservation group emphasised environmental factors most (D2Wetland, D2SeaPond, D2Mangrove) and placed less significance on factors relating to the economic feasibility of development (D2Shop, D2PrimRd, D2SecRd, Acres). Interestingly, the D2TourZone criterion was weighted highly by individuals with both pro-development and pro-conservation objectives (e.g. Deputy Director of Planning) as it encapsulates elements of a site's economic feasibility (e.g. hard service infrastructure, tourist attractiveness) for the former and consideration of previous decisions concerning where development should be confined to for the latter.

5.2. *Individual and Group Rankings*

As noted in section 4, each participant used either the weighted summation (WS) or net concordance discordance (NCD) method to produce their site rankings. Table 6 lists the final group and individual rankings for the top 25 Copeland ranked sites of the total 53 candidates. Inspection of the positions in the table reveals which sites may be suitable to the specific interests of a sub-group of participants. For example, a reasonable degree of consensus is evident for the sites that the pro-development group members ranked highly. Parcels with marked discrepancies in their ranks (e.g. 17A_14 and 17A_12, with a Copeland rank of 2.5) can be attributed to differences in the criteria sets used by individuals, particularly with respect to the highly weighted D2TourZone and Vacant criteria. In contrast, the ranks produced by the pro-conservation group members were much less similar, notwithstanding their agreement on the most preferred and the least desirable sites. Again, differences in criteria sets can account for some of the differences in their rankings, but it is also important to recall that the National Trust Scientist defined substantially greater fields of influence for the D2Mangrove and D2Wetland criteria in the data normalization process than the other group member. Finally, the ranks for the Deputy Director of Planning reflect his objective to achieve a balance between development and conservation interests although they tend to be closer to the former than the latter.

The high degree of correspondence between the Borda and Copeland rankings and some of the idiosyncratic aspects of the results produced by each method are also evident in Table 6. For instance, in addition to ranking the same parcel first, a total of 12 of the top 15 sites based on the Copeland ranks were also

Table 5. User-Specified Data Standardisation Parameters.

Participant	Criterion	Actual Min.	User Min.	Actual Max.	User Max.	Rationale
Dir. of Environ.	D2Shop	0	0	6216	1320	1/4 mile = max. walking limit to shops
	D2Mangrove	53	300	13602	1000	300 ft. = vegetation line, no > benefit after 1000 ft.
	D2Marine	0	200	5650	300	No development < 200 ft., after 300 ft. no > benefit
	D2Replent	0	200	9969	300	As above
	D2Wetland	0	200	14391	750	No development < 200 ft., no > benefit after 750 ft.
National Trust Scientist	D2Mangrove	53	300	13602	5280	No development < 300 ft., no > benefit after 1 mile
	D2SeaPond	3889	300	17450	5280	As above
	D2Wetland	0	300	14391	5280	As above
Real Estate/ Developer	Acres	1	2	279		Min. 2 acre site for hotel & services
	D2Sand	0	0	5185	1000	Tourists will walk up to 1000 ft. to a sand beach
Hotel Oper./ Developer	Acres	1	2.5	279		2.5 acres is min. preferred site size, no max. size
	D2Tourzone	0	0	9022	10*	Sites must currently be zoned for tourism
	Waterfront	0	200	6217		Min. 200 ft. needed for aesthetics and activities
Business Owner	Acres	1	0.5	279		Min. 1/2 acre site satisfactory, no max. size
	D2Sand	0	0	5185	500	Want sites on, or close to, a sand beach
	D2SeaPond	3889	100	13602	2000	No development < 100 ft., no > benefit after 2000 ft.
	D2Shop	0	0	6216	1000	Tourists will walk up to 1000 ft. to shops

CPA Member	Acres	1	0.5	279	100	Min. 1/2 acre sites, no > benefit beyond 100 acres
	D2PrimRd	5.9	0	5310	5280	Access and hard service benefits within 1 mile
	D2Sand	0	0	5185	2640	Tourists will walk up to 1/2 mile to a sand beach
	D2Shop	0	0	6216	2640	Tourists will walk up to 1/2 mile to shopping
	Waterfront	0	100	6217	500	Min. 100 ft. water frontage needed, no further benefit beyond 500 ft.
Deputy Dir. of Planning	Acres	1	0.5	279	3	Prefer 1/2 acre sites, max. of 3 acres (acres = cost)
	D2Hotel	0	0	4273	1320	Cluster accommodation due to servicing concerns
	D2Mangrove	53	50	13602	1000	No development < 50 ft., no > benefit after 1000 ft.
	D2PrimRd	5.9	0	5310	660	Easier hard service access near primary roads
	D2Shop	0	0	6216	1500	1500 ft. = max. walking distance to shops
	D2TourZone	0	0	9022	10*	Sites must currently be zoned for tourism
	D2Reef	1274	1000	20448	1500	Sites 1000 to 1500 ft. away discourage daily use but allow less frequent use

1. Except for Acres, all values are in feet.
2. Data values are rounded.
3. * indicates a User Maximum value of 0 was specified, but the researcher substituted a value of 10 to account for possible errors in the spatial data sets.
4. Criteria not listed above for a given participant were standardised using the actual values found in the case study data sets.

Table 6. Top 25 Candidate Sites Using Copeland Ranks.

Block & Parcel	Dir. of Env.	National Trust	Real Estate/ Developer	Hotel Operator/ Developer	Business Owner	CPA Member	Deputy Dir. of Planning	Borda Rank	Copeland Rank
11D_31	15	**7.5**	**5**	**4**	**2**	**5**	**6**	**1**	**1**
17A_14	17.5	18	**2**	**3**	21	**7**	**10**	**4**	**2.5**
17A_12	**3**	25.5	**3**	19	30	**2**	**4.5**	**6**	**2.5**
11B_67	19.5	15.5	**6**	**5**	**5**	12	**3**	**2**	**4**
11B_17	24	34.5	**4**	**2**	**3**	**4**	13	**5**	**5**
11B_76	13.5	**10.5**	**7**	**6**	**8**	13	**8.5**	**3**	**6**
17A_10	33,5	34.5	17	**1**	**1**	**1**	14.5	**7**	**7**
10E_35	30	39	11	**7**	17	**3**	11.5	**8**	**8**
10E_9	28.5	39	12	**8**	16	**10**	14.5	**9**	**9**
8A_3	31	43.5	20	**9**	**10**	**8**	20	11	**10**
8A_60	25	39	22	11	**7**	16	26	12	11
11B_61	**4**	28	**9**	24	20	30	16.5	**10**	12
8A_68	26.5	46	24	15	11	14	21.5	19	13
3D_110	26.5	41.5	23	13	**6**	22	23.5	17.5	14.5
3D_23	28.5	**4**	27	12	**4**	20	21.5	22	14.5
11B_62	**6**	32	19	24	24	38	**4.5**	13.5	17
17A_11	13.5	23	38	20	38	11	**7**	15	17
8A_7	35	51	28	10	12	6	18	20	17
1C_33	39.5	**3**	**1**	26	28	9	41	13.5	19.5
8A_2	32	43.5	21	17	**9**	23	23.5	24	19.5
11D_49	19.5	9	46	22	41	17	**1**	17.5	21
8A_16	43.5	49	25	14	15	18	11.5	27	22
11D_4/1/5	**1**	**1**	13	31	25	31	50	16	23
8A_14	49	51	26	16	14	21	16.5	31	24
8A_13	45	53	29	18	13	19	19	32	25

1. The top 10 ranked sites are shown in bold text for each participant.
2. A Block and Parcel number is a unique identifier the C.I. Land Registry assigns to properties. It is comprised of a grid-like Block number and a Parcel number within that specific Block.

ranked within the top 15 with the Borda method. Further, although the sites that ranked lowest on a group level are not shown in Table 6, it is worth noting that 9 of the 10 least preferred sites were consistent between the two methods.

The greater propensity for the Copeland technique to produce tied ranks due to its "win minus loss" structure of comparing each alternative with every other alternative is also shown in Table 6. This characteristic is not necessarily a disadvantage in a multi-party decision context, since it can serve as a reminder to the participants that the results of individual, group and overall ranking procedures may not sufficiently differentiate between some choice alternatives without other supporting evidence.

If the purpose of the evaluation stage of group decision making is to minimise conflict by promoting options that are highly ranked on average over alternatives that are highly favoured by some participants but ranked poorly by others, the Borda social preference function is the preferred method to use (Jankowski and Stasik, 1997). However, as Cox (1997) has noted, the Borda method can promote mediocre candidates if participant's ranking vectors are either highly diverse or if one or more individual deliberately assigns low ranks to alternatives that may threaten their own most favoured choices. The problem of strategic voting can be discounted in this research given the independence and anonymity of participants. Despite this, the averaging effects of the Borda function can be seen in Table 6, as the site that tied for second on the Copeland rank received a sixth place Borda rank due to the influence of the National Trust Scientist and Business Owner's ranks (25.5 and 30th).

A sense of the spatial distribution of each participant's ranks can be obtained by cross-referencing Table 6 with Figure 4 in which the Copeland consensus rankings are mapped. The classification scheme in Figure 4 uses progressively darker shades to signify increasing overall preference such that the highest ranked sites have the darkest shading. As Figure 4 illustrates, the most preferred tourism development sites on a group-wide basis are found on the waterfront near the southern extent of West Bay District. These sites are close to Seven Mile Beach, located within a hotel/tourism zone and are near to commercial areas. The sites also satisfy environmental concerns, although to a lesser extent, by being comparatively far from the wetlands, reefs, and replenishment zones found in the northern portions of West Bay (Figure 1) and mangroves that fringe the North Sound coastline. These same factors account also for the least preferred sites being concentrated in the north and north west parts of the study area.

Visualisation of ranks in map form supplement the tabular results and can assist participants in exploring the implications of altering criteria sets and/or priority weights. Moreover, mapped output provides insight into the spatial dimensions of consensus and conflict in the selection and evaluation stages of decision outcomes.

To test the significance of the association between pairs of participant site rankings and between the ranks for individuals and subgroups, Spearman's rank correlation coefficients (r_s) were calculated for the ranks listed in

Table 6. The test statistic assumes that at least 5 pairs of observations are present and that the observations are ranked from 1 to n with any tied ranks being represented by average ranks (Silk, 1979). Assuming that the number of pairs of tied ranks does not exceed 25 percent of n, the statistic has the following form (Siegel, 1956):

$$r_s = 1 - \left(\frac{6 \sum_{i=1}^{N} d_i^2}{n^3 - n} \right) \tag{7}$$

where: d_i^2 is the squared difference between the ranks for alternative i
n is the number of observations.

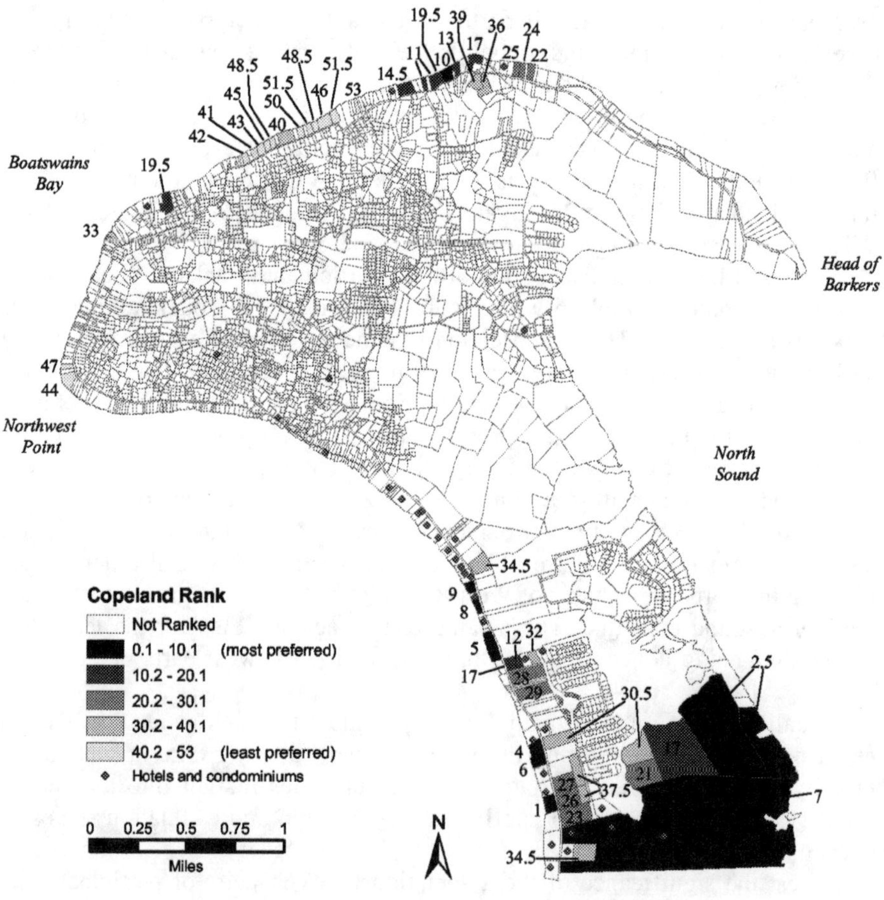

Figure 4. Copeland Ranks For The 53 Common Sites.

When n is greater than 10, the distribution of r_s approaches the t distribution allowing the significance of r_s to be tested against critical values of t as a value of t with $n - 2$ degrees of freedom (Siegel, 1956). The transformation of r_s values to Student t values is calculated by:

$$t = r_s \sqrt{\frac{n-2}{1-r_s^2}} \tag{8}$$

Pair-wise r_s values confirmed significant positive correlations between the ranks of the pro-development group members at the 99% confidence level as well as between the Director of Environment and two of the pro-development group members (Business owner, Hotel operator/Developer). Several other pairs of ranks exhibited positive association at the 95% level (CPA member and Deputy Director of Planning, Real estate/Developer and Deputy Director of Planning, Hotel operator/Developer and Director of Environment). In contrast to the other conservation-minded participants (Deputy Director of Planning, Director of Environment), the National Trust Scientist displayed only statistically significant negative association with the ranks of both the Business owner (99% level) and the Hotel operator/Developer (95% level). These results are largely due to the composition of his criteria set, which only included factors related to environmental impacts.

The correlations between participant rankings and the two methods of producing aggregate ranks are listed in Table 7. The effects of the different methodologies underlying the Borda and Copeland functions are apparent when the range of r_s values generated for each of the two methods is considered. In particular, the r_s values listed for the Borda function are higher for the "minority", pro-conservation participants who emphasised environmental factors in their criteria sets and lower for the "majority" participants who focused almost exclusively on economic development considerations (Real Estate Developer, Business Owner, Hotel Operator/Developer, CPA Member). This confirms the earlier observation that the Borda method is more likely to promote compromise alternatives than the Copeland method. In contrast, the latter is less susceptible to "extreme" minority positions and will produce aggregate ranks more closely reflecting the majority view, if one exists, as shown by the strong associations between its ranks and the four *de facto* pro-development participants.

6. SUMMARY AND CONCLUSION

This paper has presented an integrated set of methodologies that unite GIS technology with MCA-based evaluation of decision alternatives. Together the unified approach seeks to uncover and allow balancing of conflict and consensus in decision outcomes. To accomplish this, custom software was developed to accommodates input from and analysis of the views of multiple non-expert

Table 7. *Spearman Rank Correlation – Individual and Aggregate Ranks.*

Group Ranking	Participant Ranking	Rs	Student's t-statistic	Significant at 95%	Significant at 99%
Borda Rank	National Trust Scientist	0.1247883	0.89818721	No	No
Borda Rank	Director of Environment	0.581401	5.10317843	Yes	Yes
Borda Rank	Deputy Director of Planning	0.64722377	6.063363288	Yes	Yes
Borda Rank	Business Owner	0.6998065	6.99620761	Yes	Yes
Borda Rank	CPA Member	0.7567328	8.26671493	Yes	Yes
Borda Rank	Hotel Operator/Developer	0.8256592	10.4514504	Yes	Yes
Borda Rank	Real Estate/Developer	0.8388969	11.006851	Yes	Yes
Copeland Rank	National Trust Scientist	–0.0661354	–0.47333735	No	No
Copeland Rank	Director of Environment	0.461419	3.71422027	Yes	Yes
Copeland Rank	Deputy Director of Planning	0.6789115	6.603466578	Yes	Yes
Copeland Rank	Business Owner	0.7786226	8.86142543	Yes	Yes
Copeland Rank	Real Estate / Developer	0.8233136	10.3590068	Yes	Yes
Copeland Rank	CPA Member	0.8263387	10.4785456	Yes	Yes
Copeland Rank	Hotel Operator / Developer	0.9148263	16.1773126	Yes	Yes
Copeland Rank	Borda Rank	0.9792914	34.5435695	Yes	Yes

participants and overcome the complexity of use and intrinsic single user perspective that characterise most commercial GIS products.

Although the demonstration problem in land development was contrived, it mirrored many of the complexities of real world decision making. Analysis of the problem illustrated that the relatively simple site selection and evaluation models used work rather well in exposing individual and sub-group dimensions of commonality and help to identify generally acceptable group decision outcomes.

In the approach and the software that was developed for the research, a considerable amount of freedom was granted to participants with respect to choices of criteria and methods (both MCA and weighting) that were available to them to evaluate spatial selections of developable sites for tourism accommodation in a district of Grand Cayman. This is somewhat unconventional and made the process less tractable for the facilitator, but it did provide indications not only of the criteria but also the extent of their 'value' for each participant in their decision process. Moreover, Monnikhof and Bots (2000) illustrate that there is some risk that decision participants will reject both evaluation methods and outcomes if method and criteria choices are not provided. More importantly, the approach was sufficiently flexible to allow individual commonalties and differences on the issue at hand to be melded together into flexibly defined 'group' views. The resultant outcome of the decision process (i.e. the 53 sites selected by 2 or more participants from the available set of land parcels in the study area) is considerably larger than is usually reported in the MCA-based decision support literature (Joerin *et al.*, 2001). Despite the complexity this added to the evaluation stage, the paper shows that it is possible using the methods implemented to produce individual site ranks, as well as group and overall ranks, that allow acceptable options to be identified for all participants.

Cursory analysis of the site ranks shown in Table 6 reveals that the parcels with the highest degree of development consensus for tourism accommodation, as reflected by the number of times a parcel ranks in the top ten for any participant, are sites 11D_31 (6 of 7); 17A_14, 17A_12, 11B_67, 11B_17 (all 4 of 7); 11B_76, 17A_10, 10E_35, 10E_9, 8A_3 and 11B_61, 1C_33 (all 3 of 7). All of these sites also figure in the top ten for both the Borda and Copeland group ranks except 1C_33, which is located on the coast to the north of the study area (Figure 4 – Copeland rank of 19.5). Further inspection of Table 6 reveals that this site is split in its ranking pattern across participants with two ranking it as highly suitable for tourism development, two ranking it as unsuitable and the other three ranking it as of intermediate suitability. This clearly split pattern, which is a basis for decision conflict, can be better understood and potentially ameliorated with the decision approach presented in the paper. If users were to iterate through the evaluation process and the same pattern were to emerge, after reconsidering and modifying evaluation criteria and weights, such a contentious site could be removed from development consideration using an appropriate decision rule.

While both the Borda and Copeland aggregate ranks for the sites in Table 6 could be used as input for the initial stages of a tourism development plan, it is not clear from the research to what degree the ranks are susceptible to small changes in data standardisation parameters, to the numerical sensitivity of the specific MCA method used, to evaluative criteria choices, or criteria weight settings. Preliminary investigations suggest that the sensitivity of the rankings to the latter two factors is relatively low, however each of these influences must all be examined rigorously in future research.

In this context, two factors are known to influence the criteria weights shown in Table 4. First, the importance of a given criterion as a standard to judge the suitability of different options depends to a large extent on the range of the scores for that criterion in the choice set (Voogd, 1983). For example, a criterion that would usually be assigned a high weight should have its relative importance in the decision process reduced if it is discovered that all of the alternatives under consideration have similar scores for that criterion. Although not discussed in this paper, a dialog for displaying simple descriptive statistics on any data field is incorporated into the software used in the research. Few participants in the case study took advantage of this functionality during the weighting process since it required that they execute an additional menu step external to the main MCA dialog. This potentially impacted on the significance attached to the D2PrimRd, D2SecRd and Vacant criteria, since there was little difference in the distance from any land parcel to a road and most of the sites available for tourist accommodation were undeveloped.

Second, it is important to recall that criteria weights are only approximations of an individual's priorities and, as such, they can vary with time, with the set of criteria available, and with respect to the specific set of choice alternatives (Malczewski, 1999). For example, several participants noted evaluative factors that, had they been available, would have influenced their decision calculus. These factors include, for example information on current land use, the character of the sea bed, land costs, infrastructure availability and capacity, soil capacity and vegetation cover. Most, if not all of these factors, with more time, could have been added to the data set available for site evaluation and their presence would have, without question, modified the results reported above. The fact is that inclusion of these (and potentially other) factors would not necessarily change the approach or the techniques used in the approach used, although it would tend to make the results, all other things equal, more embracing in their scope. As noted above, robustness of the results between different MCA techniques is another matter, and this an important topic for future work.

It is important to note that this research was designed initially to examine how GIS-based multi-participant decision support could be used in the early stages of longer-term plan generation to explore issues of consensus and conflict. The results of this pilot exercise are promising enough to consolidate the research by expanding its application to related application areas while delving further into some of the technicalities in process and method that require further refinement.

ACKNOWLEDGEMENTS

This research was funded through a grant from the International Development Research Centre, Ottawa awarded to Dr. Brent Hall and Dr. Arthur Conning. We would like to acknowledge the input of Dr. Robert Bowerman in software development. Local arrangements in Grand Cayman were co-ordinated with Mr. Kenneth Ebanks, Director of Planning in the Cayman Islands. John Bebb of the Cayman Government Lands and Surveys Department supplied us with most of the GIS data used. We gratefully acknowledge their contributions and those of the research participants.

Department of Geography,
School of Planning,
Faculty of Environmental Studies,
University of Waterloo,
Waterloo, Ontario N2L 3G1, Canada

REFERENCES

Arrow, K.J. (1963), *Social Choice and Individual Values.* Wiley, New York.

Banai, R. (1993), 'Fuzziness in Geographical Information Systems: Contributions from the analytic hierarchy process.' *International Journal of Geographical Information Systems*, **7**(4), 315–329.

Bodini, A. and G. Giavelli (1992), 'Multicriteria analysis as a tool to investigate compatibility between conservation and development on Salina Island, Aeolian Archipelago, Italy.' *Environmental Management*, **16**(5), 633–652.

Bojórquez-Tapia, L., S. Díaz-Mondragón and E. Ezcurra (2001), 'GIS-based approach for participatory decision making and land suitability assessment.' *International Journal of Geographical Information Systems*, **15**(2), 129–151.

Borda, J.C. (1781), 'Mémoire sur les Élections au Scrutin.' *Histoire de l'Académie Royale des Sciences.*

Carver, S.J. (1991), 'Integrating multi-criteria evaluation with geographical information systems.' *International Journal of Geographical Information Systems*, **5**(3), 321–339.

CIA World Factbook (2002) (http://www.cia.gov/cia/publications/factbook/geos/cj.html). Last accessed April 15, 2003.

Couclelis, H. and J. Gottsegen (1997), 'What maps mean to people: Denotation, connotation, and geographic visualization in land-use debates.' In S. Hirtle and A.U. Frank (eds.), *Spatial Information Theory: A Theoretical Basis for GIS. International Conference COSIT '97*, Lecture Notes in Computer Science, No. 1329. Springer-Verlag, Berlin, pp. 151–162.

Cox, G.W. (1997), *Making Votes Count: Strategic Co-ordination in the World's Electoral Systems.* Cambridge University Press, Cambridge.

Densham, P.J. (1991) 'Spatial decision support systems.' In D.J. Maguire, M.F. Goodchild and D.W. Rhind (eds.), *Geographical Information Systems: Principles and Applications*, Volume 1. John Wiley, New York, pp. 403–412.

Feick, R. and G.B. Hall (1999), 'Consensus-building in a multi-participant spatial decision support system.' *URISA Journal*, **11**(2), 17–23.

Feick, R. and G.B. Hall (2000), 'The application of a spatial decision support system to tourism-based land management in Small Island States.' *Journal of Travel Research*, **39**(2), 163–171.

Fishburn, P.C. (1971), 'A comparative analysis of group decision methods.' *Behavioural Science* **16**, 538–544.

Flowerdew, R. (1998), 'Reacting to ground truth.' *Environment and Planning A*, **30**(2), 289–301.

Goodchild, M.F. (1995) 'Geographic information systems and geographic research.' In J. Pickles (ed.), *Ground Truth: The Social Implications of Geographic Information Systems*. Guilford Press, New York, pp. 31–50.

Harker, P.T. (1989), 'The art and science of decision making: The analytical hierarchy process.' In G.L. Golden, E.A. Wasil and P.T. Harker (eds.), *The Analytic Hierarchy Process: Applications and Studies*. Springer-Verlag, New York, pp. 3–36.

Hendriks, P. and D. Vriens (2000), 'From geographical information systems to spatial group decision support systems: A complex itinerary.' *Geographical and Environmental Modelling* **4**(1), 83–104.

Heywood, I., J. Oliver and S. Tomlinson (1994), 'Building an exploratory multi criteria modelling environment for spatial decision support.' *Proceedings of EGIS/MARI '94*, Vol. 1., Paris, pp. 632–639.

Hobbs, B.F., V. Chankong and W. Hamadeh (1992), 'Does choice of multicriteria method matter? An experiment in water resources planning.' *Water Resources Research*, **28**(7), 1767–1779.

Hwang, C-L. and K. Yoon (1981), *Multi Attribute Decision Making*. Springer-Verlag, New York.

Hwang, C-L. and M-J. Lin (1987), *Group Decision Making Under Multiple Criteria: Methods and Applications*. Springer-Verlag, New York.

Jankowski, P. and L. Richard (1994), 'Integration of GIS-based suitability analysis and multicriteria evaluation in a spatial decision support system for route selection.' *Environment and Planning B*, **21**, 323–340.

Jankowski, P. and M. Stasik (1997), 'Design considerations for space and time distributed collaborative spatial decision making.' *Journal of Geographic Information and Decision Analysis*, **1**(1), 1–8.

Jankowski, P., T.L. Nyerges, A. Smith, T.J. Moore and E. Horvarth (1997), 'Spatial group choice: A SDSS tool for collaborative spatial decision-making' *International Journal of Geographical Information Systems*, **11**(6), 577–602.

Jones, R.M., C.V. Copas and E.A. Edmonds (1997), 'GIS support for distributed group-work in regional planning.' *International Journal of Geographical Information Systems*, **11**(1), 53–71.

Joerin, F., M. Thériault and A. Musy (2001), 'Using GIS and outranking multicriteria analysis for land-use suitability analysis.' *International Journal of Geographical Information Systems*, **15**(2), 153–174.

Kingston, R. S. Carver, A. Evans and I. Turton (2000), 'Web-based public participation geographical information systems: an aid to local environmental decision-making.' *Computers, Environment and Urban Systems*, **24**(2), 109–125.

Malczewski, J. (1996), 'A GIS-based approach to multiple criteria group decision-making.' *International Journal of Geographical Information Systems* **10**(8), 955–971.

Malczewski, J. (1999), *GIS and Multicriteria Decision Analysis*. John Wiley, Toronto.

Malczewski, J., R. Moreno-Sanchez, L.A. Bojórquez-Tapia and E. Ongay-Delhumeau (1997), 'Multicriteria group decision-making for environmental conflict analysis in the Cape Region, Mexico.' *Journal of Environmental Planning and Management*, **40**(3), 349–374

Massam, B.H. (1988), *Multi-criteria Decision Making (MCDM) Techniques in Planning*. Progress in Planning 30. Pergamon Press, Oxford, pp. 1–84.

Massam, B.H. (1993), *The Right Place: Shared Responsibility and the Location of Public Facilities*. John Wiley, New York.

Monnikhof, R.A.H. and P.W.G. Bots (2000), 'On the application of MCDA in interactive spatial planning processes: Lessons learnt from two stories from the swamp.' *Journal of Multi-Criteria Decision Analysis*, **9**(1), 28–44.

Nyerges, T.L., R. Montejano, C. Oshiro, and M. Dadswell (1997), 'Group-based geographic information systems for transportation improvement site selection.' *Transportation Research-C*, **5**(6), 349–369.

Nijkamp, P. and A. van Delft (1977), *Multi-criteria Analysis and Regional Decision Making*. Martinus Nijhoff Social Sciences Division, Leiden.

Olson, D.L. (1996), *Decision Aids for Selection Problems*. Springer, New York.

O'Reilly, C.A. (1990), 'The use of information in organizational decision making: A model and some propositions.' In L.L. Cummings and B.M. Staw (eds.), *Information and Cognition in Organizations*. JAI Press, Greenwich (Conn.), pp. 89–1255.

Pereira, J.M.C. and L. Duckstein (1993), 'A multiple criteria decision-making approach to GIS-based land suitability evaluation.' *International Journal of Geographical Information Systems*, **7**(5), 407–424.

Phaneuf, Y. (1990) *EIS and Decision Making*. Canadian Environmental Assessment Research Council.

Pickles, J. (1999) 'Arguments, debates, and dialogues: the GIS-social theory debate and the concern for alternatives.' In P.A. Longley, M.F. Goodchild, D.J. Maguire and D.W. Rhind (eds.), *Geographical Information Systems: Principles and Technical Issues*, Volume 1, 2nd Edition. Wiley, Toronto, pp. 49–60.

Saaty, T.L. (1980), *The Analytical Hierarchy Process*. McGraw-Hill, New York.

Saaty, T.L. (1989), 'Group decision making and the AHP.' In G.L. Golden, E.A. Wasil and P.T. Harker (eds.), *The Analytic Hierarchy Process: Applications and Studies*.' Springer-Verlag, New York, pp. 59–67.

Siegel, S. (1956), *Nonparametric Statistics for the Behavioural Sciences*. McGraw-Hill, New York.

Silk, J. (1979), *Statistical Concepts in Geography*. George Allen and Unwin, London.

Silver, M.S. (1991), *Systems That Support Decision Makers: Description and Analysis*. John Wiley, New York.

Simon, H.A. (1976), *Administrative Behaviour*. Free Press.

Towers, G. (1997), 'GIS versus the community: Siting power in West Virginia.' *Applied Geography*, **17**(2), 111–125.

Van Huylenbroeck, G. and A. Coppens (1995), 'Multicriteria analysis of the conflicts between rural development scenarios in the Gordon District, Scotland.' *Journal of Environmental Planning and Management*, **38**(3), 393–407.

von Winterfeldt, D. and W. Edwards (1986), *Decision Analysis and Behavioural Research*. Cambridge University Press, New York.

Vinke, P. (1992), *Multicriteria Decision Aid*. Wiley, Chichester.

Voogd, H. (1983), *Multicriteria Evaluation for Urban and Regional Planning*. Pion, London.

Yoon, K.P. and C.L. Hwang (1995), *Multiple Attribute Decision Making: An Introduction*. Sage Publications, Thousand Oaks (California).

YOSHIO ARAI AND SHIRO KOIKE

GRID-BASED POPULATION DISTRIBUTION ESTIMATES FROM HISTORICAL JAPANESE TOPOGRAHICAL MAPS USING GIS

1. INTRODUCTION

Recently, with the rapid progress in Geographical Information Systems (GIS) and their related fields, data previously unavailable can be analyzed using this powerful geographic tool. In addition, the use of GIS in economics, sociology, archaeology and human geography has expanded remarkably. There are, however, few studies which utilize GIS to analyze time-space phenomena. There is not only a difficulty in collecting the historical data due to the lack of statistical materials, but also there are high barriers to surmount in the process of transforming analog data to digital data (Knowles, 2000). On the other hand, topographical maps drawn at the end of the nineteenth century are effective for obtaining historical spatial data. Although these early edition maps do not meet the accuracy standard of contemporary survey, they do have many valuable characteristics: they cover wide areas, and they contain large amounts of information concerning regional circumstances at the time. It is also certain that the overall coordinate system of old topographical maps is somewhat biased, but this can be corrected through comparison with recent topographical maps. These observations motivated this paper, the objective of which is to estimate historical population distributions by linking topographical maps to statistical materials.

In geography few studies have attempted to reconstruct the relationship between social data and population on a quantitative basis; exceptions include Taniuchi (1995), which estimated grid population data around 1900 in Japan using various statistical materials, and Norton (1976) which reconstructed the land use of a township in Canada. Although some studies dealing with past central place systems can be identified (*e.g.*, Morikawa, 1962; Masai and Matsumoto, 1971; Kohno, 1990), quantitative analyses of population distribution are rarely found in these studies. In historical geography, many studies were devoted to micro-scale analysis of the distribution of urban places or local villages. These studies, however, cover only a small region at a point in time and do not cover a much larger region over a long time period. The limitation seems to result from the lack of standardized data over long periods.

Recently, however, studies which apply GIS to historical phenomena (so called Historical GIS) have begun to appear. Siebert (2000) reconstructed infrastructure patterns of Tokyo in pre-World War II with GIS using old topographical maps and other data sources. Lee (1996) explained the population distribution

B. Boots et al. (eds.), Modelling Geographical Systems, 235–250.

of northern Ireland in the middle of the nineteenth century with a numerical modelling formula. Murayama and Ono (1998) displayed historical data through Internet GIS. Although these studies made remarkable efforts to overcome the barrier mentioned above, most studies place stress on reconstructing the past data and do not incorporate time-space analysis as a central research objective.

While grid data are suitable for spatial analysis, they were not introduced in the studies given above. Besides, we cannot obtain grid population data covering the whole Japan until 1970. However, if we can estimate such grid data using the same unit base employed in the present standardized national statistics system, such grid data would reveal the change of spatial population distribution clearly. Based on the methodology discussed above, we propose several data estimation methods. In practice, these methods can estimate a population distribution with an allowable accuracy and efficiency.

In this paper, after mentioning briefly the history of spatial data building in Japan (Section 2), we establish the assumptions for the estimation of a Basic Grid Square Spaces (BGSS) population distribution (Section 3). Then we explain the basic method of the estimation giving an example (Section 4). Lastly, we present a population estimation result for the Kanto district based on the established method (Section 5).

2. SPATIAL DATA BUILDING IN JAPAN

In this section, we present a brief history of three categories of spatial data (topographical maps, statistical data, and grid data) compiled at the national level in Japan, with emphasis on the Meiji period (1868–1912) up to the present.

2.1. *Topographical Maps*

The development of modern map making in Japan was initiated by Ino Tadataka at the beginning of the nineteenth century. Though his survey method was rather primitive, his maps covered the whole country with surprising accuracy, and these were used for the foundation for land and sea map making through the modernization period of the country from 1868.

In the Meiji period, the demand for surveys and inventories of the country increased as part of the policy for enriching the nation and building up its defences. Responding to this demand, the Meiji military leaders planned to construct maps with the scale of 1:20,000 covering the whole country at ten-year intervals. So called *Jinsokuzu*, Quick-Mastery-Maps, and *Kaseizu*, Temporary-Maps, were the first survey results in this series of planning. Although these maps do not fulfill the research project's expectation of accuracy, they are nonetheless rare source materials which can teach us the pattern of spatial distribution of settlements at that time.

After publication of yet another set of topographical maps, *Seishikizu*, mainly covering urbanized areas, the original plan of covering the whole country at the scale of 1:20,000 was changed to 1:50,000 scale coverage because of budget limitations. These 1:50,000 scale topographical maps covering Japan's main territory were completed in 1924.

In the high economic growth period after World War II, the land use pattern began to change drastically and demand increased for detailed topographical maps. In place of 1:50,000 scale topographical maps, which were insufficient to show the details of regional circumstances, 1:25,000 scale topographical maps became dominant. 1:25,000 scale topographical maps covering the whole country including isolated islands were completed in 1983. After that, these maps were renewed frequently and are being used as the basis of various analyses and planning activities (Figure 1).

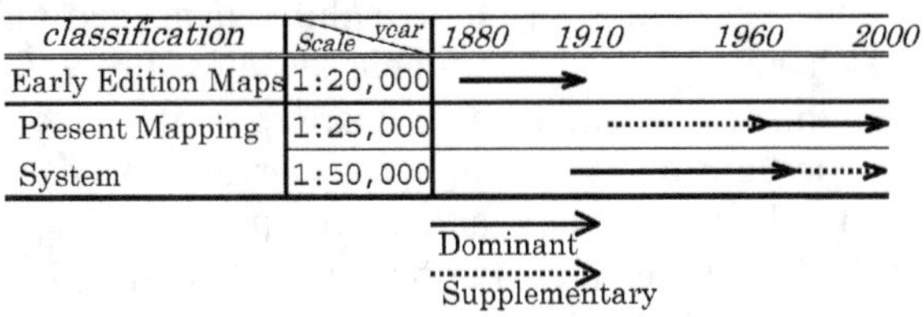

Figure 1. Topographical map making in Japan.

2.2. Statistical Population Data

Statistical population data did not exist in the Edo period (1600–1867) except for inhabitant lists according to religious sects arranged by village masters called *Shumon Aratamecho*. In the Meiji period the national government and the modernizing military establishment collected a series of secret statistics inventorying national power in preparation for a possible national emergency. These statistical reports are called *Kyobu Seihyo* and *Chohatu Bukken Ichiranhyo*. They contain not only population by sex and by village, but also military data such as tools and materials for emergent transportation needs, information typically not seen in official statistical papers. Other statistical reports by prefecture were made annually from around 1880. Although these reports are the antecedents of the modern statistical system and provide unusual insight into population distribution in the Meiji period, the accuracy and the consistency of the calculation process is rather questionable.

As for official statistical data, the first National Population Census was conducted in 1920. This census provided population information by local administrative units, *Shi Cho Son* (city, town, village) as well as other population data. From the outset, this census has provided essentially a complete enumeration of Japan's population, and the accuracy has been consistently quite high. With only one exception, the census has been regularly conducted every five years to the present. Data collection categories in the census have increased over time, making this survey the most frequently used statistical data source for clarifying population, households, and industry structure in Japan, and for establishing various policies and plans for national and local governments.

2.3. *Grid Data*

From the beginning in the National Population Census, data were basically organized by areal units, that is, municipalities (*Shi Cho Son*). It should also be noted that, from the outset, most of the municipalities were established and administered by local government systems, which were designed by Meiji government. In 1889 about 16,000 municipalities existed in the country. At this time urbanized areas roughly coincided with the areal extent of urban municipalities and, therefore, substantial urban population most likely corresponded well with administrative population figures for each *Shi.* This circumstance persisted until the period before World War II.

After the war, the annexation of municipalities happened one after another to cope with the changing local government systems. In 1956 the number of municipalities diminished by about three-quarters, that is, to about 4,000. As a result, even in the case of *Shi*, municipality areas ceased to coincide with the boundaries of urbanized areas, and municipalitites' population did not correspond with the literal urban population. Just at that time, Japan experienced high economic growth and population concentration was a remarkable characteristic of urban areas. Therefore, ever more detailed information about population distribution was required for urban planning and other purposes.

Due to these circumstances, from 1965 the concept of a zonal grid was introduced in the construction of various statistical materials in Japan. The zonal grid is a system of statistical units dividing an area into detailed rectangles. In Japan, the system called the "Standard Zonal Grid System" is commonly used. For a Standard Zonal Grid System, the area is divided by latitude and longitude. The minimum data unit which is available nationally is the "BGSS" (Basic Grid Square Spaces, approximating 1 km rectangles). BGSS population data are available continuously from the 1970 census. Unique code numbers are given for each BGSS according to the Standard Zonal Grid System.

At the beginning of the development of the system, limited statistics, such as population, were available on the basis of BGSS system. Soon after, nationally assigned statistics and various kinds of physical or environmental data became available. Recently, these grid data are being offered through large capacity media such as magnetic optical disks at a reasonable price, hand in hand with the popularization of high-spec personal computers. Since BGSS data are not affected by the change of a municipality's areal unit, they are suitable for a long span time-space analysis.

3. ASSUMPTIONS FOR THE ESTIMATION OF BGSS POPULATION DISTRIBUTION

To estimate population distribution at the end of the nineteenth century, we make two operational assumptions to utilize the BGSS framework for topographical maps. The first assumption is that almost all buildings at that time were used for residential purposes. Contrary to most Western countries, Japan started the first step of modernization in the last third of the nineteenth century. Although industrial development was gradually prioritized as national projects, agricultural circumstances were similar to those in the pre-modern age in most regions. In these regions most buildings were used as residences only. In urban areas, even though manufacturing began to grow, job places and residences were not separated for most workers. Therefore, nearly all buildings were used for residence except for those in the central area of large cities, where government and municipal offices were concentrated.

The second assumption is that population density of residences does not vary significantly across different regions. Almost all the buildings at that time were one-story or two-story wooden structures. The Meiji period modernization of buildings proceeded gradually. Few high-rise buildings appeared before World War I and these were limited to government offices or, very occasionally, upper-class people. Also, the lifestyles of people at that time were not extraordinarily varied and the composition of family structure was considered to be similar regardless of region.

In the Meiji period, population inflow to urban areas was large and a high density housing form called *Nagaya* spread rapidly. *Nagaya* is a housing form whose front is narrow and whose depth is long. On topographical maps, the buildings in these high density areas were not drawn separately, but they were drawn as spatial clusters. We treat them as different categories from ordinary houses, as discussed later in this paper.

From these two assumptions we make the hypothesis that population can be expressed as a function of the area of buildings drawn on and represented by topographical maps.

4. TRIAL WORK IN EAST-*BIWAKO* AREA IN *SHIGA* PREFECTURE

As a first step, we attempt to confirm the utility of BGSS population estimation based on the assumptions mentioned in the previous section (Koike and Arai, 2001).

The chosen area for the trial work is the so-called "East-*Biwako* Area" in *Shiga* Prefecture (Figure 2). There are three reasons for this choice. First, the *Seishikizu* version of the early edition maps covered the whole area. Second, a statistical report at the end of the nineteenth century, *Shigaken Bussanshi*, exists and we can directly confirm the accuracy of our estimations with it. Third, the area is suitable for trial estimation because of the high density of village distribution.

Based on the above assumptions, we estimate the population of a village, P, as a function of the area of buildings in the village, A, *i.e.*,

$$P = f(A) \tag{1}$$

We estimate the area of buildings by a systematic point sampling method. The systematic point sampling method is reported to be fairly suitable for calculating area accurately and easily (Okabe *et al.*, 1984). In this method,

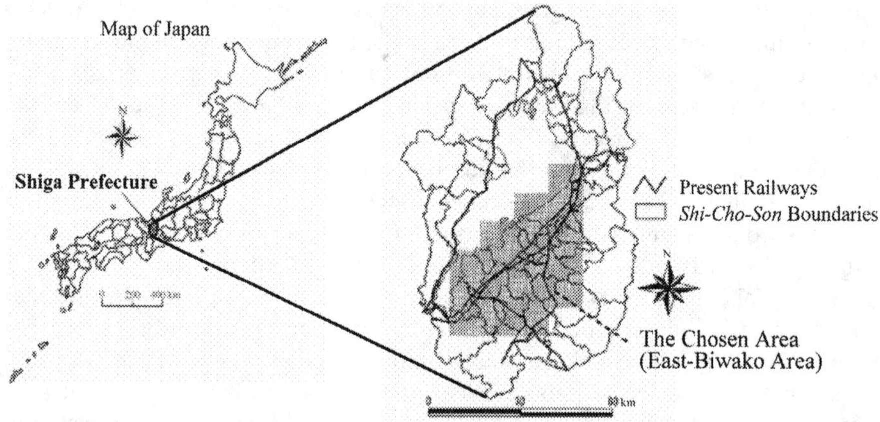

Figure 2. Location of the chosen area.

first we overlay a fine grid on scanned images of topographical maps and then extract the common points of intersection, represented by sample points, between the grid and overlapping buildings drawn on topographical maps. Let N be the number of sample points dropped on villages, then A is substituted by N.

The scattergram of N and P, where N is less than 300 (Figure 3), clearly shows high correlation between N and P. Since P tends to increase more proportionally as N increases, we applied an exponential regression function. Next, we counted the number of sample points in each BGSS cell, and calculated an estimate of BGSS population using the estimated regression formula mentioned above. The distribution of estimation errors obtained by comparing the estimated BGSS population with the actual BGSS population, which is calculated from the village population reported in the *Shigaken Bussanshi*, is shown in Figure 4. The percentage of BGSS cells whose error is within 20% is more than 60% of the total number. We may conclude from this result that the above method is accurate to a high degree.

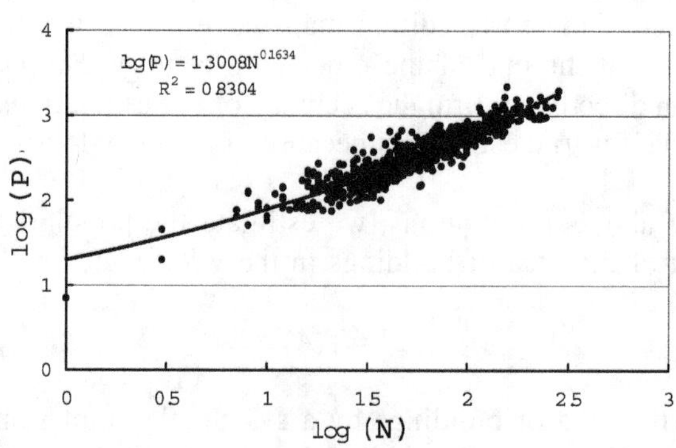

Figure 3. Scattergram of N and P in logarithmic scale.

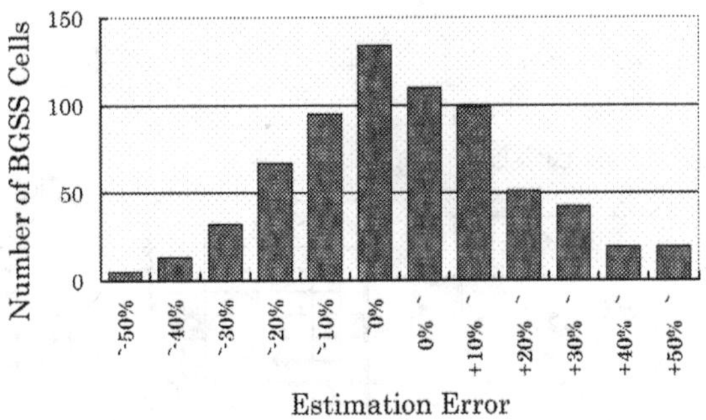

Figure 4. Histogram of error rate of estimated BGSS population.

However, we found that the spatial distribution of the error rate is fairly biased as shown in Figure 5.This bias is mainly caused by differences in the way of map drawing in each sheet. In this period a standard for map drawing had not yet been established and the manner of drawing buildings differed from drawer to drawer. It is this difference that mainly produced the spatial error bias. To exclude this bias, we tried to reduce the error by noting the *kanji* character sizes of village names written on topographical maps. Although we will not discuss this technique in detail here, it allowed us to reduce the spatial bias to some extent.

In addition, the estimation of accuracy becomes worse in highly urbanized areas. In such areas buildings were used not only for residence but also for other uses. Thus Equation (1) varies according to building types, and this brings estimation errors. To reduce these errors, we introduced three dummy variables: hospitals, public offices and law courts, which were chosen as significant variables from seven dummy variables. A multivariate regression model with N and these three dummy variables greatly improves the estimation accuracy in highly urbanized areas. It should also be noted that the populations of villages along main roads were underestimated. To deal with this bias, sample points near the main roads were treated separately in regression analysis.

As a result of these modifications, the BGSS cells in which the error rate stayed within 20% totaled over 80%, as shown in Figure 6.

As mentioned above, we confirmed BGSS population estimation using only topographical maps where significant accuracy was possible. Considering the estimation for a larger area, however, the methodology mentioned above is not always efficient, because of the need to rely on a highly labor intensive process of estimation. Therefore, we tried to reduce work time and costs through introducing GIS techniques into the estimation process. To avoid the aforementioned modified steps in the estimation process, the most time-consuming aspect of the research, we changed the building 'area' point sampling method to one well suited to GIS techniques.

As a result of the GIS-based estimation method for the same example area, automation of some steps of the estimation process reduced work time per

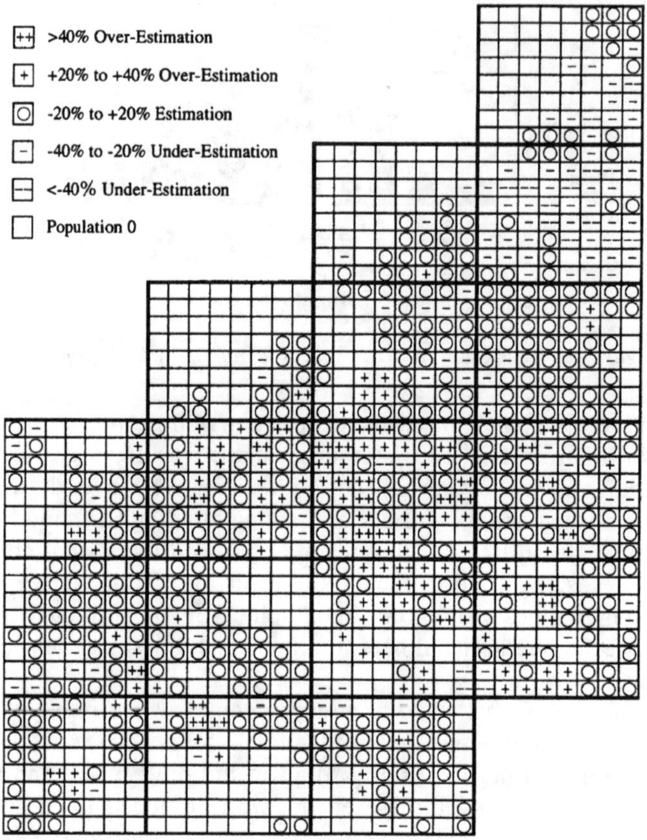

Figure 5. Spatial distribution of the error rate (Bold lines nearly represent the boundaries of topographical maps).

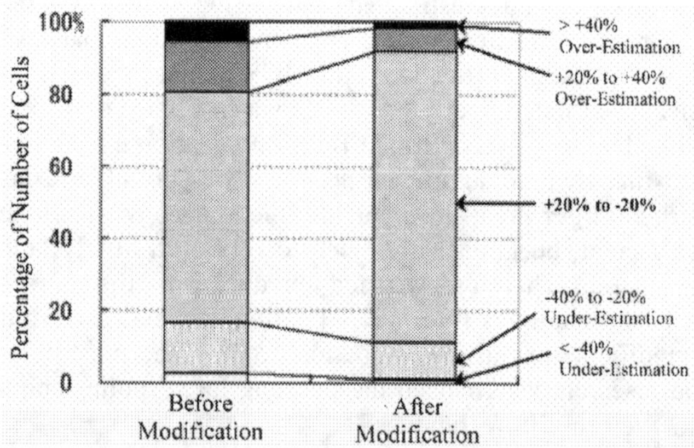

Figure 6. Comparison of estimation accuracy before and after modifications.

map by about half. Although the accuracy dropped slightly, we could obtain good estimation results for the BGSS cells, in the form of either an error rate of less than 20% or an error value of less than 100 persons, for approximately 85% of the tested cases.

We conclude, therefore, that BGSS population estimation with high accuracy is possible from the information limited to topographical maps, and that utilization of GIS as a main tool for estimation acts to greatly improve work efficiency. Although the detailed estimation method explained in this section will not be suitable for all other cases, because the manner of village drawings varied widely among the different kinds of topographical maps, estimation through the measurement of building areas on these topographical maps is thought to be generally applicable.

In the next section, we will try to estimate the BGSS population in Kanto Plain in the modern period based on the results obtained in this section. The Kanto Plain is the region of Japan associated with Tokyo, Japan's largest metropolis.

5. THE ESTIMATION OF BGSS POPULATION DISTRIBUTION IN KANTO PLAIN IN MEIJI PERIOD

Clarifying the population distribution change in the Tokyo metropolitan area, the largest metropolis in Japan from the beginning of the modern period, provides significant insight into various analytical processes such as time-space modelling of urbanization or the assessment of the impact of industry or transportation development in Japan. Due to the lack of statistical materials, however, many of the existing studies which have dealt with urbanization have been constrained to observation of short time-spans and qualitative examinations. If the BGSS data available from the beginning of the Meiji period can be estimated at certain time intervals, various new kinds of studies such as analysis of the spatial aspects of the urbanization process with time-space modelling techniques become possible. In this section, as a first step, we attempt to estimate the population distribution in the Kanto Plain at the end of the nineteenth century, utilizing the BGSS technique, incorporating brief comments based on the estimation result.

In the Kanto Plain, the *Jinsokuzu*-type maps mentioned above were published roughly between 1880 and 1886. These maps cover the main Kanto Plain area except for a pair of cities in North-Kanto as shown in Figure 7 (Ono, 1956). The digital data files for these maps, scanned and saved as JPEG format images in a CD-ROM, were recently released. We used these files as the basis of the work for this section. In addition, we used the *Chohatsu Bukken Ichiranhyo*, edited in 1891, with data from 1890, as the basis for statistical comparison. In this report, village population and other social data of the time are recorded. Although the accuracy is slightly unstable, and there is a short-lag between the publication of the *jinsokuzu*-type maps and the *Chohatsu Bukken Ichiranhyo*, these data are considered the most suitable statistical material for our purposes.

Figure 7. Covering area of Jinsokuzu-type maps.

The estimation process was slightly modified from the established GIS-based method described in the previous section. The process is summarized in Figure 8. Here we provide a short explanation for each work step.

5.1. Extraction of Map Frames

First, we extracted the frames of all the maps recorded in the CD-ROM files using a graphic handling application. This process allows all the maps to be arranged without included blanks during later coordinate transformation.

5.2. Import to GIS Application

Then, we imported the extracted frames of the topographical maps into ArcView. In our work we divided the area under observation into 46 blocks and dealt with each block as a *project* in ArcView in order for the *project file* size not to be too large.

5.3. Input Buildings Data

Next, we overlaid the vector layer of buildings on the images of the maps. Specifically, we inputted points on the objects recognized as separated buildings, and assigned polygons to the concentrated building's area on the buildings layer. In the process, a few buildings seemed to be non-residences and were therefore excluded from the data input. Then, the inputted points were converted to circles with a radius corresponding to average building size using GIS buffering application. In this way, all the building data can be saved as polygon

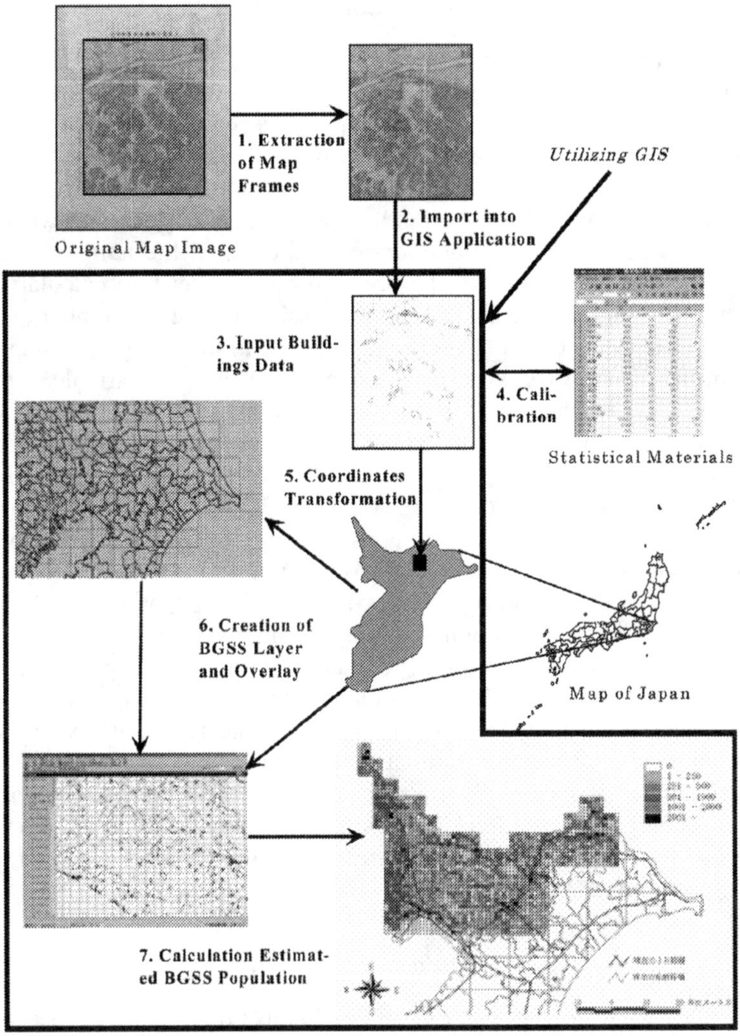

Figure 8. Work process adopted in the estimation.

data. In the center of Tokyo and in a part of Yokohama, non-residential buildings (e.g., governmental and military facilities) were dominant. These building data can be excluded by utilizing other large-scale maps and by distinguishing the boundaries of the residential areas. For the next step, we transformed these polygons to grid data by means of a systematic point sampling method. The sample points in the grid data are divided into two types depending on whether they were overlapped on originally input polygons or on polygons buffered from points. These two types of sample points are dealt with as separate variables in the next process.

5.4. Calibration

In the East-*Biwako* area drawn in the *Seishikizu*-type maps, mentioned in the previous section, village boundaries were clear in almost all cases. In contrast, in many cases the villages in *Jinsokuzu*-type maps contained blurred boundaries. Therefore, we tried to pursue calibration of the data set for selected villages where the boundaries were clear. The selected villages totaled approximately 700 villages, approximately one-tenth of all villages. In practice we selected randomly villages with clear boundaries and counted the number of sample points established according to the previous steps. Then, we calculated parameters for the multiple linear regression analysis utilizing the number of sample points for both types of explanatory variables, with the village population presented in the *Chohatsu Bukken Ichiranhyo* as a dependent variable.

However, the result of the calibration for all the selected villages indicates a biased error distribution. This bias is mainly produced because of the type of building depiction, a problem similar to the case discussed in the previous section. According to the commentary of the CD-ROM (*Nihon Chizu Sentar et al.*, 1998), there were several survey parties working for *Jinsokuzu*-type maps. Therefore, we divided the area under observation into three parts according to the territories of these survey parties to reduce the spatial error bias, and calculated population estimation formulas separately for each.

By comparing estimated village population with actual population, as contained in the *Chohatsu Bukken Ichiranhyo*, the rate of villages with an error rate less than 30% can be estimated to be over 80%. The accuracy fell slightly when compared with the trial estimation in the previous section. This appears to be caused by the lack of concrete standards for map drawing in *Jinsokuzu*-type maps compared with *Seishikizu*-type maps made later. On the other hand, the spatial bias of the error ratio distribution remained weak as a result of dividing the estimation formulas.

5.5. Coordinates Transformation

Since ArcView, the GIS application applied to this project, utilizes its own coordinate system, coordinate transformation of building grid data to some generally used coordinate system is necessary. We chose the UTM coordinate system, which is employed in recent topographical maps published by the Japanese government, to overlay and analyze comparisons with other types of the digital data.

5.6. Creation of BGSS Layer and Overlay

In this step we created a BGSS layer in ArcView and overlaid it on the transformed and merged building grid data. The BGSS layer and the identification numbers for each BGSS cell can be created utilizing ArcView script.

5.7. Calculation of Estimated BGSS Population

Next, applying the spatial merge procedure of the GIS application to the building grids data layer and the BGSS layer, we assigned an identification number for BGSS cells to building grid data. Then we counted up the sample points presenting buildings in each BGSS cell according to the BGSS identification system. Lastly, we estimated the BGSS population by substituting the calculated numbers of sample points of each BGSS cell for the estimation formulas.

Figure 9 shows a spatially smoothed trend surface calculated from the estimated BGSS population in 1890. We can point out from Figure 9 that agglomerations of population were remarkable around the present-day railways stations, though these railways had not been constructed yet. We can observe that the railways were built along high-density population areas at that time, rather than that these agglomerations developed as a result of railway construction. Although the density of city places tends to be high in the west and low in the east, many agglomerations are distributed at an approximately constant distance along the main roads at that time. We can see some characteristics of the central place system of a period when walking was the main transportation means.

Overlaying the 1890 estimation data on topographic data shows a high relationship between the population distribution at that time and land elevation. The distribution of the population occupation rate by land elevation for three points in time, 1890, 1970, 1995, is shown in Figure 10. This figure reveals that the

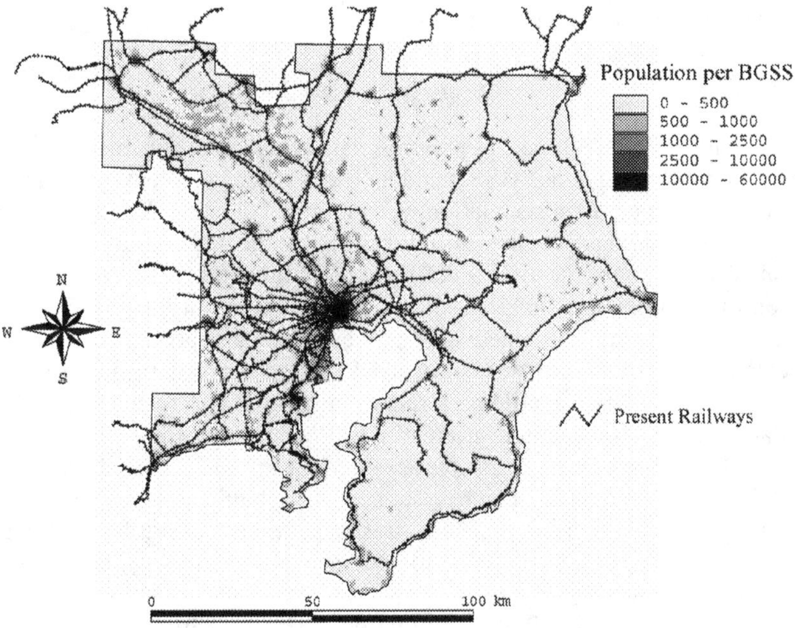

Figure 9. Trend surface based on the estimated BGSS population in 1890.

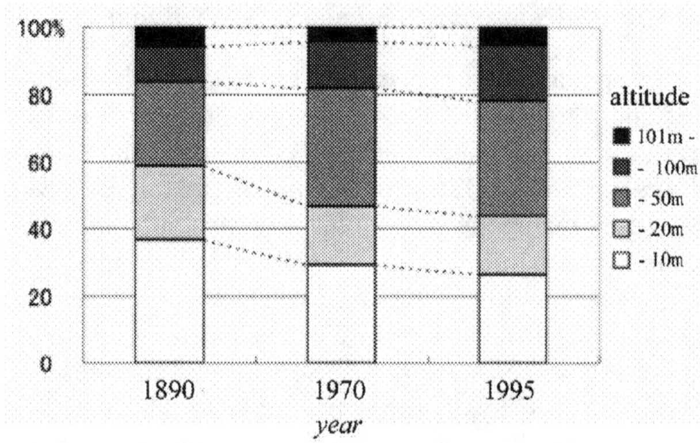

Figure 10. Change of population occupation by land height.

land use occupation rate in elevated areas has been gradually increasing. In 1890, while population concentration in lowlands was noticeable, population density was low in uplands and hilly areas. Generally, it was difficult to live on the uplands or hilly areas because of the lack of water in the pre-modern age. Although a number of small canals were built there from the seventeenth century to the nineteenth century, and new type villages called Shinden villages were formed in these areas, there was a limited ability to support larger populations. These land conditions seems to cause the differences of population density from west to east, as shown in Figure 9.

6. CONCLUSIONS

In this study, we developed population estimates based on historical topographical maps based on a BGSS method. The accuracy attained by the research was sufficient to confirm population distribution accurately. Employing a time and cost reduction procedure utilizing GIS as the main working tool, we attempted to estimate the BGSS defined population for the Kanto Plain. This method of population estimation allows for the possibility of various quantitative analyses being combined with prevailing data sources. Thus, we have presented a new direction for historical GIS-based analysis in this paper.

There are two main challenges remaining. First, the result of the BGSS population estimation method in the Kanto Plain demonstrates the possibility for spatial analysis of the central place system in the Meiji period. In the nineteenth century, the central place system was probably determined naturally according to non-vehicular, i.e., foot, traffic patterns. To overlay and analyze topographical information allows us to understand more clearly the central place system at that time.

Second, applying the BGSS population estimation method around 1930 enables time-space analyses such as the modelling of population distribution change. Supplementing the data based on estimates from 1890 and the confirmed population for 1970, we will to be able to construct a database with a nearly constant time-span from the beginning of the modern age to the present. Fortunately, since 1:50,000 scale topographical maps were published for almost the whole of the country around 1930, estimation work utilizing them is possible. However, we will need a slight modification of the estimation method because the variety of ways of life and the variety of shapes for residential buildings became more diverse as compared to the Meiji period. Because the village population data are not available, calibration should be done by incorporating municipality population data found in the National Population Census.

The BGSS population estimation method from 1930 enables various time-space analyses together with prevailing BGSS data. We would like to try such analytical methods in subsequent research utilizing the new data sets created by the estimation techniques in this study.

ACKNOWLEDGEMENTS

We would like to thank Gil Latz for editing on an earlier manuscript and to thank Barry Boots, Atsuyuki Okabe and Richard Thomas for valuable comments. We are also grateful to an anonymous referee for a lot of helpful suggestions.

Department of Human Geography,
University of Tokyo,
3-8-1, Komaba, Meguro-ku, Tokyo 153-8902, Japan

REFERENCES

Arai, Y. and S. Koike (2000), 'BGSS Population estimation along BGSS system from Jinsokuzu Maps in the middle of the Meiji-Era in Chiba Prefecture.' *Komaba Studies in Human Geography*, **14**, 1–34 (in Japanese).

Haga, T. and M. Okabe (1992), *Shashin de miru Edo Tokyo* (Edo and Tokyo through photographs). Sinchosha, Tokyo (in Japanese).

Koike, S. and Y. Arai (2001) 'BGSS population estimation from the topographical maps in the Meiji-Era.' *Theory and applications of GIS*, **9**(1), 1–8 (in Japanese).

Knowles, A.K. (2000), 'Introduction.' *Social Science History*, **24**, 451–470.

Kohno, K. (1990), 'The Change of Central Place Systems in Nagano Basin from the Meiji-Era.' *Geographical Review of Japan Ser. A*, **63**, 1–28 (in Japanese).

Lee, J. (1996), 'Redistributing the population: GIS adds value to historical demography.' *History and Computing*, **8**, 90–104.

Masai, Y. and S. Matsumoto (1971), 'Cities and towns of Kanto district, Japan – Distributional appraisal of urban settlements in the Meiji, Taisho and Showa periods.' *Geographical Review of Japan*, **44**, 1–13 (in Japanese).

Morikawa, H. (1962), 'Urban Areas distribution in the beginning of the Meiji-Era.' *The Human Geography*, **14**, 377–395 (in Japanese).

Murayama, Y. and H. Ono (1998), 'Development of internet GIS – Map information system of regional statistics in the Meiji period.' *Tsukuba Studies in Human Geography*, **22**, 99–128 (in Japanese).

Nihon Chizu Sentah (Japan Map Center) and Kiinokiniya Shoten (1998), *Meiji zenki sokuryo ni*

man gosen bun no ichi furansushiki saishoku chizu CD-ROM ban kaisetsusho. (Commentary of CD-ROM edition of French painted 1:20,000 scale maps in the beginning of the Meiji period).
Norton, W. (1976), 'Constructing abstract worlds of the past.' *Geographical Analysis*, **8**, 269–288.
Okabe, A., H. Yomono and T. Kaneko (1984), 'The accuracy of govarage estimation methods in relation to their amount of work.' *City Planning Review Extra Number*, **19**, 265–270.
Ono, M. (1956), 'Again on the Quick-mastery and Temporary surveymap of Japan.' *The Human Geography*, **8**, 104–117 (in Japanese).
Siebert, L. (2000), 'Using GIS to document, visualize, and interpret Tokyo's spatial history.' *Social Science History*, **24**, 537–574.
Takasaki, M. (ed.) (1988), *Chizugaku (Cartography)*. Asakura Shoten, Tokyo (in Japanese)
Taniuchi, T. (1995), 'Distribution of urban population in the Tokyo and Osaka metropolitan areas, 1883–1985.' *The Proceedings of the Department of Humanities College of Arts and Sciences*, **101**, 99–118 (in Japanese).

Shipeng Zhao and Toshikazu Tamura

GIS Modelling for Rain-Induced Debris-Flow Hazards in a Small Watershed

1. Introduction

Debris-flows, which are a type of mass movement initiated on hillslopes or in channels by either intense rainfall, snowmelt, or rapid runoff, can cause substantial damage and loss of life, particularly in mountainous and hilly areas. Therefore, assessment of debris-flow hazards is imperative for disaster prevention and watershed management in montane environments.

Using field survey and analysis of debris-flow deposits, geomorphologists have devised many methods for identifying areas susceptible to debris-flow hazards (e.g., Cook, 1984; Costa, 1984; Jakob, 1996; Bovis and Jakob, 1999). While these methods work well on the specific area, they are empirical in nature. During the past twenty years, there have been great advances in theoretical or parametric approaches to understanding the occurrence of rain-induced debris flow (e.g., Anderson and Sitar, 1995; Iverson, 1997a, b; Takahashi, 1977, 1981, 1991). Various process models have been used to simulate debris-flow hazard (e.g., Mizuyama and Ishikawa, 1990; Takahashi, 1991; Jin and Fread, 1997), in which the simulation was specified in a fan or a gully and performed by complex finite-element or finite-difference computation. Morgan *et al.* (1995) argued that conducting such analyses for numerous drainages on large areas was not feasible because these methods require a large amount of data along with numerous assumptions about the characteristics of the debris flow. Recently GIS has been used to conduct assessment and automatic mapping of debris flow hazard. Zhao *et al.* (1996) integrated GIS and an expert system to assess hazard of debris flow on a mountainous area of Northern China. Saczuk and Gardner (1998) coupled the recurrence interval model with a comprehensive GIS database to identify potentially hazardous areas. Although these GIS-based methods are suitable for large areas, they are actually the weighted and statistical reasoning of factors influencing debris-flow and lack a consideration of debris-flow processes.

In order to construct a process-oriented approach to the assessment of debris-flow hazard at a regional scale, we develop a method of integrating existing process models with GIS. First, we construct a physically-based process model to compute spatially distributed hydrological conditions, slope stability and debris supply to gullies corresponding to heavy rain with different return periods. Then we apply a debris-flow mobilization theory to determine the triggering

B. Boots et al. (eds.), Modelling Geographical Systems, 251–271.

of debris-flow and its possible scale. Finally, we combine the results with rainfall of different return periods to create an index of debris-flow hazard. We test the results in a hilly watershed where many shallow landslides and debris-flows were induced by heavy rain. Through modelling debris-flow hazard, we expect not only to provide a map showing spatial difference of debris-flow hazard but also to provide an understanding of the interplay of rainfall, runoff and debris flow processes.

2. METHOD

There are generally two ways of integrating GIS and process models: one is loose integration; another is tight integration (Abel *et al.*, 1994). Loose integration is pair-to-pair integration in which the GIS and process model are run independently and exchange data by means of a shared file. Tight integration is embedded-type integration in which either the process model is embedded in GIS or GIS components are embedded in a process model system. Considering the characteristics of rainfall-induced debris-flow process and the spatial analysis capacity of current GIS , we implemented all debris-flow process models in ArcView Spatial Analyst using Avenue language. A flow chart of system integration is shown on Figure 1. Slope and aspect were computed by surface functions; flow accumulation, flow direction, flow length, stream link and stream order were computed by hydrological functions. All computation of process models was transformed into operations within a grid theme or among grid themes. The implementation of each process model is explained in following sections.

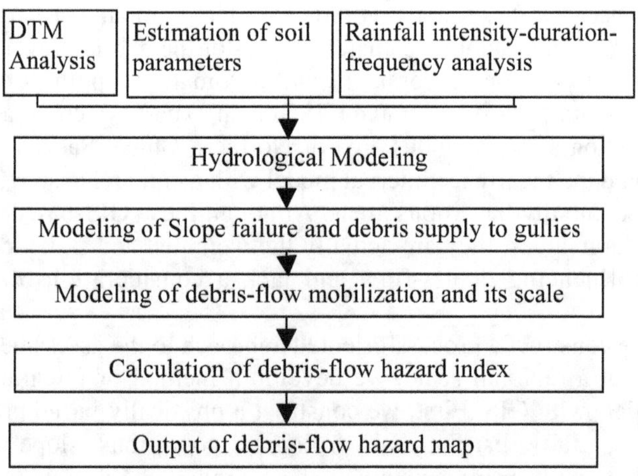

Figure 1. Flow chart of GIS modelling for rain-induced debris-flow hazards.

2.1. *Simulation Storm-Runoff Processes*

2.1.1. *Analysis of Seepage Flow on the Hillslopes*

We applied a simplified seepage flow analysis to calculate depth of saturated throughflow (Iida, 1999). The analysis is based on the assumptions that (1) the saturated throughflows occur parallel to bedrock surface, and (2) the recharge rate of the saturated throughflow is equal to the rainfall intensity. Then maximum depth ($H(N)$) of saturated throughflow with return period N can be given by the following equation:

$$(H(N)) = \max_{t} \frac{I_0(t, N)\, S_{max}(t)}{k \sin\beta \cos\beta\; CellSize} \tag{1}$$

where: $I_0(t, N)$ is the rainfall intensity-duration-frequency curve with return period N; $S_{max}(t)$ is the maximum covered area with travel time t of the saturated throughflow; k is the saturated conductivity coefficient; β is slope angle; *CellSize* is the cell size of the grid. If $(H(N))$ is greater than or equal to the depth of regolith soil, the surface will reach saturation.

The maximum covered area $S_{max}(t)$ with travel time t of the saturated throughflow can be explained as follows. For a cell (the cell at the lower right corner of Figure 2), we can get its flow contribution area (the shaded part in Figure 2) by Jenson and Domingue's (1988) method. Then we can calculate the travel time of saturated throughflow from each cell within its contribution area to the pouring point. Finally, we can obtain the maximum covered area (the lightest part in Figure 2) of the saturated throughflow in a particular time, such as an hour, by classifying and statistical method.

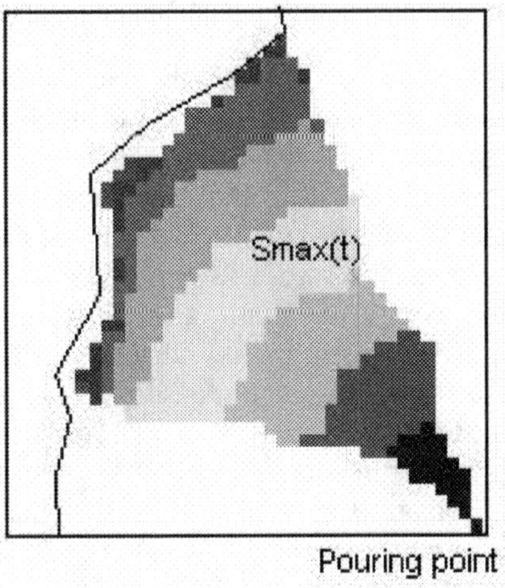

Figure 2. Diagram of the maximum covered area $S_{max}(t)$ within travel time t of saturated throughflow.

2.1.2. *Calculation of Peak Discharge in the Gullies*

According to the variable source area concept, peak discharge consists of two components: saturated-excess overland flow and subsurface storm flow, that is,

$$Q_p(N) = Q_o(N) + Q_s(N) \tag{2}$$

where: $Q_p(N)$ is peak discharge with return period N; $Q_o(N)$ is the component of saturated-excess overland flow in the peak discharge; $Q_s(N)$ is the component of subsurface storm flow in the peak discharge.

The component of saturated-excess overland flow is a mixture of direct runoff on the saturated zone and return flow. We calculate using the following equation:

$$Q_o(N) = I_0(td, N)A_{sat}(N) + \sum_t F_i(N) \tag{3}$$

where: td is the lag time of peak discharge which depends on the type of ground cover, the slope of land, and the distance along the flow path. $A_{sat}(N)$ is the area of saturated zone with return period N; $F_i(N)$ is the balance of throughflow in ith cell within the saturated zone with return period N.

We generate F_i by grid-based spatial analysis techniques. First, we establish an orthogonal coordinate system for the grid theme. Second, we divide the outflow of throughflow for each cell into two parts according to aspect and calculate them respectively by Equations (4) and (5):

$$Outflow\ X = kH(N) \sin\beta \cos\beta \sin\alpha\ CellSize \tag{4}$$

$$Outflow\ Y = kH(N) \sin\beta \cos\beta \cos\alpha\ CellSize \tag{5}$$

where: α is the aspect of slope; k, $H(N)$, β and $CellSize$ are defined as in Equation (1). Finally, we calculate inflow, outflow and balance of throughflow for each cell by means of a neighborhood operation. If the balance is positive, return flow exists in the cell. If the balance is negative, direct runoff is equal to the difference between rainfall intensity and the balance.

The component of subsurface stormflow represents the output of subsurface flow to the stream from grids abutting the stream. Subsurface flow may provide stormflow in three ways: pipeflow, macropore flow and flow through the soil matrix. It is difficult to calculate subsurface stormflow accurately because we cannot know the distribution of macropores or pipes in the soil. We simplify subsurface flow into Darcian flow through the soil matrix. Thus the Darcy velocity of seepage flow will be $k \sin\beta$ if the direction of seepage flow is parallel to the ground surface. We calculate the component of subsurface stormflow in peak discharge by Equation (6):

$$Q_s(N) = \sum_t (2k \sin\beta H_j(N) \cos\beta\ CellSize) \tag{6}$$

where: $H_j(N)$ is depth of throughflow in jth cell abutting the stream; k, β and *CellSize* are as defined in Equation (1).

2.2. *Analysis of Slope Stability*

2.2.1. *Infinite Slope Stability Model*

An infinite slope stability model is applied to determine the potential of shallow landslide triggered by heavy rain. This model described is supported by many researchers of shallow landsliding (Iida, 1999) and is described by

$$F = \frac{c + \{r_t(D - H) + (r_{sat} - r_w)H\}\cos^2 \beta \tan \phi}{\{r_t(D - H) + (r_{sat} - r)H\}\cos \beta \sin \beta} \tag{7}$$

where F is the factor of safety; c is the soil cohesion coefficient; ϕ is the soil internal friction angle; D is depth of soil; H is depth of throughflow; β is the slope angle; r_t is the unit weight of wet soil; r_{sat} is the unit weight of saturated soil; r_w is the unit weight of water.

The factors of safety can be directly calculated for each cell on the basis of the depth of throughflow and the above defined soil parameters.

2.2.2. *Supply of Landslide Mass to The Gullies*

Shallow landsliding will cause the downhill motion of a debris mass. According to the principle of sliding friction, the debris mass will move downhill until the following equations are satisfied:

$$\tan \beta \leq \frac{r_t - mr_w}{r_t} \tan \phi \tag{8}$$

where: m is the coefficient of wetness, which is equal to H/D. r_t, r_w, β and ϕ are defined as in Equation (7).

The greater the coefficient of wetness, the further the debris mass can move. When the coefficient of wetness is close to 1, the debris mass may even be mobilized into debris flow (Iverson, 1997a). The process of downslope movement of debris mass can be simulated by a neighboring operation of grid cells. According to the aspect, we determine which cell the debris mass goes to. According to the slope angle, we determine if the debris mass will stop on a cell.

2.3. Determination of Debris-Flow Mobilization AND Its Scale

2.3.1. Debris Flow Mobilization Model

According to Takahashi (1981), the condition for the occurrence of debris flow is $a_L \geq \kappa\, h_0$. Here a_L is the thickness of mobilized debris; h_0 is the depth of water in the gully; κ is a coefficient related to density of debris flow. If a_L is considerably less than h_0, the flow will be highly concentrated water flow.

Because the cohesion bond is broken when landslides occur, the thickness of mobilized debris is calculated as follows in case of cohesion coefficient $c = 0$:

$$a_L = D \qquad \text{if} \quad \theta \leq \theta_1; \tag{9}$$

$$a_L = \frac{\rho \tan\theta}{C_*(\sigma - \rho)\tan\phi - \{C_*(\sigma - \rho) + \rho\}\tan\theta} \qquad \text{if} \quad \theta_2 \leq \theta \leq \theta_1; \tag{10}$$

$$a_L = 0 \qquad \text{if} \quad \theta \leq \theta_2; \tag{11}$$

where: D is the depth of debris in the gully; h_0 is depth of water in the gully; θ is the slope angle of the gully; ϕ is the internal friction angle; C_* is the grain concentration by volume in the static debris bed; σ is the density of solid parts of debris; ρ is the density of water. θ_1 is the slope at which the whole debris layer will be mobilized; θ_2 is the minimum slope at which sediment gravity flow will occur. The slopes θ_1, θ_2 were calculated as follows:

$$\tan\theta_1 = \frac{C_*(\sigma - \rho)}{C_*(\sigma - \rho) + \rho(1 + h_0/D)}\tan\phi \tag{12}$$

$$\tan\theta_2 = \frac{C_*(\sigma - \rho)}{C_*(\sigma - \rho) + \rho(1 + h_0/d)}\tan\phi \tag{13}$$

where: D, h_0, θ, ϕ, C_*, σ and ρ are as defined previously; d is the diameter of a representative particle of the debris.

2.3.2. Scales of The Debris Flows

According to Takahashi's (1981) study, the discharge of debris flow can be estimated by

$$M(N) = \frac{C_*}{C_* - C_d}Q_p(N) \tag{14}$$

where $M(N)$ is the discharge of debris flow induced by a storm with return period N; C_d is the volume concentration of debris flow; C_* is defined previously; $Q_p(N)$ is defined in Equation (2).

The mobilization of debris flow is simulated on each segment of the stream channel, which is composed of cells for which flow accumulation is larger than the threshold. If the thickness of mobilized debris is large enough to cause the debris flow, we will estimate its scale according to Equation (14).

2.4. *Calculation of the Debris-Flow Hazard Index*

After getting the results on scale of debris flows with different return periods, we can draw a chart on the magnitude-frequency relationship of debris flows (Figure 3). Here the frequency is represented by the reciprocal of the return period. Then the debris-flow hazard index is defined as the area between the curve and the *X* axis, which is approximated by the discrete method:

$$H_{index} = \sum_i \left[\frac{M(N_i) + M(N_{i+1})}{2} \left(\frac{1}{N_{i+1}} - \frac{1}{N_i} \right) \right] \quad (15)$$

where, H_{index} is the debris-flow hazard index; $M(N_i)$, $M(N_{i+1})$ are the scale of debris flows with return period N_i, N_{i+1}, respectively.

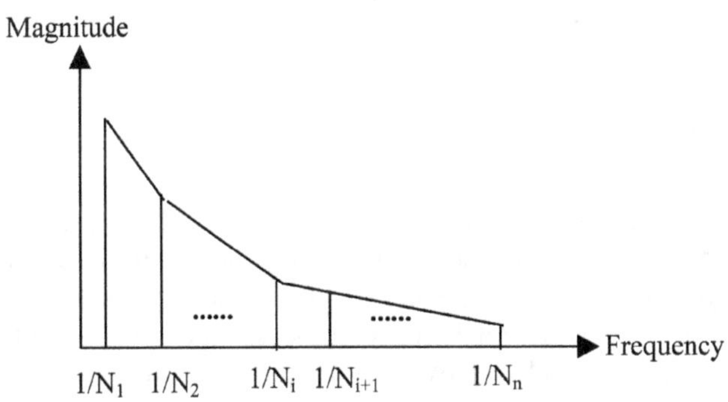

Figure 3. Diagram for magnitude-frequency relationship of debris flows.

3. TRIAL WATERSHED

3.1. *Geographical Description*

The trial was made in the Kawauchizawa watershed which covers an area of approximately 7 km^2 in Takadate Hills, about 10 km south of Sendai, Northeastern Japan (Figure 4). The watershed is mainly underlain by andesite (88 percent of the area) and tuff-breccia which belong to the Takadate Formation of the Miocene age. Volcanic rock of the Takadate Formation is weathered into thick clayey material, which provides parent material of red soil considered to be a paleosol. A part of the red soil is covered by weathered pumice beds of the Mid- and Late Pleistocene. The slope lacking red soil is covered by brown forest soil which moved in part by soil creeping (Tamura, 1999).

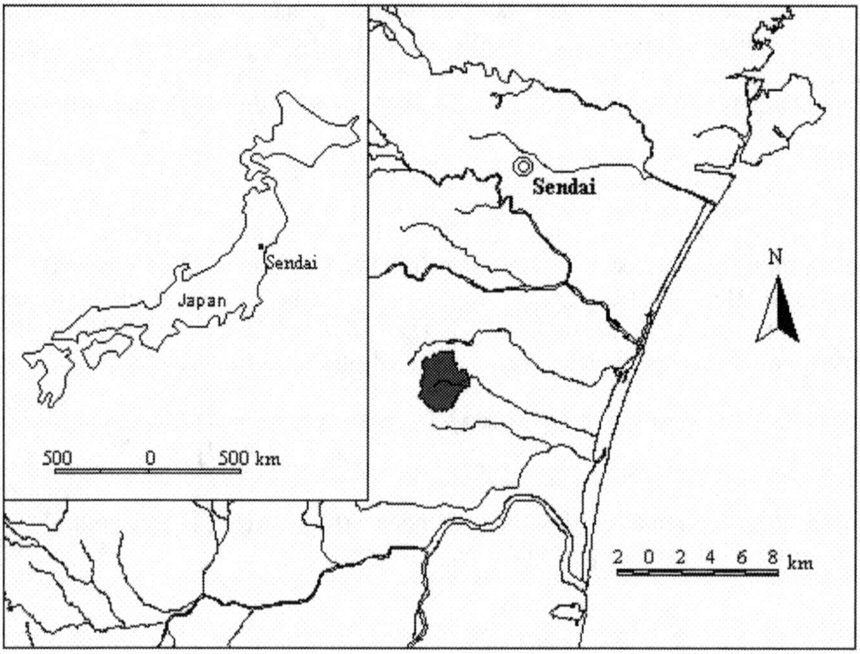

Figure 4. The location of the Kawauchizawa watershed.

The watershed has a narrow crest and broad hillslope. Altitude ranges from about 300 m a.s.l. Steep slopes are concentrated on the western part of the watershed and some lower slope segments of the eastern part. Drainage density in the Kawauchizawa watershed is 10.4 km/km^2. Some river segments in the watershed show slope gradients over 20 degrees, providing a condition for the occurrence of debris-flows. Most of the area in the watershed is covered by forest which consists mainly of secondary forests dominated by broad-leaved deciduous trees such as *Quercus serrata* and plantations of *Cryptomeria japonica*. The proportion of secondary forests to *Cryptomeria japonica* plantations is about 6:4. Farmland and settlements are distributed on alluvium-filled narrow valley-bottoms.

The Kawauchizawa watershed has a temperate monsoon climate with a mean annual precipitation of about 1,200 mm and an annual mean temperature of about 12 °C. About 70 percent of the annual precipitation is concentrated in the period from May to October. Almost every year a storm occurred with more than 50 mm daily precipitation. On 22–23 September, 1994, a big storm, with total rainfall of about 500 mm in 20 hours and maximum hourly rainfall of about 85 mm, induced about 400 surface regolith slides and many debris-flows in the watershed. A series of hydrogeomorphic phenomena in the event was surveyed intensively (Tamura, 1999). The distribution of shallow landslides in relation to hillslope form and regolith profile was analyzed by Chatterjee

(1999). Radiocarbon dating of former slide and debris-flow deposits indicated that the average return period of shallow slide swarm in the hills including the Kawauchizawa watershed is about 400 years (Li and Tamura, 1999).

3.2. Data Preparation

3.2.1. Rainfall Intensity-Duration-Frequency Curve on the Sendai Region

Rainfall intensity-duration-frequency curve on the Sendai region was taken as Talbot type:

$$I_0(t, N) = \frac{a(N)}{t + b(N)} \tag{16}$$

where t is rainfall duration (hours); N is rainfall return period (years); I_0 is rainfall intensity (mm/h). After Ishiguro (1961), the parameters, $a(N)$ and $b(N)$ can be approximated by the following empirical equations,

$$b(N) = \left(24 - \frac{I_{1N}}{I_{24N}}\right)\left(\frac{I_{1N}}{I_{24N}} - 1\right); \quad a(N) = [b(N) + 24]I_{24N} \tag{17}$$

where I_{1N} is the maximum hourly rainfall and I_{24N} is the maximum daily rainfall with N-year return period.

Recurrence interval of various rainfall intensity on the Sendai region was summarized and analyzed by Sano (1999) on the basis of 68-year meteorological data. Using the data, parameters $a(N)$ and $b(N)$ were calculated for an appropriate return period N (= 5, 10, 15, 20, 30, 40, 50, 60, 80, 100, 120, 140, 160, 180, 200, 250, 300, 400, 500, 600, 800, 1000) and fitted as power functions of N. Eventually, the rainfall intensity-duration-frequency curve on the trial watershed was expressed as follows and shown in Figure 5.

$$I_0(t, N) = \frac{112.874N^{0.1854}}{t + 2.3007N^{-0.037}} \tag{18}$$

3.2.2. Digital Terrain Analysis on The Kawauchizawa Watershed

We constructed a 10 m × 10 m DEM of the Kawauchizawa watershed using the following procedures. First, we traced contours, stream lines and labeled spot elevations from a 1:25,000 Digital Map Image published by **Japan Map Center**. Second, we interpolated elevation of some discrete points along main streams. Third, we built a TIN model using the contours and these elevation points. Finally, we converted the TIN model into a grid digital elevation model (Figure 6). Then we conducted detailed terrain analysis: (1) derivation of slope angle from the DEM (Figure 7); (2) derivation of slope aspect from the DEM; (3) calculation of flow accumulation for each cell; (4) calcu-

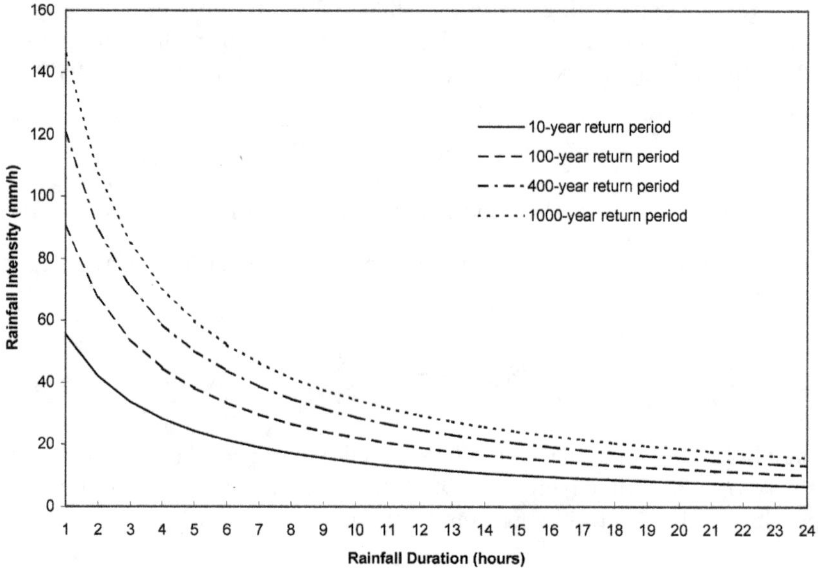

Figure 5. The rainfall intensity-duration-frequency curve in the Sendai area.

lation of maximum upstream flow length for each cell; (5) calculation of mean slope within a catchment for each cell; (6) extraction of drainage network by Tarboton's method (1991).

3.2.3. Estimation of Soil Depth on the Kawauchizawa Watershed

Soil depth is considered to be influenced not only by systematical factors such as slope and specific catchment area but also some random factors. Some researchers (e.g., Iida and Tanaka, 1997) discovered that the frequency distribution of soil depth could be expressed approximately by a log-normal function. Based on surveying data at over 20 points within the watershed using portable penetration test apparatus, soil depth on the Kawauchizawa watershed was simulated by a log-normal distribution function with mean 0.82 m and standard deviation 0.5.

3.2.4. Generalization of Channels in the Kawauchizawa Watershed

Through field survey, we found that most channels in the Kawauchizawa watershed initiate from contributing areas bigger than 2,500 m^2. Comparing field measured cross-section of channels with distribution of flow accumulation, we generalized the parameters of channels on the Kawauchizawa watershed as shown in Table 1.

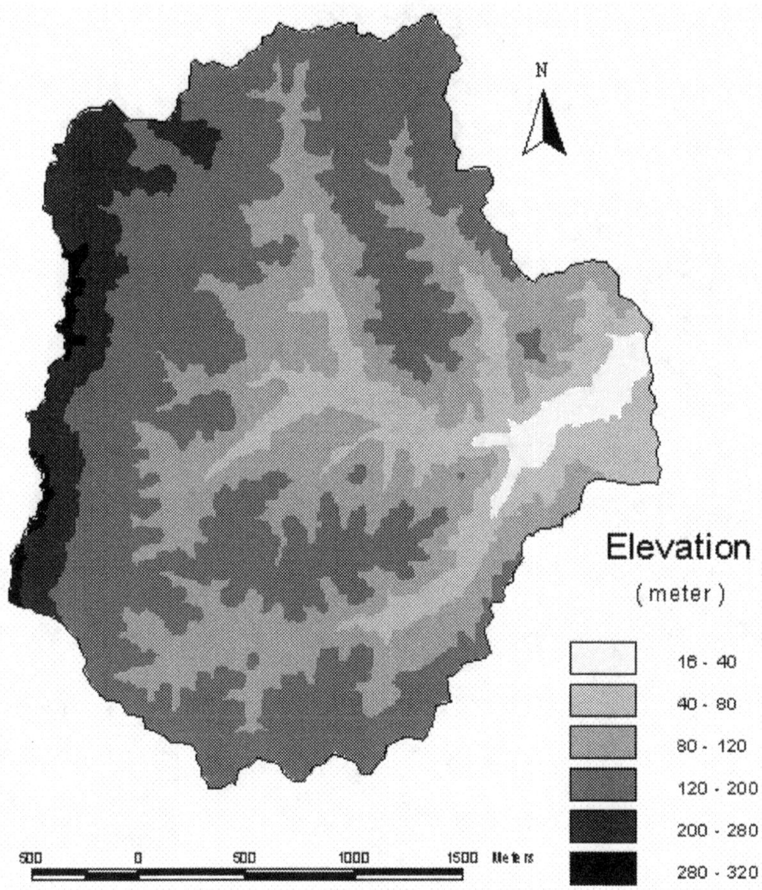

Figure 6. Relief map of the Kawauchizawa watershed.

Table 1. Classification of Channels in the Kawawuchizawa Watershed.

Type of channel	Area of catchment (km^2)	Width of channel (m)	Manning's *n*
1	0.0025–0.0999	1	0.080
2	0.1–0.9999	2	0.060
3	1–1.9999	3	0.048
4	2–3.9999	5	0.040
5	4–6	8	0.035
6	> 6	10	0.030

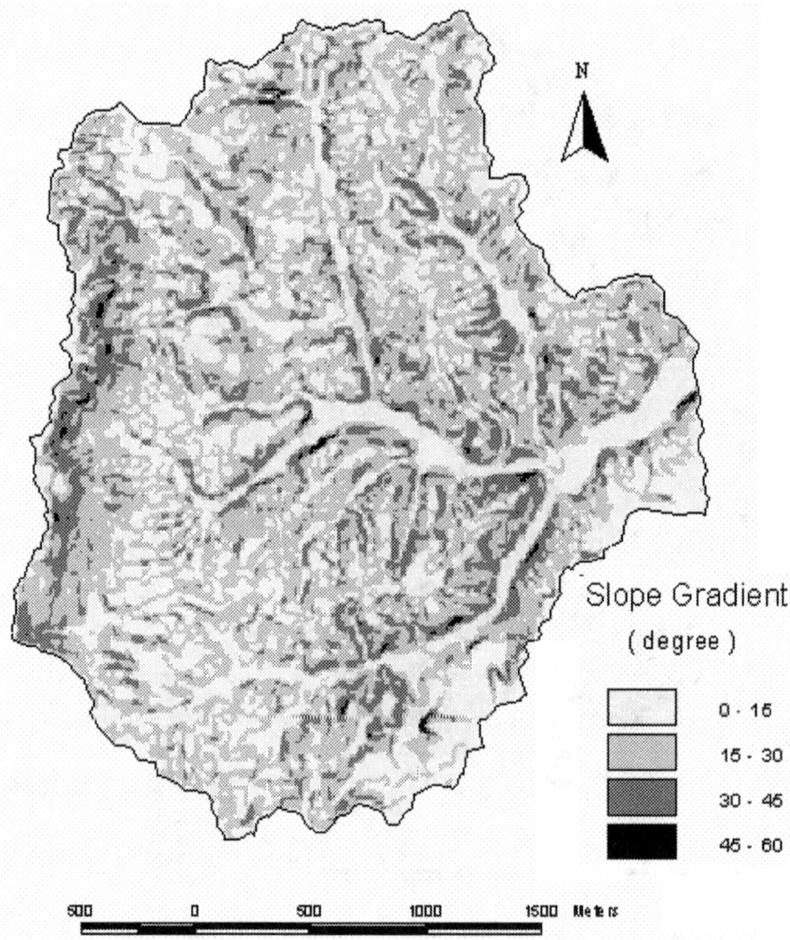

Figure 7. Distribution of slope gradient in the Kawauchizawa watershed.

3.3. Results and Analysis

3.3.1. Distribution of Hydrological Conditions Under Heavy Rain With Different Return Periods

Setting the soil saturated conductivity coefficient as 6×10^{-5} m/s based on the measurement of soil samples at a spot within the Kawauchizawa watershed and analysis of measured data from other researchers (Shindo, 1983), we conducted analysis of seepage flow under heavy rain with 10-year, 50-year, 100-year and 400-year return periods. Distribution of saturated zone under heavy rain with a 10-year return period is shown on Figure 8. It illustrates that while hillslopes close to channels have been saturated, crest- and nose-slopes have

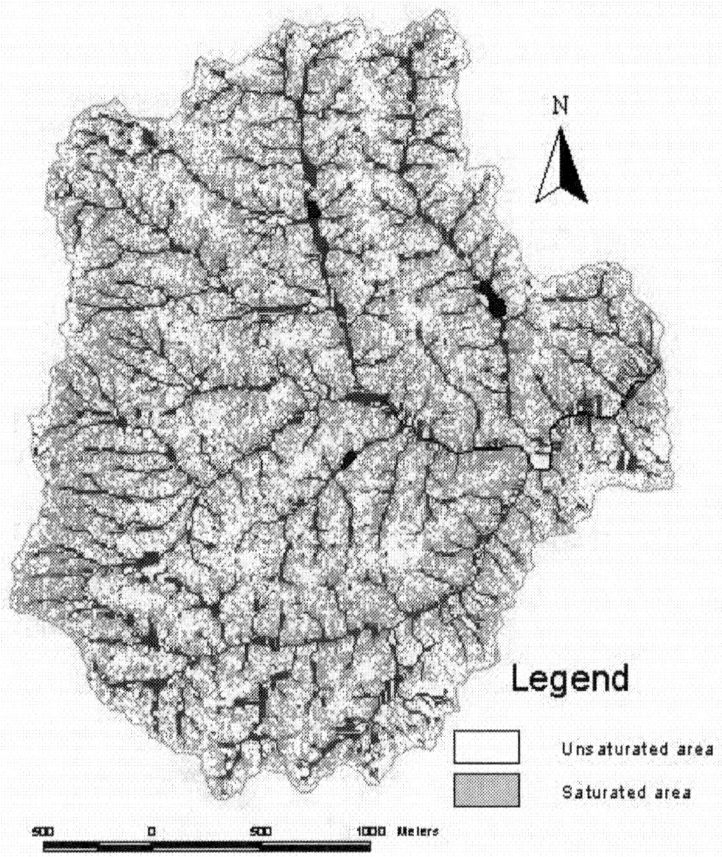

Figure 8. Distribution of saturated zone under heavy rain with 10 year return period.

not. As rainfall intensity and duration increase, saturated zones expand upslope from valley-bottoms.

By determining the lag time of peak discharge using the empirical formula (Sugawara and Ozaki, 1988): $t_d = 0.15\sqrt{A}$; where A is catchment area (km^2), we calculated peak discharge on the Kawauchizawa watershed under heavy rain with 10-year, 50-year, 100-year and 400-year return periods (Table 2). Distribution of peak discharge under heavy rain with a 400-year return period is shown on Figure 9, which demonstrates that peak discharge increases downstream from source heads.

3.3.2. Distribution of Source Areas for Debris Flows and Debris Supply to Gullies Under Heavy Rain With Different Return Periods

Based on the analysis of soil samples, the following soil parameters were used: the unit weight of saturated soil and wet soil, 1,600 kg/m^3 and 1,400 kg/m^3,

Please supply original

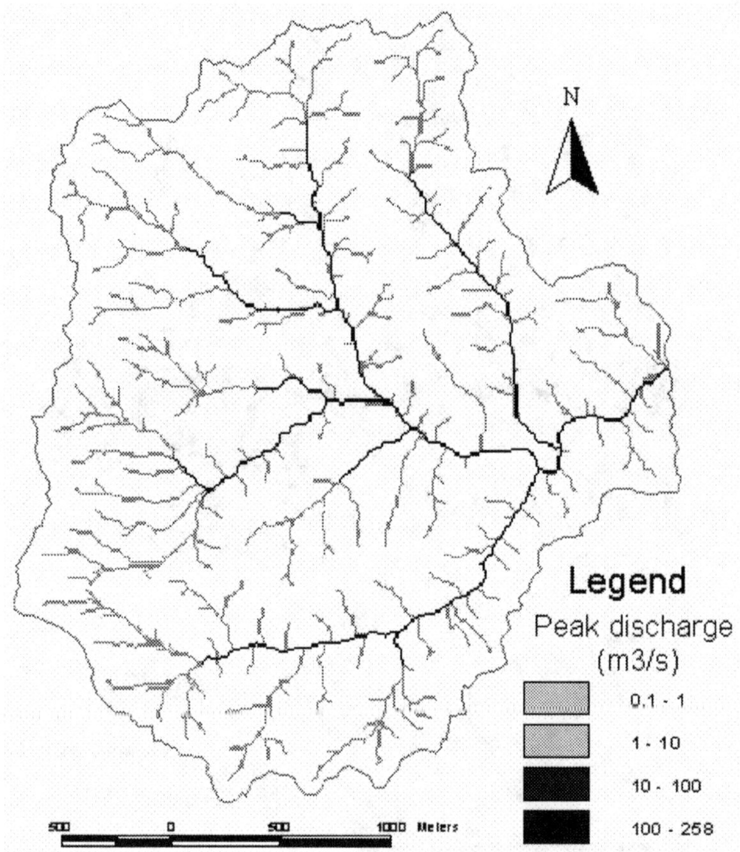

Figure 9. Peak discharge under heavy rain with 400-year return period.

Table 2. Results of Flood Hazard Assessment on the Kawawuchizawa Watershed.

Return Period (years)	10	50	100	400
Percent Area of Saturated Zone (%)	63.2	76.1	80.7	88
Peak Discharge at Outlet Section (m^3/s)	84.5	143	175.6	257.2

respectively; soil internal friction angle, 35 degrees; soil cohesion coefficient, 500 kgf/m^2. Then we calculated Factors of Safety (FS) for landslides in the watershed under heavy rain with 10-year, 50-year, 100-year and 400-year return periods. Distribution of FS under heavy rain with a 400-year return period is shown in Figure 10. Theoretically speaking, the area where SF is less than 1 will be susceptible to shallow landsliding. Comparing shallow landslide scars in the 1994 event with the spatial distribution of calculated FS under heavy

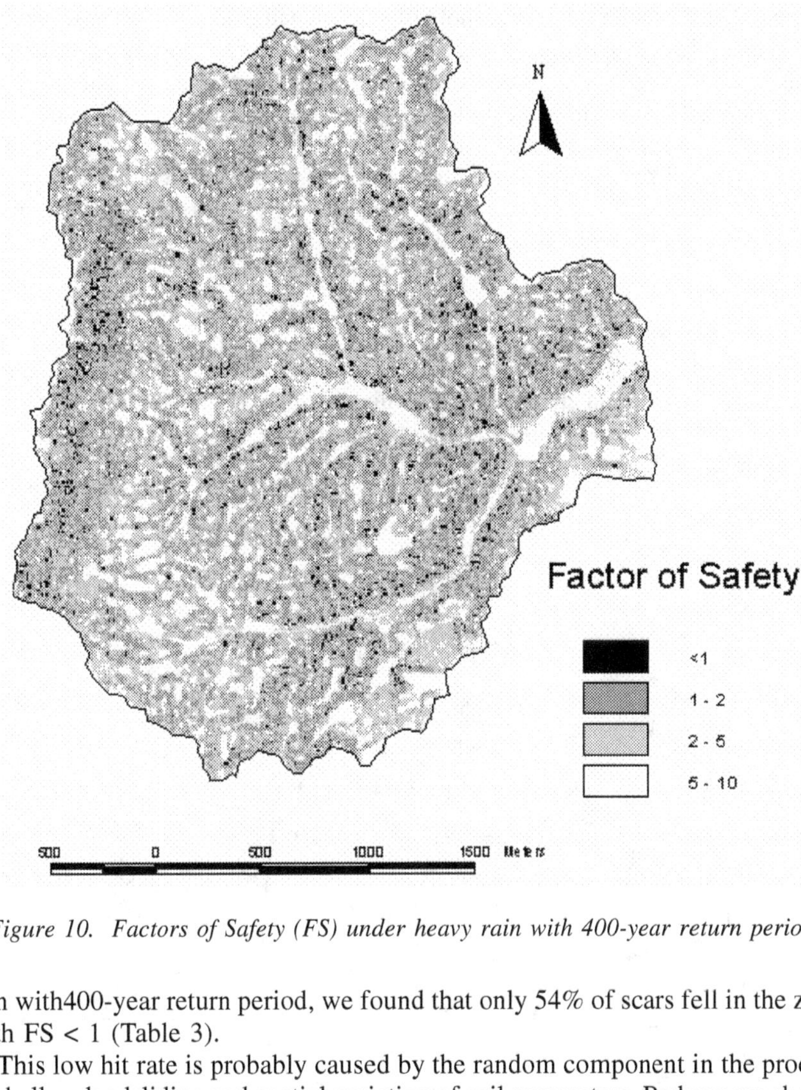

Figure 10. Factors of Safety (FS) under heavy rain with 400-year return period.

rain with400-year return period, we found that only 54% of scars fell in the zone with FS < 1 (Table 3).

This low hit rate is probably caused by the random component in the process of shallow landsliding and spatial variation of soil parameters. Perhaps we should

Tables 3. Comparison between Distribution of the Scars in 1994 and the Factors of Safety Calculated under Heavy Rain with a 400-year Return Period.

The calculated Factors of Safety	Distribution of scars in 1994
< 1	54%
1–1.9999	40%
2–4.9999	3.4%
5–10	2.6%

take a higher FS value as the threshold of shallow landsliding in order to increase the hit rate. The adoption of a higher FS value causes many more cells to be predicted as unstable than those where slides occurred actually. Therefore, we maintained FS = 1 as the threshold of landsliding. Assuming the whole layer of regolith soil on the grids with FS < 1 would slide downslope, we calculate the depositional position of the sliding mass. Figure 11 shows the distribution of sliding mass due to heavy rain with a 400 year return period. It tells us that sliding masses stop on gentle slopes and in channels.

3.3.3. Debris-Flow Hazard Assessment

The spatial units for assessment of debris-flow hazard are the links, that is the sections of a stream channel connecting two successive junctions, a junction and the outlet, or a junction and the source of the stream. All links are assigned an unique identifier and an order by Strahler's method (1957) so as to facili-

Please supply original

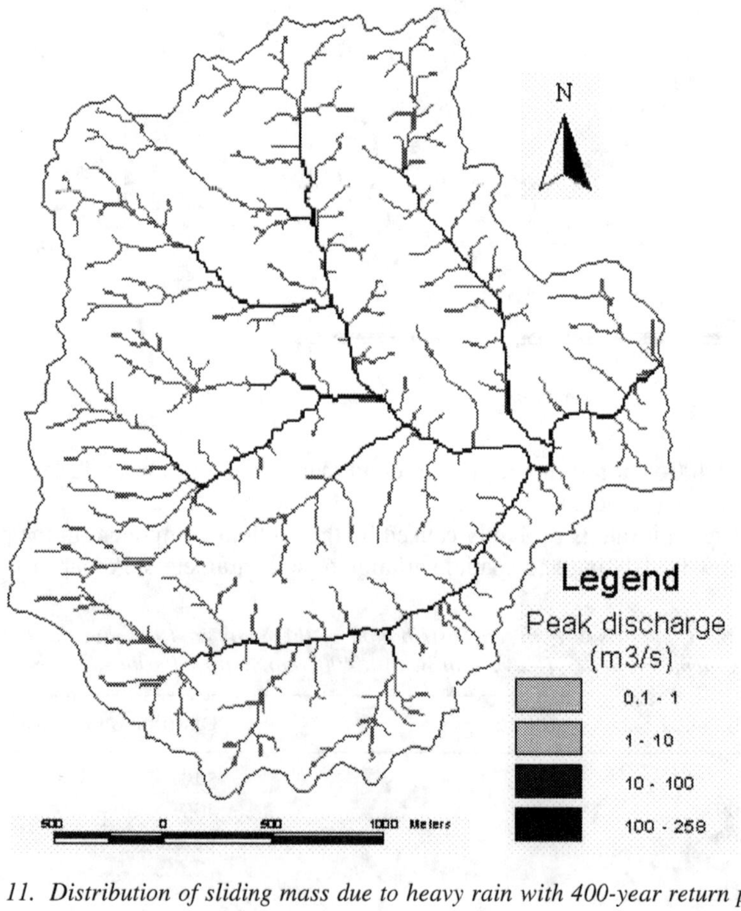

Figure 11. Distribution of sliding mass due to heavy rain with 400-year return period.

tate spatial analysis and operation in the assessment of debris-flow hazard. Hazards in the alluvial fan are excluded from the assessment because almost all debris flows on the Kawauchizawa watershed enter the main stream and wash out by flood and only very small debris-flow fans exist.

In order to judge whether debris flows will occur and to estimate their possible scale in a link due to heavy rain with 10-year, 50-year, 100-year and 400-year return periods, some information on hydrological conditions, slope gradient and debris supply must be transformed and calculated on the link-base. First, the maximum peak discharge in a link was sought and the maximum water depth was calculated using Manning's formula. Second, the slope gradient of a link is calculated by the length of a link divided by the maximum difference of elevation within the link. Third, the thickness of mobilized debris due to heavy rain with 10-year, 50-year, 100-year and 400-year return periods was calculated using Takahashi's (1981) method by assuming the volume concentration of the solid particles as 0.65, the density of solid particles as 2.6, and the medium diameter of particle size as 0.1mm. On the basis of above information, we judge whether debris flow will occur in a link by setting κ as 0.7 and comparing the thickness of mobilized debris with water depth in the link. We determine whether the link downstream to the debris-flow link is a conveyance path for debris flows based on two simple rules: the deviation of flow direction between two links is less than 45 degrees and the slope gradient is greater than 6 degrees. Otherwise debris flow will be deposited on the junction of the stream or gentler channel. Then we calculate the possible scale of debris flows in these links due to heavy rain with 10-year, 50-year, 100-year and 400-year return periods.

The index of debris-flow hazard on the Kawauchizawa watershed was calculated using data on the magnitude of debris flow due to heavy rain with 10-year, 50-year, 100-year and 400-year return periods. Considering the practical situation in the Kawauchizawa watershed, hazard of debris flows was classified according to standards in Table 4. Figure 12 is the resulting hazard map for debris-flows. Field survey proved that debris-flow deposits exist at the outlets of the high-hazard gullies and some medium-hazard gullies presented in Figure 12.

Table 4. Classification of Calculated Index for Debris-Flow Hazard.

The Index for debris-flow hazard	Hazard class
0	No Hazard
0.0001–0.2499	Low Hazard
0.25–0.5	Medium Hazard
> 0.5	High Hazard

Please supply original

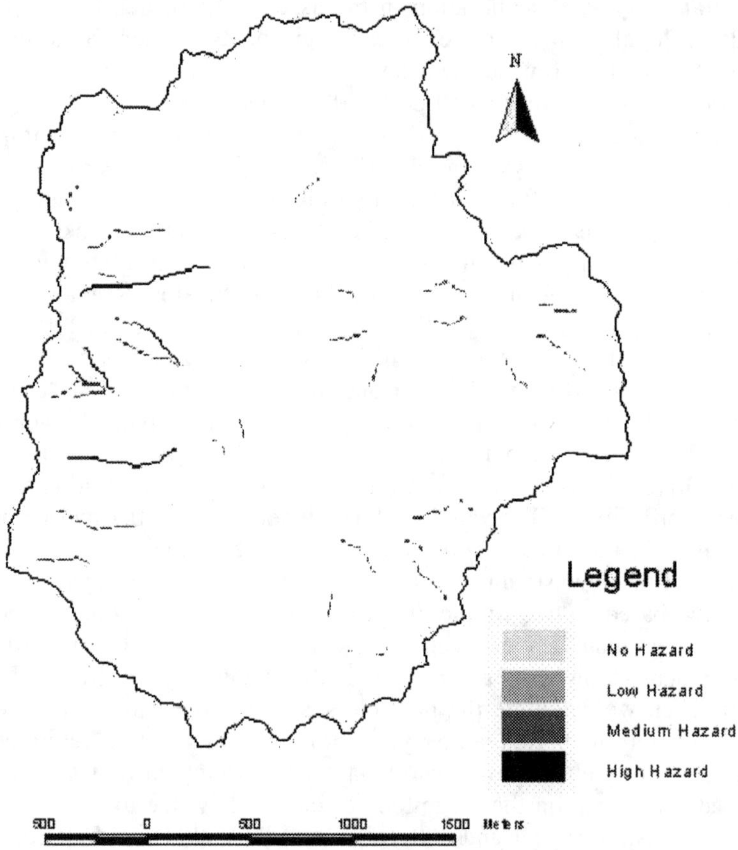

Figure 12. Map of debris-flow hazard on the Kawauchizawa watershed.

4. Discussions and Conclusions

A method integrating GIS and process models for the assessment of debris flow hazard in a small hilly watershed was put forward and tested on the Kawauchizawa watershed of Northeastern Japan. The debris flow hazard map was shown to be reliable by field survey. The research indicates that coupled hydrologic, slope stability, debris-flow mobilization models and GIS provide a powerful tool for assessment of debris flow hazards on a region. By this method, we not only got a final assessment of debris flow hazard, but also many intermediate results on rainfall-runoff, slope failure and debris flow. It is very useful for us to understand debris flow processes by this GIS-based simulation.

The complexity of debris-flow processes means that any evaluation result contains a considerable amount of uncertainty. The fact that hit rate for shallow landslide occurrence was 54% implies the role of model simplification and random elements. The affect of spatial uncertainty of shallow landsliding to debris supply to a gully may be reduced by spatial synthesis within a sub-watershed. This is true for the situation where debris flow is induced by many landslides within a watershed as in the Kawauchizawa watershed. This is because the occurrence of debris flow depends on debris supply from many shallow landslides and debris supply from other areas with FS < 1 can make up debris supply scars within a sub-watershed. The affect of uncertainty in the process of triggering debris flow on hazard assessment may be reduced by temporal averaging over a long period because the hazard map combines results with different return periods.

Many variables and parameters were simplified when we applied the GIS-based process model to the trial watershed. Some variables were assumed to be spatially uniform such as the soil saturated conductivity coefficient, the soil cohesion coefficient and the internal friction angle. As Montgomery and Dietrich (1994) concluded, assumptions of spatial uniformity for strength and hydrological properties are convenient but not necessary. Should such data as the spatial variation in transmissivity and frictional strength be available, they can be incorporated into model simulations. Perhaps this will help to improve the accuracy of site-predication of shallow landslides. Soil thickness was simulated by a random distribution rather than assumed to be constant. This may closer to the real situation from the viewpoint of watershed scale.

It should be noted that the method presented here is mainly suitable for landslide-type debris flow which is formed by many rain-induced shallow landslides entering gullies. For bed-type debris flow it is difficult to measure thickness of debris accumulated on the bed of gullies within a watershed. For debris flow mobilized from deep-seat landslides, great uncertainty exists in the prediction of its occurrence. It is important to remember that the method presented here serves for mapping the spatial difference of debris flow hazard on a watershed scale.

ACKNOWLEDGEMENTS

Financial support for the work was provided by the Grant-in-Aid for Scientific Research of the Ministry of Education, Science, Sports and Culture of Japan, Project Number: 98070 (1997–1999). We would like to express my appreciation to Dr. K. Sakaida, Institute of Geography, Tohoku University, for providing meteorological data and Dr. B.P. Gyawali, presently at Nepalese Ministry of Forests, for his assistance in the field work.

Institute of Geography,
Graduate School of Science,
Tohoku University,
Sendai 980-8578, Japan

REFERENCES

Abel, D.J., P.J. Kilby and J.R. Davis (1994), 'The system integration problem.' *International Journal of Geographical Information Systems*, **8**, 1–12.

Anderson, S.A. and N. Sitar (1995), 'Analysis of rainfall-induced debris flows.' *Journal of Geotechnical Engineering*, **121**, 544–552.

Campbell, R.H. (1975), Soil slips, debris flows, and rainstorms in the Santa Monica Mountains and vicinity, Southern California: *U.S. Geological Survey Professional Paper 851*, 51 pp.

Chatterjee, D. (1999), 'A GIS-assisted analysis of landslide distribution on hillslopes divided by slope breaks.' *Science Reports, Tohoku Univ., 7th Ser. (Geography)*, **49**, 69–80.

Cooke, R.U. (1984), *Geomorphological Hazards in Los Angles: A Study of Slope and Sediment Problems in a Metropolitan County*. London, 206 pp.

Costa, J. E. (1984), 'Physical geomorphology of debris flows.' In J.E. Costa and P.J. Fleisher (eds.), *Developments and Applications of Geomorphology*, Springer-Verlag, Berlin, pp. 268–317.

Hubbard, B. and M.F. Sheridan (1999), *GIS Model of the 1920 Seismogenic Debris Fow in Veracruz, Mexico*. UCGIS Buffalo HomePage.

Iida, T. (1999), 'A stochastic hydro-geomorphological model for shallow landsliding due to rainstorm.' *Catena*, **34**, 293–313.

Iida, T. and K. Tanaka (1997), 'The relationship between topography and soil depth measured with the portable penetration test apparatus.' *Transactions, Japanese Geomorphological Union*, **18**, 61–78, (in Japanese with English abstract).

Ishiguro, M. (1961), 'On the estimation of probable rainfall intensity formula by specific coefficient methods.' *Transactions, Japan Society of Civil Engineers*, **74**, 19–26 (in Japanese with English abstract).

Iverson, R.M. (1997a), 'Debris-flow mobilization from landslides.' *Ann. Rev. Earth Planet. Sci.*, **25**, 85–138.

Iverson, R.M. (1997b), 'The physics of debris flows.' *Reviews of Geophysics*, **35**, 245–296.

Jenson S. K. and J.O. Domingue (1988), 'Extracting Topographic Structure from Digital Elevation Data for Geographic Information System Analysis.' *Photogrammetric Engineering and Remote Sensing*, **54**, 1593–1600.

Jin, M., and D.L. Fread (1997), 'One-dimensional routing of mud/debris flows using NWS FLDWAV MODEL.' In C.L. Chen (eds.), *Debris-flow Hazards Mitigation: Mechanics, Prediction, and Assessment: Proceeding of First International Conference*, San Francisco, USA, pp. 687–698.

Li, Y. and T. Tamura (1999), 'Regolith slide recurrence history indicated by radiocarbon ages in the hills around Sendai, Northeastern Japan.' *Science Reports, Tohoku Univ., 7th Ser. (Geography)*, **49**, 55–67.

Mizuyama, T., and Y. Ishikawa (1990), *Prediction of Debris Flow Prone Areas and Damage: American Society of Civil Engineers, Proceedings of International Symposium Hydraulics/Hydrology of Arid Lands*. San Diego, California, pp. 712–717.

Montgomery, D.R. and W.E. Dietrich (1994), 'A physically based model for the topographic control on shallow landsliding.' *Water Resources Research*, **30**, 1153–1171.

Morgan, B.A., G.F. Wieczorek, R.H. Campbell and P.L. Gori (1997), *Debris-Flow Hazards in areas Affected by the June 27, 1995 Storm in Madison County, Virginia*, USGS open file report, pp. 97–438.

Saczuk, E.A.R. and J.S. Gardner (1998), 'GIS-based mapping and modeling of debris flows in Banff National Park.' *Canadian Journal of Remote Sensing*, **24**, 53–59.

Sano, T. (1999), 'Analysis of two storm events triggering shallow landslides.' In T. Tamura (in Japanese).

Shindo, S. (1983), 'Occurrence of landslides at the head of stream and its relation to subsurface flow. Report of Grant-in-Aid for Scientific Research of the Ministry of Education, Science, Sports and Culture of Japan, Project No.: 5702509 (1982), pp. 76–77 (in Japanese).

Strahler, A.N. (1957), 'Quantitative analysis of watershed geomorphology.' *Transactions, American Geophysical Union*, **38**, 913–920.

Sugawara, M. and E. Ozaki (1988), 'On the analysis of runoff in Yangtze River.' *Journal of Japan Society of Hydrology & Water Resources*, **1**, 9–47.

Takahashi, T. (1977), 'A mechanism of occurrence of mud-debris flows and their characteristics in motion.' *Annual Reports, Inst. Disaster Prevention, Kyoto Univ.*, **20**, 1–31.

Takahashi, T. (1981), 'Estimation of potential debris flows and their hazardous zones: soft countermeasures for a disaster.' *Natural Disaster Science*, **3**, 57–89.

Takahashi, T. (1989), 'Debris flow hazard zone mapping.' In *Proceedings of the Japan-China (Taipei) Joint Seminar on Natural Hazard Mitigation*, Kyoto, Japan, pp. 363–372.

Takahashi, T. (1991), *Debris Flow*. A.A. Balkema, Rotterdam, 165 pp.

Tamura, T. (1999), Geomorphic position and recurrence history of regolith slides and their relation to hillslope development, Report of Grant-in-Aid for Scientific Research (B) of the Ministry of Education, Science, Sports and Culture of Japan, Project Number: 09480021 (1997–1998), 127 pp. (in Japanese with English abstract).

Tarboton, D.G., R.L. Bras and I. Rodriguez-Iturbe (1991), 'On the Extraction of Channel Networks from Digital Elevation Data.' *Hydrological Processes*, **5**, 81–100.

Zhao S.P., C. Zhou, Y. Xie and X. Xu (1996), 'Study on Integration of GIS and Expert System for Mudflow Hazard Assessment.' *Remote Sensing of Environment (China)*, **11**(3) (in Chinese with English abstract).

Section C

The Internet

NARUSHIGE SHIODE

A GEOGRAPHICAL INTERPRETATION OF CYPERSPACE: PRELIMINARY ANALYSIS ON THE SCALING TENDENCY OF INFORMATION SPACES

1. INTRODUCTION

Rapid development of information and communication technologies over the last two decades has brought a significant impact upon many aspects of our societal and economic activities. With an estimate of over 120,000,000 host computers and 300,000,000 users from round the world (ISC, 2001), the Internet has grown to conceive, within its own network, electronically generated information spaces collectively referred to as cyberspace. The variety exhibited by such information spaces ranges from the telecommunication network of satellites and fibre-optics to the metaphorical space of the World Wide Web and further to a full 3D virtual environment. Each space has its own structure and distinctive functionality yet is closely linked to one another to comprise cyberspace as a whole, and its rapid growth is indeed beginning to generate a new geography within the wider structure of our societal and economic grounds (Batty, 1993).

As well as providing an efficient service for the global and local markets, information technologies (IT) have invariably merged to the existing framework of our society, forming almost a ubiquitous network environment of information spaces (Ludwig, 1996). They are increasingly becoming crucial to our urban life, making it vital to interpret their spatial and geographical features to maintain and improve the quality of their services.

Various attempts at measuring and interpreting the structure, size and connectivity of such spaces have been made but their growth and evolution generate a constant need for new measurements and interpretations (Albert *et al.*, 1999; Janelle and Hodge, 2000). In fact, the vast extent of information spaces and their rapid rate of expansion are making it difficult to perform a comprehensive review on the various information spaces from the geographical viewpoint. The fact that these spaces do not exactly follow Euclidean metric but are largely dependent on the topology makes it even more difficult to apply quantitative analysis. It would be of particular benefit to provide a way to treat these spaces as an objective of spatial modelling and to seek a method of quantitatively representing different types of information space so as to provide a coherent structure for measuring all the spaces under a single framework.

This chapter aims to provide three things. First, it identifies different types

B. Boots et al. (eds.), Modelling Geographical Systems, 275–293.

of information spaces and those emerging alongside the IT deployment. It then extracts some examples from different types of information spaces, which not only illustrate the distinctive geography within them but also attempt to provide an insight on the coherence between each information space. Finally, it explores the possibilities of measuring information space as a measurable entity and the ways to model the dynamically evolving structure and contents of information spaces.

2. GEOGRAPHICAL COMPONENTS OF INFORMATION SPACE

Increasingly, we hear the use of terms combining information and space such as IP address *space* and virtual *worlds*, Web *space* or, more generally, *cyberspace*. Their notion of space is somewhat ambiguous in that some of the information spaces have practically no entity in the real space, and whether they should be addressed as geographical places within which people can live part of their lives remains controversial.

Some argue that it contains nothing more than an archive of digital information; others see an electronically invented Heaven or the Utopia in cyberspace (Wertheim, 1999). The legend of 'the death of distance' also persists, suggesting that the efficiency of cyberspace has virtually overcome all the problems of distant communication and interaction by providing a near instantaneous access to the relevant information (Cairncross, 2001).

Most studies, however, assume that information space has a firm ground of spatiality, if perhaps different from the conventional interpretation of geography, and they demonstrate this in various ways including the visualisation of its structure (Dodge and Kitchin, 2000), quantitative analysis (Abraham, 1996; Bray, 1996; Pirolli *et al.*, 1998; Huberman and Adamic, 1999) as well as the establishment of cyberspace geography (Batty, 1993, 1997; Murinon and Healey, 1998; Sheppard *et al.*, 1999). Yet most of them take only a subset of one particular type of information space as the object of their study, as the typology and the extent of information spaces are so diverse and volatile. These spaces comprise elements that are interdependent with those of other spaces but nonetheless can be defined and extracted as part of a single type of space. Assuming that each information space is discrete from one another, we can provide a definition and quantitative measures to help categorise them.

Figure 1 illustrates a suggested typology of information spaces with respect to their spatial attributes. It is by no means exhaustive or conclusive but nonetheless provides a picture of the range found in the current information spaces. The boundaries between each group are flexible and may be adjusted as the spaces evolve or when a new type of space emerges.

At the most physical end of the spatial sequence is the conventional geography of the real world that gives the basis for spatial reference to all other spaces. At the same time, this world of physical places is reoriented by the development of the new information spaces, forming a feedback loop (Batty, 1997). The second layer is the physical aspect of IT space comprising the infrastruc-

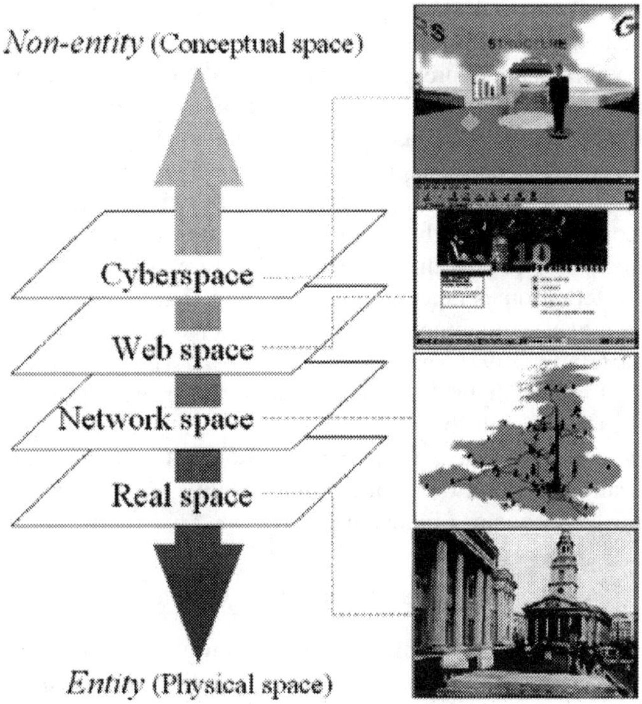

Figure 1. Layers of different types of information spaces.

ture of the Internet, the fibre backbones and satellite networks, and the distribution of servers and clients represented by IP addresses. The fact that the components of such networks are embedded within the real world binds it to the conventional geography to some degree (Malecki and Gorman, 2001), but traffic flows observed amongst them follow a unique spatial order of its own, thus forming a unique telecommunication space (Cairncross, 2001; Goodchild, 2001).

The third tier consists of metaphorical space such as multimedia contents and hyperlinks of the Web. Their existence depends on the physical network identified above, but the structure of these Web spaces is in fact determined by their topological frameworks, thus having much less relevance to the real geography. With regards to the recent transitions in societal and economic activities, the newly emerging economy of e-commerce, for instance, there has been a persistent debate on if the rise of information space has overcome the physical distance. In fact, both arguments can be valid, provided that they are looking at different types of information space, i.e. the network space and the Web space, which are strongly connected to each other, thus making it difficult to distinguish the different elements.

Finally, a variety of 3D cyber cities and other 3D-like virtual environments are located on the top layer, representing the most conceptual of the invented environments. Compared to the previous three groups, they are still small in number and to assume the popularity of a fully immersive 3D environment may sound far fetched at present. In fact, the standard 3D technology is yet to be established and, currently, they range from a flat map representation to a 3D cyber city with the full navigation feature; making it somewhat controversial to form a category of its own. However, their presence is being rapidly enhanced by their wide applications including urban simulation and e-commerce that are becoming increasingly important (Batty *et al.*, 2001).

Some papers also review the geographical nature of information space from the human-interaction perspectives, suggesting that concepts of space may depend on the way we interact with the computer resources and their network environment (Batty, 1997; Sheppard *et al.*, 1999). These are primarily represented in the forms of hybrid and intermediate spaces which inherit the spatial characteristics from both the real world and cyberspace but are by and large an extension to the real space, where the ubiquitous information structure realises an electronically supported world of enhanced reality. These spaces are explained in depth in a later chapter (Takeyama, 2001).

3. Examples of Different Cyberspaces

While each of the spaces identified above follows its unique spatial order, they have a common feature in that the distribution pattern of their entities has a scaling tendency which will be discussed in later sections. Also, with some extension, we may infer a distance metrics within each space. These points will be illustrated in the following by elaborating on a portion of each space or one of its numerous sub-spaces that comprises it. Examples will be drawn from each of the three top layers, namely the network space, Web space, and cyber cities.

3.1. Network Space: The Internet and Fibre-Optics

The physical structure and components of computer networks essentially comprises three types of elements. First, there are a number of 'nodes' or the computer terminals that are placed in front of the users – these can be further classified into servers and clients. Then comes the 'edges' or the hardware that links these nodes together; e.g. fibre backbones, satellite networks and telephone wires. Finally, the network becomes visible and active to the users when 'traffic' or the information exchange between these nodes occurs.

Shiode and Dodge (1999) investigated the geographical distribution of the registered address, within the United Kingdom, for IP address, a unique numerical sequence allocated to each computer linked to the Internet. Figures 2(a) and 2(b) respectively show the population density and the IP address block density of the greater London area. The difference between the two maps is evident in that the IP address blocks are registered only at a limited number of places

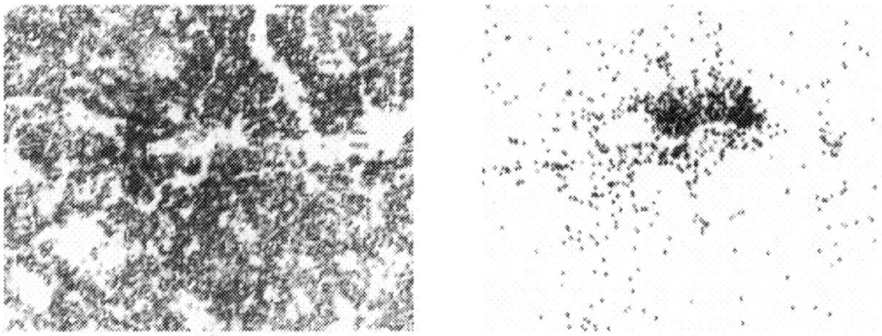

(a) (b)

Figure 2. (a) Population density of the Greater London, and (b) IP address registry distribution within the same area.

as contrast with the wide distribution of population, and the focal points of the two maps are also clearly different.

In fact, as Sheppard *et al.* (1999) and Malecki and Gorman (2001) suggest, there is IT networks are continuously converging, and they are still based on the real geography; nevertheless, as seen in Figure 2, they possess their own momentum and geography which separate these spaces from the real world. Figure 3 shows the rank-size plot for the IP address blocks registered at each postcode zone within the United Kingdom. The flat sections show that more than several locations have the same capacity of IP address blocks registered within its area, suggesting that its geography is uniquely different.

3.2. Metaphorical Space: Topological Space of WWW

The Web is increasingly becoming popular since its dissemination in the early 1990s. Although the volatile and rapid change of its contents makes it difficult to grasp the structure of the Web, several novel attempts have been made to capture and measure this metaphorical space and its transitions.

For instance, a research group at University of Notre Dame has predicted that two randomly chosen documents on the Web are 19 clicks away from each other on average (Albert *et al.*, 1999). Also, Huberman and Adamic (1999) observed that there exists a power-law between the distribution of the number of Web sites and the size of their contents. Their investigations suggest that the Web is dominated by a relatively small collection of large sites.

Murnion and Healey (1998) showed that there might be some correlation between the geographical location of users and that of online contents in that a certain degree of distance decay effect was to be found, especially in relation to the extent of cultural zones. Generally speaking, however, nodes, edges and the information exchanged at this level all exists within the electronic space, unlike those in the physical network which are indeed strongly connected to the geography of the real world (Goodchild, 2001). In other words, the geographic location of servers is less significant, and the spatial order is dependent

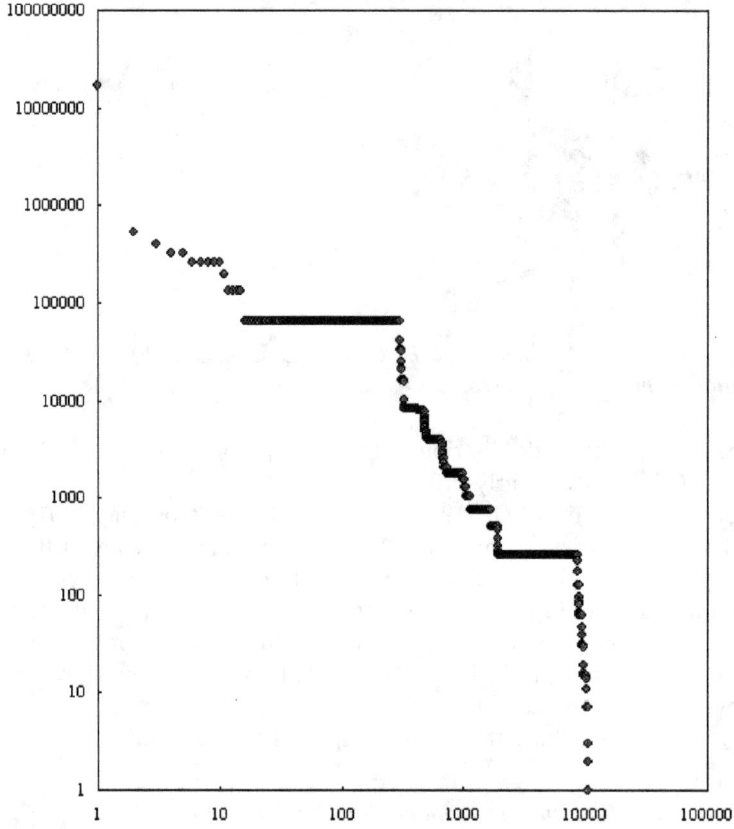

Figure 3. *Rank-size plot for IP address blocks within the United Kingdom.*

on arbitrary hyperlinks. In this sense, the Web space is a topological world, virtually free from any constraints on distance or direction in the conventional terms. The most decisive factors are the number of hops and the overall accessibility to the relevant information.

Thus, it is critical to know the most connected, focal point, and to measure the connectivity of the other points from that centre. Shiode and Dodge (2000) measured the connectivity of the global domains and visualised the Web space in terms of the number of links found amongst the domains. The same data set can be used for interpreting the Web in the context of real geography. Assuming that the size of each domain reflects the amount of Web resource within its corresponding country, we can overlay the connectivity graph on top of a world map (Figure 4). The difficulty lies in that super-national level domains such as *.com* and *.org* require careful estimation of the extent of their contribution by the US firms and those based in other countries. In Figure 4, these domains are treated as external factors where *.com* is placed in the Atlantic Ocean, while *.edu, .net*, and *.org* are positioned on the Antarctic continent. It

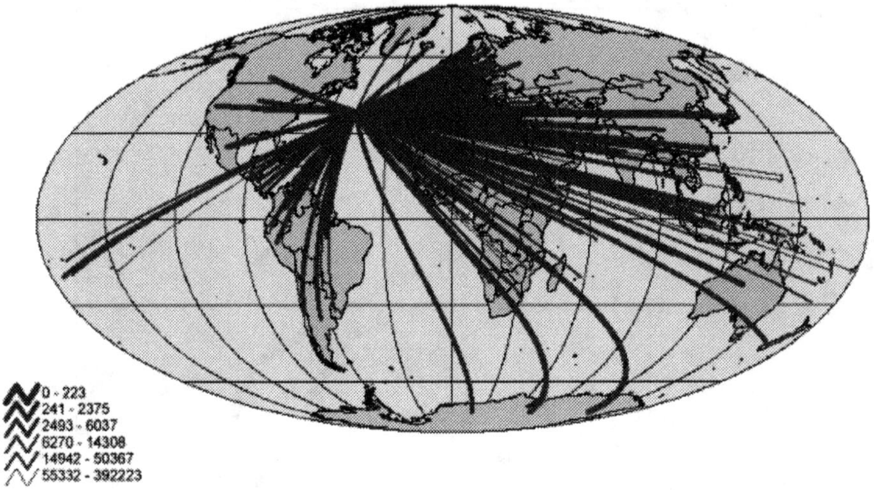

Figure 4. An atlas overlaid with the connectivity of the global domains illustrating the virtual distance from .com *site placed in the centre of the Atlantic Ocean (Shiode and Dodge, 2000).*

is clear from the map that the majority of the links are found amongst the more developed countries (Shiode and Dodge, 2000).

Take a step further and we can compare the statistical distribution of the Web with those of demographic and economic data. This is measured respectively by the number of Web sites, the number of hyperlinks associated with each domain (i.e. incoming, outgoing and intra-domain links), real GDP, and national population of each nation. Figure 5 shows the distribution of the rank size of the four indices on logarithmic scale in descending order where the value of each nation-states unit is reflected in the vertical direction and its frequency of occurrence on the horizontal axis. As the overall rank-size patterns of the Web, hyperlinks and GDP are similar, it is perhaps reasonable to conclude that the distribution of Web domains and their links broadly reflects existing economic activity patterns, despite the difference in the distribution pattern of population and Web services. Shiode and Batty (2000) verify this by measuring the correlation between each distribution.

3.3. Virtual Space: Cyber Cities and Cyber Places

The third and final category focuses on cyber cities and places. Despite its young history and lack of the standard technology, this group is increasing in its number as well as its significance. The 3D technologies involved are still under development, and the rules properties found in these spaces are vary to a great extent. However, these are starting to converge around a couple of leading technologies and protocols such as Web3D (including VRML and Java3D). In one sense,

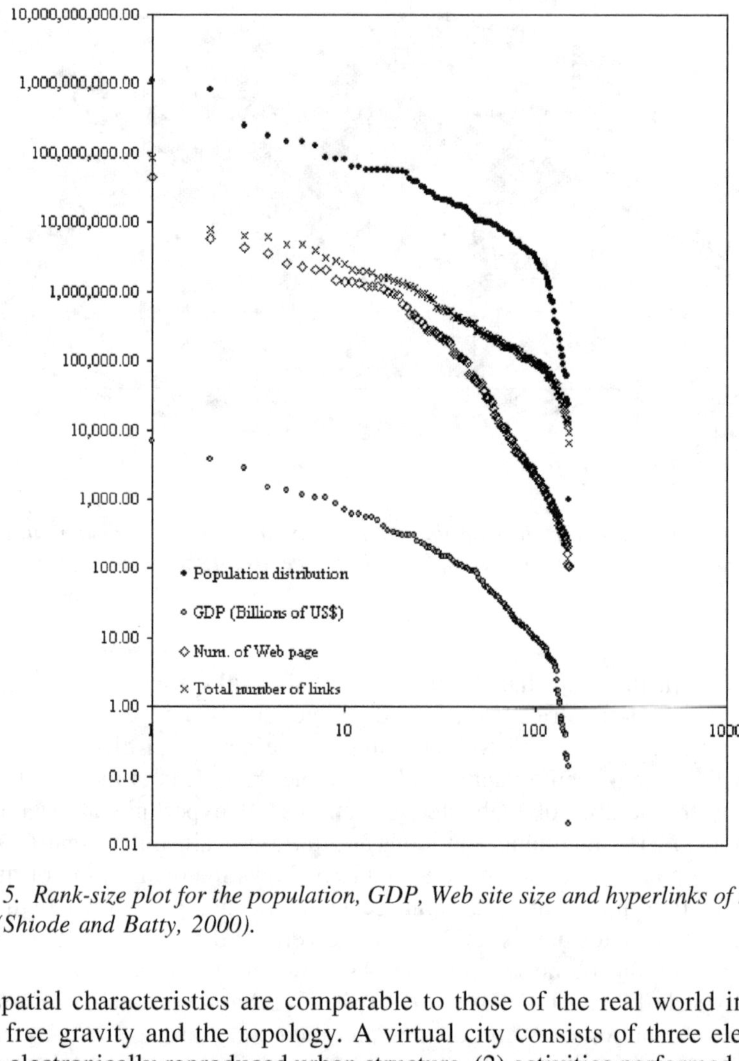

Figure 5. Rank-size plot for the population, GDP, Web site size and hyperlinks of nation-states (Shiode and Batty, 2000).

their spatial characteristics are comparable to those of the real world in spite of the free gravity and the topology. A virtual city consists of three elements (1) the electronically reproduced urban structure, (2) activities performed within these cities, and (3) participants who perform activities. Typically, it is a space where, on the surface, holds the normal Euclidean metrics. The two factors distinctively different from the real world are (1) the flexibility of the gravity that allows users to float and navigate the city, and (2) hyperlinks that provide instant access to the remotest locations.

Unlike the real cities, cyber cities are electronically constructed and maintained within a server, and users can enter them via computer networks. It seems then to be virtually free from financial constraints, apart from the running cost of the server. However, a new price system for the land value of 'cyber estates' is emerging; and they are dependent on their topological position in relation to the major portal sites. As accessibility has become valuable commodities in the information space, a site or a location with heavy network traffic such as the entrance gate to that 3D world is considered to be valuable which is exposed to a larger number of potential customers.

Once a new cyber city emerges, it develops at a much faster rate than a real city would grow, continuously sprawling outwards from the centre. Shiode (1998) applied fractal analysis to measure the growth rate of AlphaWorld. It may not be based on the most representative of the technologies currently available but nonetheless provides a good example of a popular pseudo-3D environment, as it was arguably the largest 3D multi-user world of the time. The study confirms that the city follows a fractal structure and has developed remarkably within a 14-month period; in fact, in terms of the flat area coverage rate, it is now as dense as the central London area measured by Batty and Longley (1994) (Figure 6). Rank-size plot (Figure 7) also confirms that it is comparable to a real world city in terms of its distribution of the block size in the central area, thus implying that a cyber city bears an appearance similar to those of the real environment.

4. GEOGRAPHICAL MODELLING

Categorisation of information spaces proposed in the previous sections suggests that we can regard cyberspace as a collection of different types of informa-

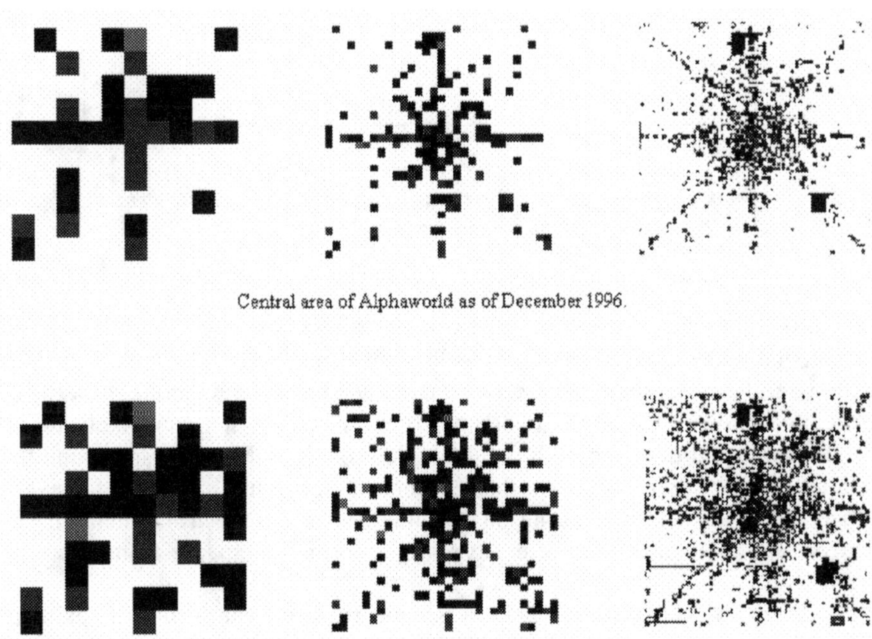

Figure 6. Visualisation of the central area of AlphaWorld at three different grid scales with pictures from December 1996 and February 1998 (Data source: ActiveWorlds).

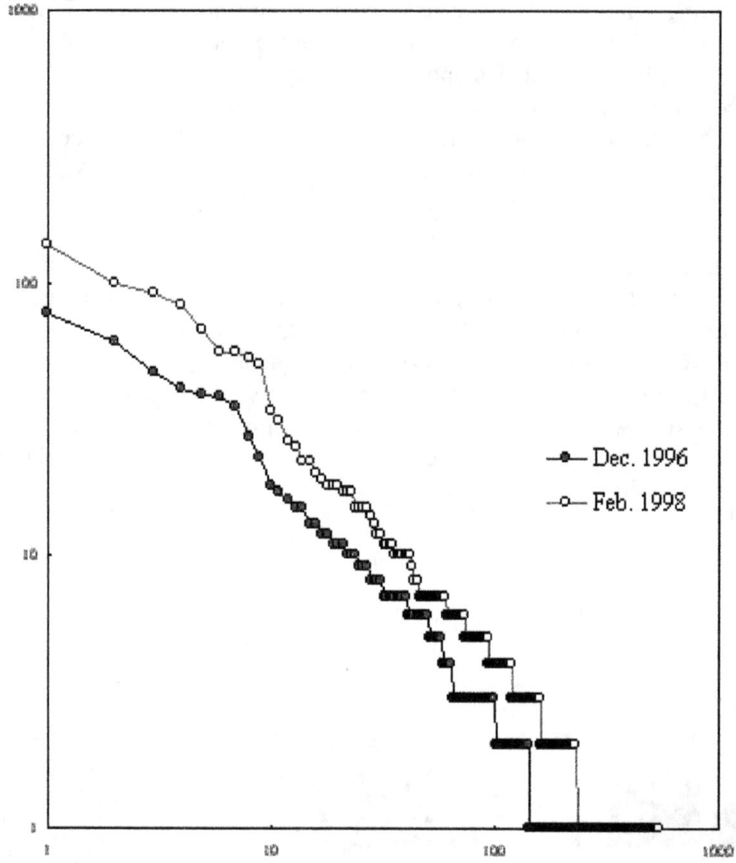

Figure 7. *Rank-size plot for the distribution of block size in the central area of AlphaWorld in December 1996 and February 1998.*

tion. Modelling such variety of spaces requires a generic framework to accommodate the wide range of spaces as well as the capacity to tolerate any changes within and between the layers, which take place once a new mode of network structure or hyperlink-like network emerges. In reality, it is not likely that a single model can describe both the dynamic growth within each space and the relationship between them. As the temporal data are not available for every space type, we will concentrate on building a framework with which we can analyse and simulate the different types of information spaces from one time point.

As far as the visual representation is concerned, we can utilise the existing GIS software packages by projecting everything onto a single layer and separating the elements from different network structures with the attribute labelling. When combined with a conventional spatial interaction model, this would suffice the demand for visualising the linkage between the different layers of information spaces. The main obstacle is to transform the various types of information space into a single layer of visible, Euclidian space.

In terms of analysing information spaces, there are already a vast amount of statistical resources as well as numerous theoretical contributions to interpreting the Internet in general and the Web in particular, and amongst the many studies conducted thus far, four approaches have been taken. These are mostly limited to investigating a single cyberspace type, but provide useful insights even when addressing the entire information space. We will summarise them by way of setting the context to our model.

4.1. Relevant Existing Studies

4.1.1. Statistical Approaches

The most obvious method for grasping the overall impression of the Internet is to collect its statistical information. A number of institutes and individuals have already attempted to capture the state of the Web through a survey on the number of various Web sites, active servers and the user groups (Bray, 1996; Coffman and Odlyzko, 1998; ISC, 2001; MIDS, 2001). However, due to the exponential growth rate of the Internet and its increasingly complex structure, most of these figures inevitably are estimated values, or the rough indicators of its scale obtained typically through a sampling method.

4.1.2. Visual Representation

Most services provided by the Internet consist of electronically generated contents and have no physical entity, but various cartographic techniques are being applied to visualise this virtual domain from a variety of perspectives. Some focus on the pattern displayed by search queries (Carriere and Kazman, 1999), while others depict the topological connectivity of hyperlinks (Shiode and Dodge 2000). Visualisation, if properly applied, can provide persuasive, intuitively comprehensible results (Dodge and Kitchin, 2000). However, they are usually confined within a single information-space type, which restricts the possibility of further exploration of the contents.

4.1.3. Data Mining

In contrast to the statistical approach, data mining methods are primarily focused on a single local spot or on a particular point of interest and carries out an in-depth analysis, possibly combined with some visualisation, to comprehend the exact impacts and effects at lower levels. Examples include local traffic distance measurement (Murnion and Healey, 1999) and IP address distribution at the district level of a country (Shiode and Dodge, 1999). Whilst these methods can be applied to a local or specific aspect of the Web, it is practically impossible to maintain such level of detail if the entire information space needs to be searched as we invariably wish.

4.1.4. Model-Based Approaches

The final approach is aimed at understanding the Internet by constructing a model of its structure. There is already an extensive collection of studies that tries to apply the existing models to the newly emerging information spaces.

For instance, there is a stream of studies on the expansion of spatial interaction model which aims to provide the accessibility or the latency measure within the changing geography of information space (Janelle and Hodge, 2000). These include the attempt to interpret the urban landscape in terms of the degrees of association or spatial clustering associated with certain aspects of information flows (Ord and Getis, 1995), as well as a more straightforward measurement of accessibility and the geography of opportunity (Shen, 1998). They are particularly useful when estimating the accessibility and opportunity within the real geographic space but, with some minor adjustment, can be also applied to the interpretation of multiple layers of cyberspaces.

Connectivity and the topological structure of the Web are other aspects of cyberspace that have been studied rigorously (Abraham, 1996; Kleinberg, 1997; Wheeler and O'Kelly, 1999). Amongst these studies is the application of the concept of social network which reflects the 'small world' assumption (Watts and Strogatz, 1998). The underlying idea is that for a variety of global network phenomena, all objects or people are connected to one another within a chain of six acquaintances, which is popularly known as the 'six degrees of separation'. As aforementioned, Albert *et al.* (1999) have applied this concept to measure the degree of connectivity of the Web, predicting that Web pages are separated by an average of '19 clicks'.

This connectivity measurement is closely linked to the idea of power laws describing networks where 'the probability of finding documents with a large number of links is significant, as the network connectivity is dominated by highly connected Web pages' (Albert *et al.*, 1999). It also confirms the scaling pattern observed in the examples drawn in the earlier section. The idea that each information space follows the power law to a certain degree gives a useful insight when overlaying the spaces, and we will expand on this aspect as we conduct our analysis.

4.2. Proposing the Model

It is clear from the above discussion that, although each method has much use in a particular context, few of them are applicable to simultaneously addressing different types of cyberspaces. This owes to the diversity of information space which, on one hand, requires a quantitative model that produces a pattern shown in each space, and a method of transformation on the other hand, that will bind the different spaces together in one layer. Based on the idea that each information space has a fractal-like and scaling pattern of spatial distribution, we will formulate a simple cellular-automaton-type (CA) model that generates a scaling distribution. As we do not know the growth rate and spatial characteristics of each space, the model will not produce any results that can be verified. But we will nonetheless expand on an existing model and aim to extract implications from the application of such model in other incidents.

4.2.1. Building a CA Model

In terms of modelling social activities, a number of morphological models have been developed to this date, many of which have simple structure yet addressing complex physical or social phenomena such as the emergence of galaxy, urban growth, and fractal and chaotic process (Fujita *et al.*, 1999). In fact, the rapid growth of information space is not dissimilar from that of the universe, despite the fact that the former is an artificially generated virtual environment. For this reason, we shall turn to astrophysical models. Initially, Groth and Peebles (1977) proposed that the distribution of galaxies could be described by a power-law correlation function

$$\xi(r) = \left(\frac{r}{r_0}\right)^{-1.8} \tag{1}$$

where r_0 is the correlation length of the two-point correlation function ξ (Groth and Peebles, 1977). This means that the fluctuation in the galaxy density has a fractal dimension of $D = 3 - 1.8 \approx 1.2$ – i.e. the degree of complexity remains statistically constant regardless of scale (Vicsek and Szalay, 1987). It was then suggested that the correlation function of the distribution of galaxy clusters also follows a power law of the same slope, but its amplitude was systematically changing with the richness or the number of galaxies per cluster. And because this change in amplitude scales in a way that the mean distance of objects and r_0 have a constant ratio, there is no way to tell a map of galaxy from the cluster representation of an urban configuration as long as their fractal dimension coincides with each other.

Taking this interpretation further, we could model the morphological structure of virtually any phenomena by selecting the right fractal dimension and cluster correlation. In the case of information spaces, we may spatially and temporally simulate the dynamic growth regardless of their actual scale or rate of growth and evolution, once the growth rate of each space is found. The difficulty lies in that cyberspace consists of multiple layers of information spaces each of which bearing different fractal dimension thereby producing different scaling curves as we will see later.

There is still no 3D-VR equivalent of such simulation model, but we can utilise a simple cellular-automaton type model for this purpose. In particular, we adopt the model proposed by Vicsek and Szalay (1987) which is shown as

$$a(i, j, k+1) = \frac{1}{5}\left[\sum_{l=-1}^{1} a(i+l, j, k) + \sum_{l=-1}^{1} a(i, j+l, k) - a(i, j, k)\right] + R(i, j, k) \quad (2)$$

where i, j ($i, j \subset N$) correspond to the 2D grid and k to the total elapsed time, respectively. Also, $a(i, j, k)$ is a number attributed to the site (i, j, k) and $R(i, j, k)$ is a random number which, for the sake of simplicity takes on the values 1, –1 with equal probability, corresponding to white noise. Equation (2) is a generalised form of the cellular-automata in that $a(i, j, k)$ can take any real number. In Vicsek and Szalay model (1987), R represents the initial random density fluctuations in the Universe whereas in our cyberspace simulation, this will be substituted with the social attractiveness of each space. As its distribution is not known, we shall leave it as a random function of white noise, but it can be fitted with any distribution later. We will also assume that site (i, j, k) will form a space if $a(i, j, k) > A$, where A is a constant threshold.

By definition, it is clear that Equation (2) 'correspond to a three-dimensional growth process which advances layer by layer, and the value of the site function $a(i, j, k)$ in the next layer is determined by the average over the five nearest neighbours in the previous layer and the value of the random number $R(i, j, k)$' (Vicsek and Szalay, 1987).

We cannot simulate the growth of cyberspace as a whole, for we need to measure the growth rate at each level of cyberspace. However, this model should be fundamentally applicable to each level of space, and we can digest it in the context of multiple information spaces by adding a suffix which indicates the level of space, provided that a distance metric can be inferred for each space. As we saw in the examples, the method of defining a distance metric within the space may vary according to the type of cyberspace. Also, as seen earlier, the scaling trend-line from two different periods shows a constant growth of cyber cities. If we could obtain temporal data for other information spaces and verify their consistency, then this model can be used to simulate the entire cyberspace.

Assuming that we will be able to overcome these obstacles in time, we proceed to build a model that will generate a simulated distribution of cyberspace which is a slightly generalised form of Equation (2):

$$a(\chi | \chi_j + 1) = c\left[-a(\chi) + \sum_{\substack{i,j \subset N \\ i \neq j}} \sum_{l=-1}^{1} a(\chi | \chi_j + 1)\right] + f(\chi) \quad (3)$$

Here, i and j are represent the different types of cyberspace, f is a residual term corresponding to a random walk, and c is a constant. The apparent problem here is that it is impossible to make one layer contagious to another unless we know the growth rate of each space and their relative positions. As such we will, in the following simulation, assume that $i = j = 1$ and carry out the experiment for a single cyberspace type.

The implications obtained from Vicsek's model are as follows. If the fractal dimension remains statistically constant, the spatial and temporal explosion simulated by the model will continuously maintain its frequency and will produce a staggering growth of power series. In other words, the above model predicts that, unlike some other social phenomena that reaches mutual equilibrium – mainly due to capacity limit – the information spaces are likely to maintain its dynamic growth rate unless a new constraint external to the system of cyberspace is introduced.

In practice, however, there are certain elements that may act as a constraint on the growth of cyberspace. These include the capacity load within networks and the maximum digit allowed by the Internet protocol of the time which will be continuously improved as the demand grows higher. To what extent these factors affect the growth of cyberspace in long-term is yet to be measured, but within a short period, there is few physical constraints that would prevent information spaces from growing.

4.2.2. *Simulating the Distribution of Information Spaces*

The sample data exhibited in Figures 3, 5 and 7 consistently indicated a slope $-q$ between 1.4 and 1.6. Using these values as the threshold A for generating the spaces, we obtain, after several iterations, a rank-size plot for the distribution of our simulated information spaces (Figure 8).

Technically speaking, the results can be visualised in a multi-dimensional grid space. An example of such projection is shown in Figure 9. However, the resulting image is quite dissimilar from the map shown in Figure 6, which has the same frequency of occurrence for the cells but may be affected by some other forces that affect its formation. Once this underlying force for cyberspace formation is revealed, we can simulate a full scale simulation of the growth and distribution pattern of information spaces.

Apparently, we need to extract the precise measurement of the scaling slope of each cyberspace and also measure their growth rate before making any decisive prediction – we will also need to take account of the fact that the number of different types of cyberspace may increase in time, and that the entire system would become even more complex; at which point the spatial variable would further increase.

5. CONCLUSION

This paper proposed a preliminary framework for modelling different types of information spaces. The spaces discussed here are essentially offsprings of the rapidly growing information and communications network, namely the physical

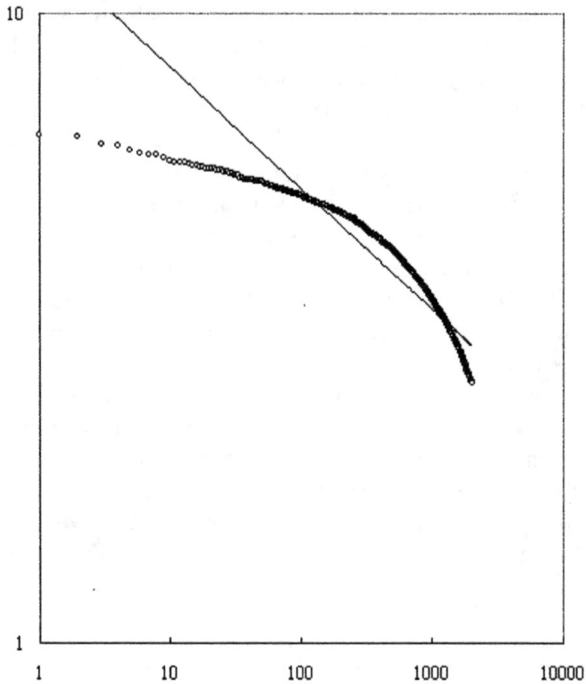

Figure 8. Rank-size plot for the distribution of simulated cyberspace.

infrastructure of the computer networks, topological structure comprising the information space, and pseudo-3D worlds of cyber cities. We took a subset of each space and observed their scaling tendency. Based on an astrophysical model, we then developed a cellular-automaton type model that generates the scaling distribution which confirms the pattern observed in the examples. The patterns obtained through the simulation, although not yet verified, shows how the distributions found in a cyberspace can be projected onto a 2D grid space. The results are still to be evaluated by comparing them against the existing spaces; in fact, the proposition of the models forms only the first avenue of a long quest for an effective model of information space which we hope to develop in due course.

At the moment, all information spaces are continuously developing at a rapid speed, becoming more and more complex as they absorb more resources and information. It is an uneven, asymmetric process of multi-polarisation rather than decentralisation, and more complication than simplification. It is conceivable that, in the very near future, a large part of urban population would start using the Internet in their urban life, consciously or not. In order to retrieve information and to express themselves, each user would need to depend more on its information services – the Web, newsgroups, online chat rooms and video conferencing and whatever virtual environments and services that will emerge from there.

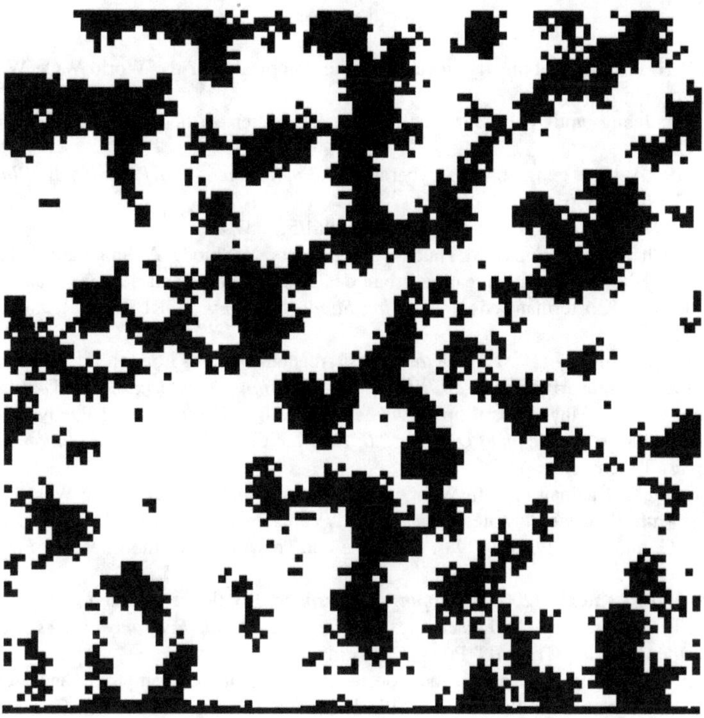

Figure 9. Projection of the distribution of a simulated Web-like space on to a grid cell of 100 by 100.

When the current generation of information spaces first emerged, they did not largely substitute for the conventional service of the entity, but complemented it and generated new types of demands. Similarly, when these spaces evolve and perhaps yield to the formation of new generation of cyberspaces, they would not completely replace the existing services, but are more likely to be added to their predecessors and provide different mode of services. We may end up with a variety of modes and spaces allotted for different activities.

In fact, if we abide by the present principle of *laissez-faire*, the Internet and its various information spaces would eventually form a massive labyrinth of miscellaneous data to which we would be forced to adjust ourselves in return for the information dependent urban lives. It is therefore vital to analyse and model information spaces to help interpreting their extent and growth which may contribute to preventing further complication of information spaces and building more efficient IT services.

Centre for Advanced Spatial Analysis,
University College London,
1–19 Torrington Place,
London WC1E 6BT, UK

REFERENCES

Abraham, R.H. (1996), Webometry: measuring the complexity of the World Wide Web, unpublished working paper.

Albert, R., H. Jeong, and A.-L. Barabasi (1999), 'Diameter of the World-Wide Web.' *Nature*, **401**, 130.

Batty, M. (1993), 'The geography of cyberspace.' *Environment and Planning B: Planning and Design*, **20**(6), 615–616.

Batty, M. (1997), 'Virtual geography.' *Futures*, **29**(4/5), 337–352.

Batty, M., D. Chapman, S. Evans, M. Haklay, S. Kueppers, N. Shiode, A. Smith and T. Paul (2001), 'Visualizing the city: communicating urban design to planners and decision makers.' In R.K. Brail and R.E. Klosterman (eds.), *Planning Support Systems*. ESRI Press, Redlands, CA, pp. 405–443.

Batty, M. and P. Longley (1994), *Fractal Cities*. Academic Press, London.

Bray, T. (1996), Measuring the Web, *Proceedings for the 5th International World Wide Web Conference*, 6–10 May, Paris (http://www5conf.inria.fr/fich_html/papers/P9/Overview.html).

Cairncross, F. (2001), *The Death of Distance 2.0: How the Communications Revolution Will Change Our Lives*. Texere, New York.

Carriere, J. and R. Kazman (1999), WebQuery: searching and visualizing the Web through connectivity (http://www.cgl.uwaterloo.ca/Projects/Vanish/webquery-1.html).

Coffman, K.G. and A. Odlyzko (1998), 'The size and growth rate of the Internet.' *First Monday*, **3**(10).

Dodge, M. and R. Kitchin (2000), *Mapping Cyberspace*. Routledge, London.

Fujita. M., P. Krugman and A.J. Venables (1999), *The Spatial Economy: Cities, Regions and International Trades*. The MIT Press, Cambridge, Massachusetts.

Goodchild, M.F. (2001), 'Towards a location theory of distributed computing and e-commerce.' In T.R.Leinbach and S.D.Brunn (eds.), *Worlds of E-Commerce: Economic, Geographical and Social Dimensions*. John Wiley & Sons, Chichester, pp. 67–86.

Groth E.J. and P.J.E. Peebles (1977), 'Statistical analysis of catalogs of extragalactic objects.' *The Astrophysical Journal*, **217**, 385–405.

Huberman, B.A. and L.A. Adamic (1999), 'Growth dynamics of the World-Wide Web.' *Nature*, **406**, 450–457.

Internet Software Consortium,(2001), Internet domain survey (http://www.isc.org/ds/).

Janelle, D.G. and D.C. Hodge (2000), *Information, Place, and Cyberspace: Issues in Accessibility*. Springer-Verlag, Berlin.

Kleinberg, J.L. (1997), 'Authoritative sources in a hyperlinked environment.' *IBM Research Report*, RJ10076.

Ludwig, G.S. (1996), Virtual reality: a new world for geographic exploration (http://www.utexas.edu/depts/grg/eworks/wie/ludwig/earthwor.html).

Malecki, E.J. and S.P. Gorman (2001), 'Maybe the death of distance, but not the end of geography: the Internet as a network.' in T.R. Leinbach and S.D. Brunn (eds.), *Worlds of E-Commerce: Economic, Geographical and Social Dimensions*. John Wiley & Sons, Chichester, pp. 87–108.

Matrix Information and Directory Services (1999), 'State of the Internet.' *Matrix Map Quarterly* **603** (http://www.mids.org/mmq/603/).

Murnion, S. and R.G. Healey (1998), 'Modelling distance decay effects in Web server information flows.' *Geographical Analysis*, **30**(4), 285–303.

Ord, J.K. and A. Getis (1995), 'Location spatial autocorrelation statistics: distributional issues and an application.' *Geographical Analysis*, **27**, 287–306.

Pirolli, P., J.E. Pitkow and R. Rao (1998), 'Silk from a sow's ear: extracting usable structures from the Web.' *Proceedings for Conference on Human Factors in Computing Systems* (http://www.acm.org/sigchi/chi96/proceedings/papers/Pirolli_2/pp2.html).

Shen, Q. (1998), 'Spatial technologies, accessibility, and the social construction of urban space.' *Computers, Environment and Urban Systems*, **22**, 447–464.

Sheppard, E., H. Couclelis, S. Graham, J.W. Harrington, and H. Onsrud (1999), 'Geographies of the information society.' *International Journal of Geographical Information Science*, **13**(8), 797–823.

Shiode, N. (1998), 'Modelling the structure of cyberspace as a fractal network.' *Proceedings of International Conference on Modelling Geographical and Environmental Systems with GIS*. Hong Kong, pp. 640–644.

Shiode, N. and M. Batty (2000), 'Power law distributions in real and virtual worlds.' *CASA Working Paper Series*, No. 19 (available from http://www.casa.ucl.ac.uk/powerlaw.pdf).

Shiode, N. and M. Dodge (1999), 'Visualising the spatial pattern of Internet address space in the United Kingdom.' In B.M. Gittings (ed.), *Innovations in GIS 6: Integrating Information Infrastructure with Geographical Information Technology*. Taylor & Francis, London, pp. 105–118.

Shiode, N. and M. Dodge (2000), 'Spatial analysis on the connectivity of information space,.' *Theory and Applications of GIS*, **8**(2), 17–24.

Takeyama, M. (2000), Geographical conceptualizations of smart places (this volume).

UCAID (2000), The Internet2 Project (http://www.internet2.edu/).

Vicsek T. and A.S. Szalay (1987), 'Fractal distribution of galaxies modeled by a cellular-automaton type stochastic process.' *Physical Review Letters*, **58**(26), 2818–2821.

Watts, D.J. and S.H. Strogatz (1998), 'Collective dynamics of "small-world" networks.' *Nature*, **393**, 440–442.

Wertheim, M. (1999), *The Pearly Gates of Cyberspace: A History of Space from Dante to the Internet*. W. W. Norton, New York.

Wheeler, D.C. and M.E. O'Kelly (1999), 'Network topology and city accessibility of the commercial Internet.' *Professional Geographer*, **51**(3), 327–339.

ROBERT G.V. BAKER

ON MODELLING INTERNET TRANSACTIONS AS A TIME-DEPENDENT RANDOM WALK: AN APPLICATION OF THE RETAIL AGGREGATE SPACE-TIME TRIP (RASTT) MODEL

1. INTRODUCTION

The mathematical description of the Internet is a new challenge facing applied modellers. There are now new spatial and temporal accessibilities to consider and new concepts emerging, such as, 'e-tailing', where commercial transactions can take place globally and almost instantaneously. This freedom of access into the Internet for consumers means issues of physical location, travel time or market area may be less relevant and the research frontier has to deal with such things as 'virtual distance' and unrestricted shopping opportunities between countries. There even appears to be some sort of time substitution for spatial interaction (particularly from time-poor affluent households). A key theoretical question is whether cyberspace is a product of what Marx described as 'time annihilating space'.

The Internet forms the physical network of connectivity (such as, optical cables and phone wires), where there are nodes or 'routers' that navigate packets of data from one computer to another (Barabasi, 2001). The Internet is therefore spatially specific in that flows occur through physical space. Conversely, in the World Wide Web (WWW) or the Web, links can be easily established arbitrarily as virtual connections between any two computers independent of spatial co-ordinates. The idea of the WWW originated from a hypertext technique where a considerable amount of multimedia is interconnected (Jiang and Ormeling, 2000). It is more content-specific and its properties are analysed by 'maps' that tell how the pages are linked together.

Both the Internet and the Web can be regarded as a network of nodes and links forming a complex graph defining what is known as 'cyberspace'. Cyberspace is a computer generated landscape which integrates these networks into a virtual space An important question therefore emerges: how does this new landscape affect the flows of information and the importance of distance? What is increasingly becoming apparent is that we have to try to understand the relationship between cyberspace and geographic space and how to develop models that recognise their distinctiveness.

Much of the current Internet research involves the application of graph theory

B. Boots et al. (eds.), Modelling Geographical Systems, 295–316.

to the study of the Internet and the Web (for example, Barabasi and Albert, 1999; Albert *et al.*, 2000; Cohen *et al.*, 2000). It is interesting that this connectivity and its theoretical descriptions are expressed in terms of time and that there is little recognition of the spatial domain. For example, The Internet Traffic Report (2001) uses a time-based index describing the round trip travel time of major paths on the Internet (also termed 'latency'). The distance factor is replaced by how much time it takes to transfer data. Further, the so-called maps plot connectivity and are essentially aspatial. Within a geographical context this is not satisfactory, because the flows of time-dependent Internet traffic around the world are passing through countries and time zones relative to a 24-hour boundary (Figure 1). This is in contrast to an aspatial view of Web traffic and connectivity using graph theory (Figure 2). The aim of this review is to look at the Internet as a geographical system in space and time and endeavour to set a modelling context for future research.

A second question concerns how retailing fits into a model of the Internet. This is not the aim of this review. However, despite the considerable euphoria and a stock market boom in the late 1990s in technology stock and 'dot com' companies, there are very few retail success stories from marketing on the Internet. Even such an e-tail 'success' story as Amazon.com recorded a \$390 million loss in 1999. Why is this the case? There is the possibility that the structure of central places (particularly in terms of cities as points of distribution) are different fundamentally to the evolution of connectivity within the Internet and the Web. Much of the efficiency of cyberspace is seen in time minimisation, but there still could be a place for distance minimisation strategies (and the gravity model) for the distribution of goods and services. The failure to

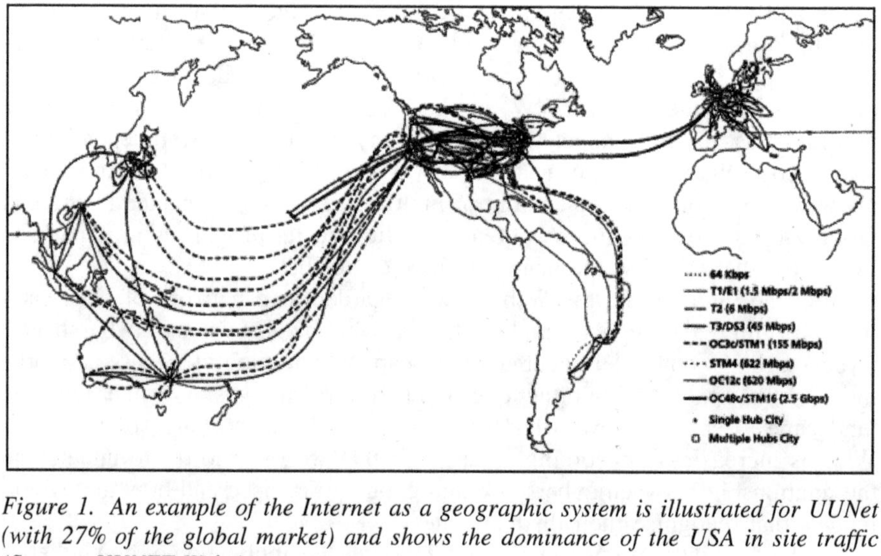

Figure 1. An example of the Internet as a geographic system is illustrated for UUNet (with 27% of the global market) and shows the dominance of the USA in site traffic (Source: UUNET Website, www.uu.net).

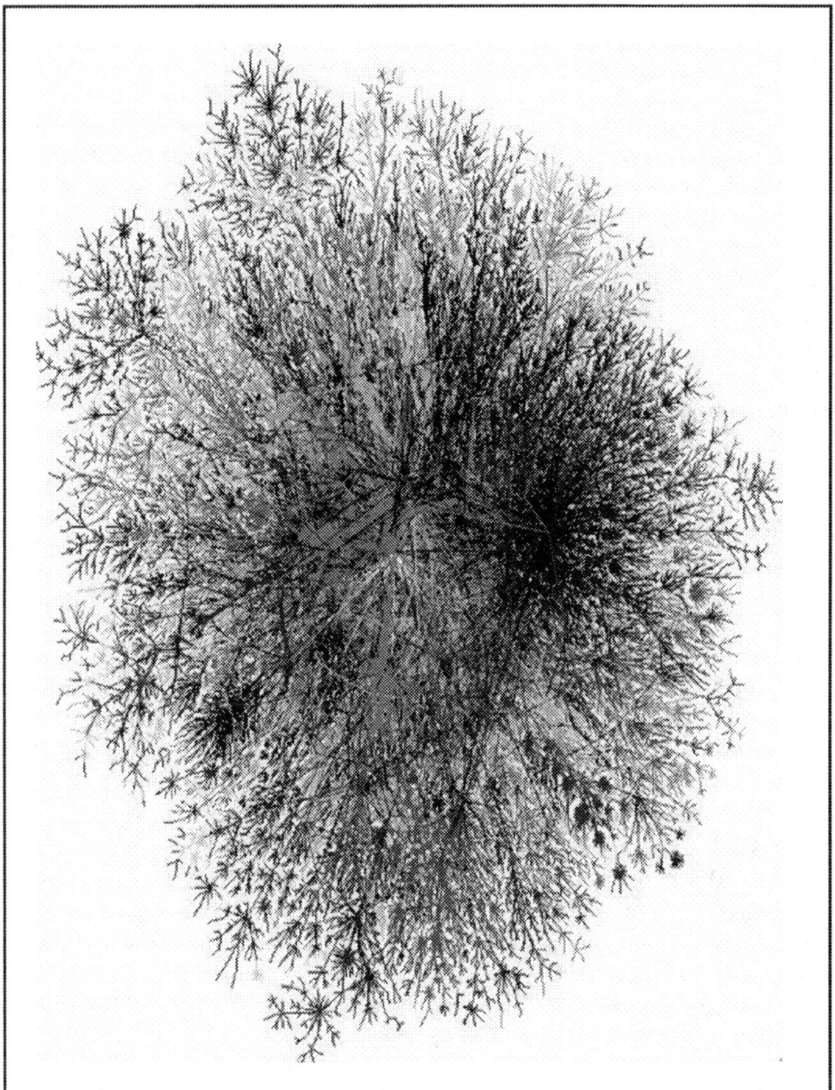

Figure 2. The generation of an Internet Web tree showing the aspatial connectivity from 100,000 Internet routers and the hierarchial structures that develop from a few highly connected nodes (Source: Cheswick, 1999).

understand this difference and its geographical underpinnings, could be a major barrier to successful marketing and profitability for Internet retailing. This suggests that there is much research needed to understand the operation of the Internet as a geographical system and as a way of receiving and distributing commercial and retail transactions.

2. MODELLING BACKGROUND

2.1. *Introduction*

The application of random graph theory to define the connectivity of the Internet and the Web is a growing research area at present in the physical sciences. This work will be briefly reviewed. An alternative is to view random processes of connectivity along a time line through differential equations. One such application is the retail aggregate space time trip (RASTT) model (for example, Baker, 1994, 2000). It has been applied extensively to study trips to and from point densities (shopping malls) along a time line, where its time-dependent solutions relative to a periodic boundary, suggests some fruitful insights into how the Internet can be modelled. Its underpinnings will also be summarised to set the context for its application to cyberspace.

2.2. *Graph Network Models of the Internet*

The analysis of complex networks can be divided into two major classes based on their connectivity distribution $P(m)$ which defines the probability that a node in the networks is connected to m other nodes (Albert *et al.*, 2000).

(1) The first type of networks is characterised by a $P(m)$ that peaks at an average $\langle m \rangle$ and decays exponentially for large m. These networks are homogeneous in that each node has approximately the same number of links. Exponential networks (such as the random graph model of Erdos and Renyi, 1960) have a connectivity that follows a Poisson distribution peaked at $\langle m \rangle$ which decays for $m \gg \langle m \rangle$.

(2) The second type belongs to inhomogeneous networks (or 'scale-free' networks) where $P(m)$ decays as a power law (or $P(m) \sim m^{-\gamma}$) free of the characteristics of scale. This network has a majority of nodes with only one or two links, but a few large nodes of links guaranteeing that the system is fully connected. An example of this type of network is the World Wide Web and this type of model can be visualised by the Internet tree simulated by Chiswell (1999) (Figure 2).

Barabasi (2001) uses geographical examples to distinguish both types of networks and these are pertinent to the development of an Internet model. An exponential network is a road map that has cities as nodes and expressways as links, because most cities are central places located at the intersection of the motorways. Conversely, an airline route map is a Type 2 network, because although most airports are served by a small number of carriers, they have a few hubs (such as, London) from which links emerge to almost all other US or European airports. The WWW is seen as an example of the latter because a majority of documents have only a few links. It appears that Type 2 networks are also hierarchical. They are also preferential, since they contain nodes that have a high probability of being connected to another node with a large number of links. For example, a new Web page is more likely to be linked to the most popular documents on the Web, since these pages are the ones we know about. Research by Faloutsos *et al.* (1999) have shown that the network behind the Internet also appears to follow the power-law distribution of inhomogeneous networks. This means that the physical wiring of the Internet is also dominated by several highly connected hubs. As Barabasi (2001) states: why do systems as different as the Internet, which is a physical network, and the Web, which is virtual network, develop scale-free networks with a power-law decline in connectivity?

This analysis is distinctly aspatial, but is still imbedded in time-dependent variables for the transfer of information. How should the Internet and WWW be viewed as a geographical system where both space and time are fundamental to interaction? The RASTT model may provide some insights into the framing of this question.

2.3. *The Retail Aggregate Space Time Trip (RASTT) Model*

The RASTT model defines 'when' and 'where' consumers enact aggregate shopping behaviour and is underpinned by the so-called 'supermarket' equation (Baker, 1994, 2000). It is constructed around a differential equation of spatial and temporal operators acting on a population function ϕ (which is standardised to ϕ_o, the equivalent calibration function per hundred shoppers divided by the size of the aggregation unit). A mathematical operator describes what has to be done on a function much like a verb does within a sentence. The 'supermarket' equation is different to classical diffusion, because time is differentiated twice and space once. This second order time operator is of immediate interest to theoretical geography, because when this is made equivalent to a first order spatial operator, the solution involves a gravity model of trip distributions in space and a periodic function of time-based demand. This relevant differential equation takes the usual form of:

$$\frac{\partial \phi_o}{\partial x} = \frac{1}{M} \frac{\partial^2 \phi_o}{\partial t^2} \tag{1}$$

where x defines the spatial coordinate, t, the trip time and M a transport constant for a calibrated population density of ϕ_o of shoppers. This linear equation (with the transport coefficient M constant) can only apply to one shopping centre, but the operators ($\delta/\delta x$ and $\delta^2/\delta t^2$) can apply equally for individual or group shopping. Equation (1) is stating there is a trade-off between trip operators through space and time. In other words, 'where' a consumer shops is dependent on the shopping cycle (time of the day or day of the week). In the particular solution, the gravity model of trip distance D is the spatial solution between residences (aggregated in concentric one kilometre bands) and the shopping centre (or $x - x_o$, where $x_o = 0$ defines the location of the shopping centre). The underlying behavioural assumption is distance minimisation in trip assignment. The corollary in the time solution of Equation (1) is that such shoppers make regular time-based trips to and from a shopping centre at $x_o = 0$. A population density of shoppers ϕ_o in this model therefore assumes populations regularly select trips that minimise distance (through exponential decay) to and from a shopping centre. A socio-economic group that best approximates this assumption are 'over 65 years', whilst for trip purpose, the weekly food and grocery trip fits this model well (Baker, 1994, 1996). Mathematically, this statement can be expressed as a particular solution of Equation (1) for one centre as:

$$\phi_o = A\ exp(-\beta D) \left\{ \begin{array}{c} sin(kt) \\ cos(kt) \end{array} \right\} \tag{2}$$

where β is the gravity coefficient and k, the interlocational trip frequency (ITF) defines how many trips are made by individuals or groups to the shopping centre. The ITF is introduced arbitrarily as the separation constant to solve the differential equation. This solution is stating that the undertaking of regular or periodic trips (*sin kx*) to a shopping centre is discounted by how far away we are to a centre (*exp* $-\beta D$), since there are increasing opportunities to shop elsewhere the further they live from this centre.

Implicit in this approach is that destinations are located along a time line of shopping opportunities. The advantage of this method is that time boundaries can be introduced as part of solving the differential equation and this will affect spatial patterns of the gravity model of trips to and from the centre. The policy corollary is that the shopping hours a mall trades will affect the extent of the surrounding market area (Baker, 2000). The RASTT model therefore deals with shopping trip distributions from a particular centre where time boundaries on a time line of destinations can change the spatial distributions. Other methods have specific problems dealing with time. For example, entropy maximising can also derive gravity spatial interaction by optimising the assignment of trip origins to destinations through statistical laws of large numbers (Wilson, 1967; Roy and Lesse, 1981). However, this method has problems dealing with time in maximising the assignment problem to and from a shopping centre. The increase in entropy only occurs between opening and closing times: when the shopping centre is shut, the complexity of the trip assignment problem disentangles and trips approach zero (contrary to the physical analogue which approach infinity). The RASTT model does not suffer from this difficulty because the solution is defined by the hours that the centre is open (0 to T) and all other possibilities are zero.

3. THE RASTT MODEL AND INTERNET TRANSACTIONS

The operators of the RASTT model ($\delta/\delta x$ and $\delta^2/\delta t^2$) are not affected by the numbers involved in interaction (applying equally to individuals or populations of billions) and are therefore classified as scale invariant. These operators are defined relative to time boundaries for movement through physical space. Yet are these mathematical operators applicable to the Internet, where there is still a real time boundary (the 24 hour rotation of the Earth) defining the movement of transactions? Consumers can also make virtual rather than real trips to retail sites and the RASTT model therefore offers the scope to explore the movement of demand through virtual space as well as physical space.

The immediate problem in the RASTT model is that relative time functions to the boundary can be either positive or negative. In previous work, the idea of negative time in the context of the process of shopping trips was thought initially to be meaningless and the time boundaries were only applied from 0 to T (and the 0 to $-T$ range discarded; see Baker, 2000). However, the idea of negative values relative to the direction from the boundary for Internet transactions is not as nonsensical as it first appears. Rather than framing the problem over 168 hours per week, we set it for trips or transactions through space over a 24-hour period (the daily cycle). The spatial origin could be located at a computer at an arbitrary location and the consumer can either go forward or backward along a time line relative to this 24-hour boundary. For example, if the individual is located at Sydney (33° S Lat and 161° E Long), that person can either go two hours forward in time to a site in Auckland (37° S Lat and 175° E Long) or two hours backwards in time to Perth (32° S Lat and 116° E Long). The RASTT model can be derived for physical trips to a mall and such trips are only viewed positively along time lines. Conversely, virtual trips on the WWW can be defined as either moving backwards or forwards relative to the 24-hour time boundary. This is a radical statement because it gives a plausible example of how relative time can exist as a corollary of virtual distance and have different properties to physical time. Boulding (1985) states that in the physical sciences, time is assumed to approach infinity in order to focus on spatially specific solutions. Alternatively here, we assume initial spatial locations and produce time-specific solutions, including solutions that can be negative. The study of the Internet as a geographical system therefore provides an opportunity to introduce a new concept and to see if it has any further properties of interest.

There is a possibility of a convergence of virtual distance into a fixed point (the computer screen) at any time. An important question is: can relative time influence the patterns of virtual space? It returns to a concept of a dynamic convergence of locations found in the geographical literature of the late 1960s, where the evolution of spatial reorganisation changes in space-time connectivity, particularly from improvements in transportation and technology (Janelle, 1968, 1969; Forer, 1978; Gatrell, 1983). The Internet is perhaps the next stage in the evolution of this space-time connectivity. Blaut (1961) argued that every empirical concept of space must be reducible by a chain of definitions to a process and Janelle (1969) states that inherent in Blaut's view is the implicit existence of a temporal pattern in each and every spatial pattern. In the RASTT model, this process is summarised as a second order time differential (or operator) that can yield positive and negative time-based solutions. This means that unlike physical time, relative time can lead to reversible time-based processes, a truly remarkable possibility. For example, it means that in the election of a US president, polling booths can be closed in the east, yet the proportion of votes counted and reported on TV can feedback simultaneously to voters on the west coast who are still voting (and can change their votes based on the east coast trends). Reversibility of a result is possible within the boundaries of relative time. It is possible to have two simultaneous sites connected by virtual distance on a computer screen in different time zones. The Internet presents a new horizon to geographical systems because we have to now distinguish between relative time (to a rotating boundary) and physical time (to infinity) and real trips (where people change their spatial co-ordinates) and virtual trips on the Internet (where the location is still at the initial spatial co-ordinate). As this stage we can only make guesses as to how this evolution in space-time connectivity develops, but there are some clues already discernible from the nature of the RASTT model. What features could be expected from a RASTT model representation of Internet transaction? There are two areas of immediate interest.

3.1. The Condition for Space-Time Convergence

The condition for space-time convergence in the solution of the supermarket equation (Equation 1) when a 24-hour boundary is applied, yields the same relationship between the gravity coefficient β and the square of the mean interlocational trip frequency k divided by the transfer constant M, namely:

$$\beta = \frac{k^2}{M}. \tag{3}$$

The interlocational trip frequency (ITF) defines the average number of trips or transactions undertaken per day by users and because it is squared it can be applied to virtual trips either forwards or backwards through relative time. The RASTT model suggests that there would still be gravity interaction of physical distance for Internet patronage, but this would be at least one order of magnitude lower than gravity coefficients computed from shopping trips to malls using concentric aggregation. Yet we would expect that it would vary for the type of transaction. For example, weekly food orders would have (with k higher) greater β values, indicating the distribution of food would be more localised than for a lower frequency consumption item such as compact disks. We would therefore expect that one feature of the Internet as a geographical system would be 'very weak' gravity interactions, but this would still be relative to the type of transaction and the limits of the distribution system.

If β is assumed to be very small and the frequency of patronage the same order of magnitude as shopping trips to malls (one visit to a site per day), the transfer coefficient M for Internet usage would have to be very large compared to its value for physical trips to a mall. This is not hard to visualise, with Internet traffic at least one order of magnitude higher than the physical trips to particular shopping malls. The other key question is whether M remains a constant, or a variable, making the differential equation non-linear.

3.2. Space-Time Distributions of Internet Demand

The type of space-time distributions that could apply to Internet patronage are simulated in Figure 3 for $\beta = 0.0001$, $T = 24$ hours, $x_0 = 0$ to $x_0 = 10{,}000$ km and an arbitrary population density $\phi_o = 10$ for a sequence of k values where $k = 0.1, 0.2, 0.3, 0.4, 0.5 \ldots 1.0$. The simplest distribution of spatial demand for Internet patronage at a site, receiving both positive and negative flows of transactions, is a gaussian-type distribution between $k = 0.1$ and 0.2 (Figure 3). This is not surprising since a gaussian distribution is an equally valid solution to Equation (1) for a time-based random walk problem. The solution has some advantages in this probabilistic form, because variables can be expressed as average quantities, such as, 'distance' and 'number destinations per visit'. A gaussian distribution can be expressed as probability distribution $P(t, x)$ for a density of web transactions at a site $\phi_o = \phi_{os}$. If this site receives n_a transactions per unit distance d, with total transactions $\Phi = n_a\, d$, the probability distribution is defined as:

$$\frac{\phi_o}{\phi_{os}} = P(t,x) = \frac{1}{2(\pi Mx)^{1/2}}\, exp(-t^2/4Mx) \tag{4}$$

where t is equal to the time for each transaction to travel to the site. The transfer coefficient M can be defined alternatively as:

$$M = \frac{1}{2}nt^2. \tag{5}$$

The transfer constant is then the number of transactions per unit distance multiplied by the relative time t taken to reach the site. Equation (4) is the type of distribution that has been simulated in the $k = 0.1$ to $k = 0.2$ range in Figure 3. It is an unbounded gaussian time distribution, where transaction densitics can be plotted for ln ϕ versus t^2 and the slope of the straight line is $(4Mx)^{-1}$. The average time taken by the transaction is defined by the mean square displacement $(\overline{\Delta} t^2)$, namely:

$$\overline{\Delta} t^2 = 2Mx. \tag{6}$$

The RASTT model can therefore define the possibility of a number of distinctive features of Internet patronage relative to traditional spatial interaction modelling. First, the gravity model of spatial interaction would have very small β coefficients compared to a regional shopping mall;. Second, technology allows for a space-time convergence to occur on the computer screen rather than

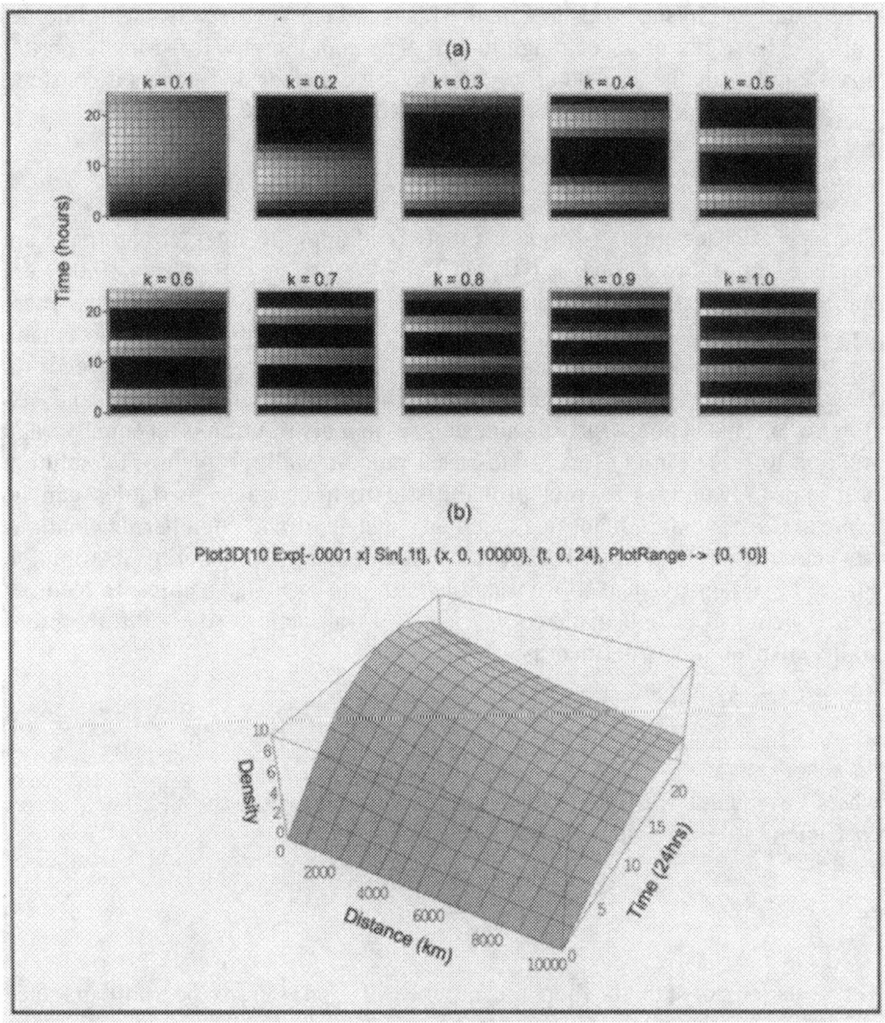

Figure 3. (a) A range of possible space-time distributions that could apply to Internet demand are simulated for $\beta = 0.0001$, *$T = 24$ hours,* $x_0 = 0$ *to* $x_0 = 10{,}000$ *km and a scaled* $\phi_{o\ max} = 10$ *for a sequence of k values where* $k = 0.1, 0.2, 0.3, 0.4, 0.5 \ldots 1.0$. *(b) A three dimensional plot visualising a likely form of the demand wave for* $k = 0.1$.

shopping malls and virtual distance allows for the possibility of simultaneous connections both forwards or backwards in relative time. Third, such connections can have implications for activities in different time zones, such as the US presidential elections or stock market activity. Last, transactions to sites should be represented in their simplest form as time-based gaussian distributions.

This type of model (and differential equation) is not found in traditional applications of applied mathematics because of the problem of dealing with positive and negative time-based functions. In the case of the Internet, such difficulty is an advantage because transaction flows can be modelled globally relative to a time boundary. It means that time has to be viewed differently at this particular scale (defined by the rotation) and has different properties to physical time (such as reversibility).

The next step is to look more formally at the derivation of the 'supermarket' equation in this Internet context for transaction interaction between a number of web sites.

4. DERIVING THE RASTT MODEL FOR INTERNET TRANSACTIONS (AFTER GHEZ, 1988)

Consider a network of web sites linked by a time line with an arbitrary origin at W, where these sites are designated through integers $i = 0, \pm 1, \pm 2, \pm 3, \ldots$. For example, a household at Sydney could have a choice of other sites at $i = \pm 1$ at Auckland or Perth (Figure 4). Each web site serves a number of households at a particular locality and there are Φ_i households linked to each site i. Assume that each of these households can jump to adjacent web sites with a frequency Γ that does not depend on the characteristics of i. These households can access sites forward in time or backwards in time. It is assumed the movement forwards or backwards are equally likely. Therefore, movement from site i to site $i + 1$ per unit distance occurs at a rate of $1/2\ \Gamma\ \Phi_i$. Likewise a household or web page at site $i + 1$ can reply to the household or web page at i at a rate of $1/2\ \Gamma\ \Phi_{i+1}$. The resulting rate of exchange is:

$$E_{i+1/2} = \frac{1}{2}\Gamma(\Phi_i - \Phi_{i+1}) \tag{7}$$

and for $i - 1$ into the site i, the flux is

$$E_{i-1/2} = \frac{1}{2}\Gamma(\Phi_{i-1} - \Phi_i). \tag{8}$$

The change in web traffic into and out of the ith site at an origin or hub (such as Sydney) is given by a definition of all possible transitions:

$$\frac{d\Phi_i}{dx} = -\frac{1}{2}\Gamma\ \Phi_i + \frac{1}{2}\Gamma\ \Phi_{i+1} - \frac{1}{2}\Gamma\ \Phi_i + \frac{1}{2}\Gamma\ \Phi_{i-1}. \tag{9}$$

The space discounting equation (or rate equation) in terms of the distribution of users in and out of the ith site is (by collecting terms)

$$\frac{d\Phi_i}{dx} = -\frac{1}{2}\Gamma\ (\Phi_{i+1} + \Phi_{i-1} - 2\Phi_i). \tag{10}$$

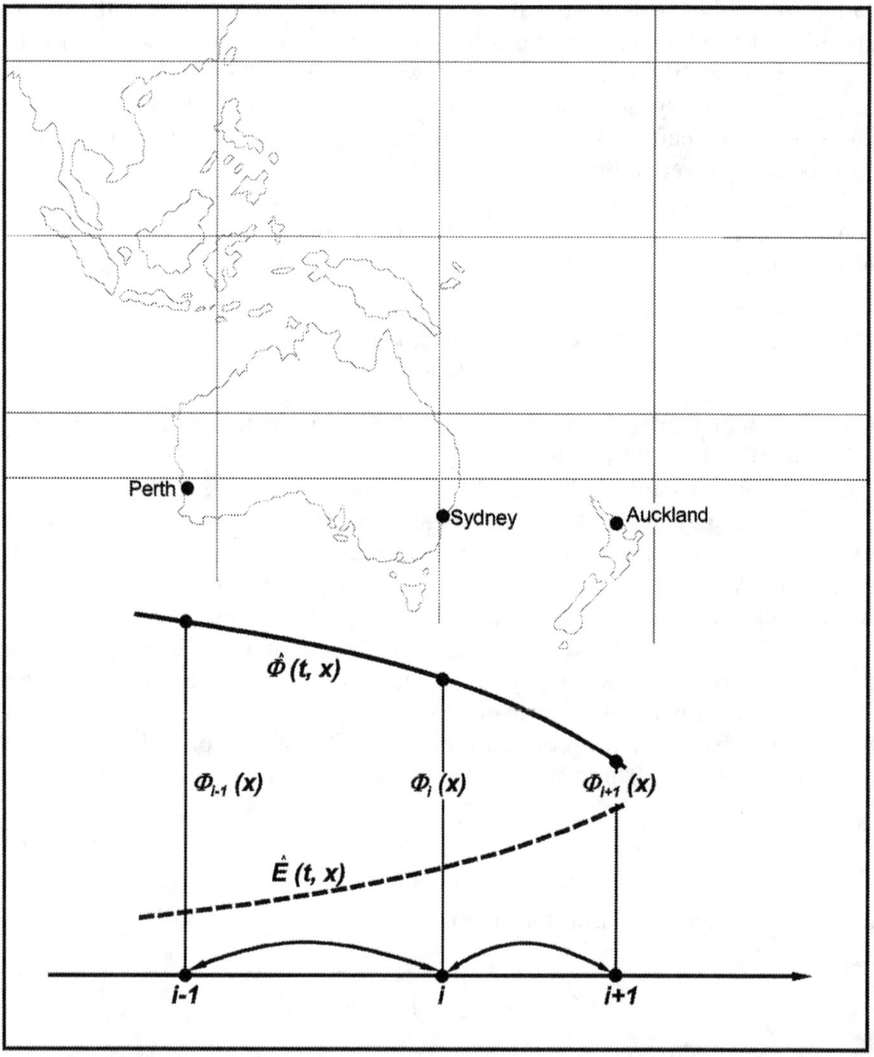

Figure 4. *The equal likelihood of jumping forwards in time to sites in Auckland or backwards to Perth from the ith Sydney site defines the underpinnings of the type of differential equations in Equations (18) to (20).*

This exchange rate of web or Internet traffic is between nearest-neighbour destinations on the time line around site i and this can be expressed in terms of the exchange rate between sites (using Equation 1 and 2)

$$\frac{d\Phi_i}{dx} = -(E_{i+1/2} - E_{i-1/2}). \tag{11}$$

The change in the web site content is defined by the difference between flows in and flows out of transactions within the connectivity.

4.1. Comments

(1) The jump frequency of transactions between sites is constant and it is assumed independent of the site index i and its location in space.

The data signal should not change its frequency within the network and does not depend on the location of the computers. This appears a reasonable assumption and agrees with the aspatial nature of the graph theory approach.

(2) This frequency of movement does not depend on the distribution of households or users in the neighbourhood of the ith site.

The distribution does not have to be homogeneous. This also appears to be a good approximation and parallels the assumptions of graph theory.

(3) The time distance between sites and the type of transfer network does not influence the process, the only thing that is important is the time-based ordering of the points.

The receipt of transaction does not depend on physical location, but on a time-dependent ordering of site hits. Once again this is a reasonable assumption for the Internet or Web.

Equations (7 to 11) define a time line between sites where the distance between points and the hierarchical network of sites is not relevant, rather, what counts is the ordering of hits to the site. Equation (11) states a conservation law where the transactions in and out defines the content of a web site. The time distance p between web sites is assumed to be equal between the origin (such as Sydney) and the ith site and has the co-ordinate of $t_i = ip$ on the time line of transaction flows. The transaction density $\phi_o(x, t)$ is assumed to interpolate the previous function at site i, with the co-ordinate $i\phi_o(x)$ by the following assumption:

$$\phi_o(t, x) = \Phi_i(x) \tag{12}$$

at destinations located at $t = t_i$. but is arbitrary elsewhere. The assignment of this transaction density function around this web site at t_i can be expanded by a Taylor series:

$$\phi_o(t_{i\pm 1}) = \Phi_o(t_i) \pm p\frac{\partial \phi_o}{\partial t}\bigg|_{t_i} + \frac{1}{2}p^2\frac{\partial^2 \phi_o}{\partial t^2}\bigg|_{t_1} + \text{terms of order } p^3 \tag{13}$$

and using the condition in Equation (12), the expansion becomes when substituted into Equation (10) (noting that we are interested in both forward and backward motion relative to the 24-hour cycle):

$$\frac{\partial \phi_o}{\partial x} = \frac{1}{2}\Gamma p^2 \frac{\partial^2 \phi_o}{\partial t^2} + \text{terms of order } p^4. \tag{14}$$

4.2. *Comment*

The condition for this approximation is that $p \times k \ll 1$, or in other words, the time distance between sites is very much smaller than the smallest significant wavelength and the A amplitude of the demand wave (A sin kt) must be insignificant outside $|k_{max}|$.

With the definition of latency between sites expressed in milliseconds, such a condition appears reasonable for the Internet of Web traffic.

The continuous exchange function between sites with transaction densities can be written as:

$$E(x,t) = -\frac{1}{2}\Gamma p \frac{\partial \phi_o}{\partial t} \tag{15}$$

and the conservation law can be rewritten from Equation (10) as:

$$\frac{\partial \phi_o}{\partial x} = -p\frac{\partial E}{\partial t} + \text{terms of order } p^4. \tag{16}$$

Now, the average transaction density $\dddot{\phi}_o$ is equal to = $\dddot{\phi}_o/p$ where p is equal to the average transaction time between sites and Γ equal to the average number of web sites visited per unit trip distance. The transport constant M to the centre is defined in the previous context as:

$$M = \Gamma p^2. \tag{17}$$

The rate of exchange, the conservation law and supermarket equation for web traffic on the Internet becomes, respectively:

$$E(x,t) = -\frac{1}{M}\frac{\partial \phi_o}{\partial t} \tag{18}$$

$$\frac{\partial \phi_o}{\partial x} = -\frac{\partial E}{\partial t}. \tag{19}$$

$$\frac{\partial \phi_o}{\partial x} = \frac{1}{M}\frac{\partial^2 \phi_o}{\partial t^2}. \tag{20}$$

Equation (18) defines a condition for the conservation of information on the Internet. The minus sign in Equation (18) is a point of debate, because it implies that information flows from sites of high densities to low densities (akin to classical diffusion) with the analogy to graph theory that the high point densities are the hubs in the network. This is a reasonable assumption in the context of inhomogeneous scale-free networks in graph theory with a power-law decline in connectivity.

5. EMPIRICAL EVIDENCE FOR A RASTT MODEL FOR INTERNET TRANSACTIONS

There are a number of hypotheses that are consequence of the application of the RASTT model to Internet transactions. These will be briefly reviewed in the context of recent geographic-specific Internet experiments and data sites (Padmanabhan and Subramanian, 2001; Internet Weather Report, 2001).

5.1. Is Weak Gravity Interaction a Feature of Internet Flows?

The spatial solution to Equation (2) would suggest that gravity spatial interaction is still relevant to Internet traffic flows. Padmanabhan and Subramanian (2001) show an example of the gravity relationship between client and proxy sites for the America On-line network. The cumulative probability for clients follows the gravity distribution of a regional shopping mall except there is a difference in the order of magnitude for distance (Figure 5a and Figure 5b). This strongly suggests that the Internet exhibits very weak gravity interactions.

5.2. Is Pause Time (Δt) a Surrogate Measure for Distance?

In Equation (2), one of the predictions of a time-based gaussian distribution is that the average transaction time (or delay or latency) taken by the transaction is defined by the mean square displacement $\overline{\Delta t^2} = 2Mx$. In other words, one of the conditions of space-time convergence is that the transaction time (to and from a client from a site) is a function of the distance from the site. Padmanabhan and Subramanian (2001) undertook an experiment with a probe machine at Seattle, USA, measuring transaction delay in four categories (5–15 ms; 25–35 ms; 45–55 ms; 65–75 ms) relative to geographic distance (Figure 5c). The results support the prediction of a linear relationship between latency and distance (as suggested by Equation 6). Over 90% of small delay values (under 10 ms) come within 300km from the source. The RASTT model therefore provides a clear mathematical relationship to test these experiments and suggest the existence of time-based gaussians.

5.3. Is the Transport Coefficient M a Measure of Congestion?

Congestion in the network is one of the problems in using time delay as a surrogate selecting minimums in delay samples (10 to 15 samples) between hosts

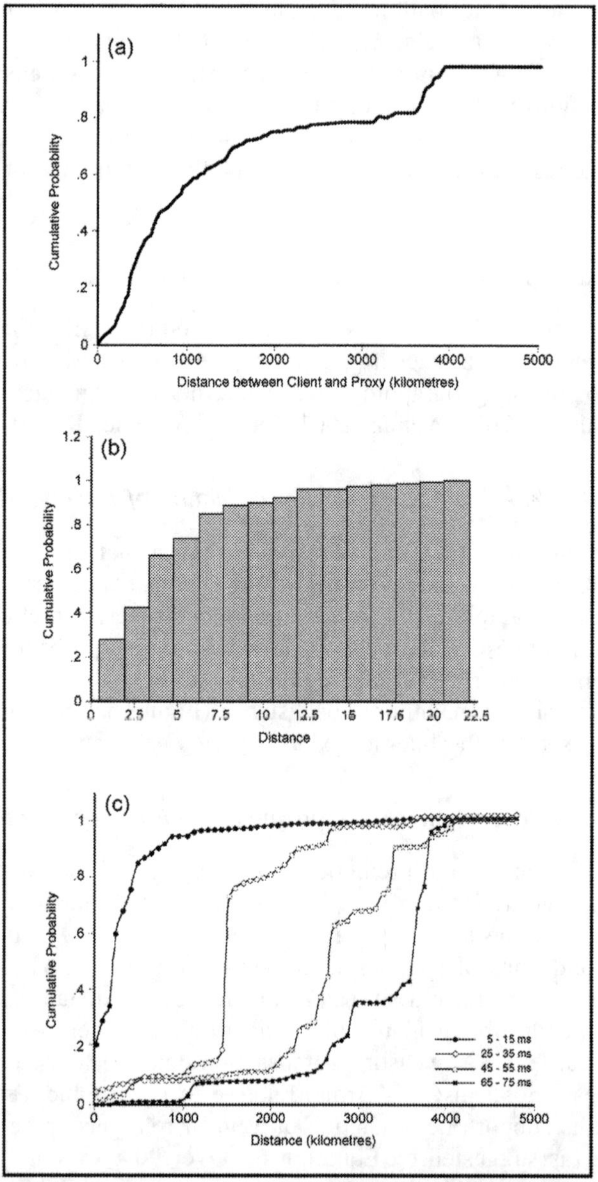

Figure 5. *(a) The cumulative probability for a gravity-type distribution for the distance between client and proxy for America-Online (Source: Padmanabhan and Subramanian, 2001); (b) The cumulative probability for a gravity-type distribution for a regional shopping mall (Bankstown Square, 1998 afternoon distribution; Baker 2000); (c) The results of a probe machine at Seattle, USA, measuring transaction delay in four categories (5–15 ms; 25–35 ms; 45–55 ms 65–75 ms) relative to geographic distance. The results support the prediction of a linear relationship between latency and distance suggested by Equation 7 (Source: Padmanabhan and Subramanian, 2001).*

to eliminate the effect of congestion. Equation (5) from the time-based gaussian suggests that the transfer constant M will be a function of the number of web sites and the transaction time. This transaction time can vary and so M can be a variable and the differential equation non-linear. What we could do is to seek M_{min} from a substantial sample size and set this as a constant in the differential equation and $M - M_{min}$ would then represent a measure of congestion in the system.

5.4. Are Spatial Demand Waves of Internet Traffic Observable?

Equation (2) predicts that there is a spatial demand wave circumnavigating the globe on a 24-hour cycle which, in our simulation (Figure 3), would most probably take the form of a time-based gaussian between $k \sim 0.1$ and $k \sim 0.2$. The Internet Weather Report (2001) provides an animation of this type of event, where variables of latency (the bigger the circle, the slower the return time trip) and the number of hosts at a given location (defined by a colour spectrum) varies over a 24-hour cycle.. The slowness of the latency (the time taken for a return trip between a host and a client) may be a function of demand and if this is the case, then the arrival of the Internet demand wave can be visualised in the USA. For example, on October 1, 2001 at 1.00 am (CDT), there are few congestion nodes and smaller latencies compared to 4.00 pm (CDT) when there are peak demand times for both east and west coast interactions (Figure 6). The latency changes through a 24-hour cycle, where there are a few small guassian-type distributions in the early morning, but as the demand wave envelops, these grow in the peak afternoon period when the network comes under heavy load.

5.5. Is the Nearest Neighbourhood Assumption Justifiable?

In the derivation of a time-based random walk for Internet traffic (Equations 7 to 11), the exchange rate assumes nearest neighbour sites within the jumping between hosts. Is this justifiable? The WWW is a small-world network, which is a sparse network where nodes are connected to other nodes in their neighbourhood but otherwise the average distance between nodes is high (Watts and Strogatz, 1998). Padmanabhan and Subramanian (2001) in their experiments found their geographic clustering algorithm (GeoCluster) produced the best results over varied data sets. This would be expected if time-dependent guassian distributions are generated at different scales (either for countries or individual hosts such as a university campus). This nearest neighbour assumption appears to have some credence, at least in mapping information generated from Web traffic.

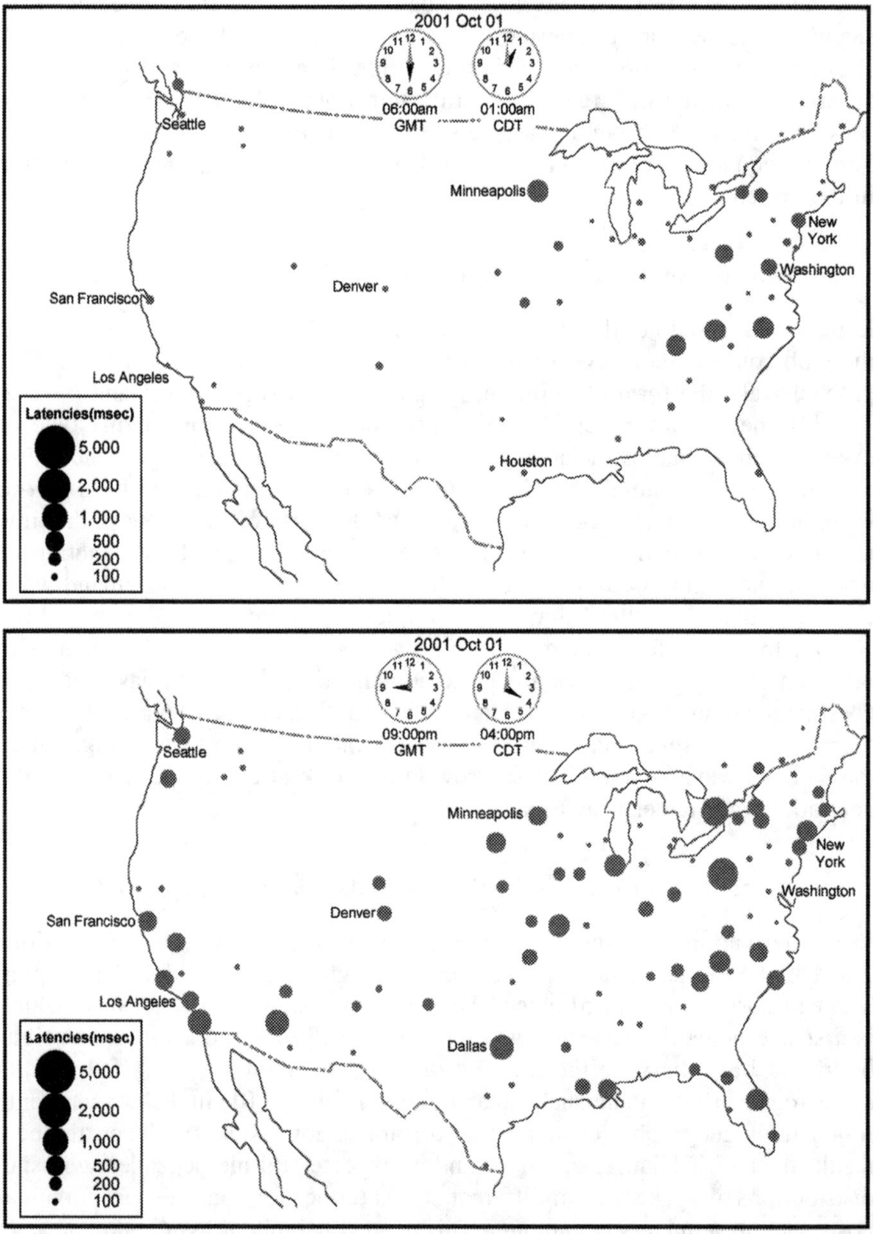

Figure 6. Internet demand wave can be visualised in the USA on October 1, 2001, when at 1.00am (CDT) there are few congestion nodes and smaller latencies compared to 4.00 pm (CDT) when there are peak demand times for both east and west coast interactions (Source: Internet Weather Report; 2001).

6. SOME FURTHER THEORETICAL THOUGHTS

One of the basic assumptions in the time-dependent random walk is that it is equally likely that a transaction can move either forwards or backwards in relative time. This might be a satisfactory assumption for the internal USA situation or the Sydney-Perth-Auckland domain, but for a global model, there is going to be a higher probability of moving in either direction towards the high connectivity of USA sites. The jump frequencies can no longer be assumed to be isotropic (namely, the jump probabilities to the right or left have the same value $1/2\Gamma$). Therefore, the jumps forwards or backwards in time are therefore not equal for Internet sites because of the influence of USA hubs. In the derivation, the exchanges now do not vanish because of the asymmetry in jump frequencies between sites. A constant distribution is no longer a necessary condition for equilibrium, but the conservation law remains, as well as other assumptions in the derivation (such as, nearest neighbour and data ordering; Ghez, 1988). This global Internet equation that can account for the asymmetry in transaction flows can be defined if the jump time distance is now $t_{i\pm 1/2} = (i \pm 1/2)a$ and there is introduced what is termed a drift velocity $v = \Delta a$ into the model. Equations (18–20) can now be rewritten:

$$E(x,t) = -\frac{1}{M}\frac{\partial \phi_o}{\partial t} + v\phi_o \tag{21}$$

$$\frac{\partial \phi_o}{\partial x} = -\frac{\partial E}{\partial t} \tag{22}$$

$$\frac{\partial \phi_o}{\partial x} = \frac{1}{M}\frac{\partial^2 \phi_o}{\partial t^2} - v\frac{\partial \phi_o}{\partial t}. \tag{23}$$

The drift velocity v through the network is proportional to the difference in the jump frequency between sites and is thus a measure in the bias that the USA sites introduce into the system, even if the households are evenly distributed among the sites. This drift velocity could simply be interpreted as the Internet's mean velocity field (which can fluctuate depending on the congestion or distance between sites within the network). Equation (22) represents a new differential equation which describes the Internet, where the first term may describe the hub exchange of transaction flow (the reverse diffusion), whilst the second is the overall movement throughout the network. It now appears to be non-linear and its solutions are beyond the current scope. Nevertheless, it allows a global view of the Internet and data flows.

There is an interesting connection with this Internet equation to the telegrapher's equation:

$$\frac{\alpha}{M}\frac{\partial^2 \phi}{\partial t^2} + \frac{\mu}{M}\frac{\partial \phi}{\partial t} = R\frac{\partial^2 \phi}{\partial x^2} + s\frac{\partial \phi}{\partial x} \tag{24}$$

where α, μ, R and s are appropriate constants. In the early days of telegraphy, the signal diffusion reduced the data rate in long cables such as the early Atlantic cables (Montroll and West, 1979). The wave propagation over time was replaced by a diffusion packet. For the Internet equation, the time operators are the same as the telegrapher's equation (apart from the sign of the first time operator). However, in the Internet equation, there is no second order space operator ($\partial/\partial x^2$) and spatial interaction is exclusively defined by the gravity operator ($\partial/\partial x$). The mathematical relationship between both differential equations is also an area for future inquiry.

7. Concluding Remarks

The application of a RASTT-type model (time-dependent differential equations with a second order time operator) to Internet transactions is explored and the results suggest that this type of model has some relevance to understanding the dynamics of the Internet and WWW as a geographical system. Such a model has a number of implications that warrant further consideration.

Firstly, this model involves the idea of relative time where flows of data can go either forwards or backwards in time relative to the rotation of the Earth. Such second order time operators in the RASTT model are a peculiarity and there is a corollary of the possibility of reversibility within time functions. The example of this process is in the USA elections where voters in California (because of different time zones) can change their vote according to events broadcast after the closure of polling in the eastern states. Relative time, therefore, has different characteristics to physical time where events are irreversible and assumed positively to infinity.

Secondly, in the graph network model of the Internet, the power-law evolution of the structure is not the same as a random-walk generation of linkages, where in the latter, the gaussian distribution (and negative exponential functions) is an integral part of the solutions of the evolution of connectivity. Perhaps this difference is significant, because the evolution of supply points might still follow random-walk generation and the gravity model might limit the physical accessibility of much of the power-law evolution of the WWW or Internet. It might be easy to get connected to the WWW, but to receive goods and services from e-tailers still relies on central places and points of distribution. This might be one reason for the difficulty of e-tailers in making profits from their WWW networks.

There are a number of encouraging results empirically that support predictions from the RASTT model.

1. There is an example of very weak gravity interaction in the relationship between the distance between client and proxy sites for the America On-line network.
2. The prediction from a guassian random walk that pause time is a function

of distance to the site has been replicated in experiments by Padmanabhan and Subramanian (2001).

3. The postulated spatial demand wave can be observed across the USA and globally in the Internet Weather Report (2001).

Such results should provide motivation for further experimental work. The RASTT model provides a mathematical framework to look at the dynamics of the Internet and WWW. The bias from USA sites can be introduced into the model and this form has some similarity to the telegrapher's differential equation, which itself is an interesting area for further inquiry. The idea of a mean velocity field of Internet traffic flows has immediate applicability to geographical information systems and visual representations of the Internet.

The RASTT model has been successful at looking at distance trip behaviour to spatial centres of demand (shopping malls) between 10^0 and 10^1 orders of magnitude. There is also some indication that virtual trips with physical distances at 10^3 and 10^4 orders of magnitude through the Internet can be described by the same operators in the RASTT model. The implication is that these operators are scale invariant and supports the idea that operator-based modelling is a way to overcome problems of scale.

Finally, modelling the Internet is a new frontier for spatial interaction modelling. The equations from the RASTT model have no physical analogy. They are essentially geographic, where time zones and distance decay are fundamental to the dynamics. Operators are the key to this view of space-time processes. They allow 1960s concepts, such as, the space-time convergence at central places, to be equally applicable to the computer screen in the 2000s. Such opportunities occur because of the mathematical tractability of using partial differential equations to describe geographic processes.

ACKNOWLEDGEMENTS

I would like to thank Barry Boots for encouraging this research and to Mike Roach for his cartography.

Division of Geography and Planning
School of Human and Environmental Studies
University of New England, Armidale, N.S.W. 2351
Australia

REFERENCES

Albert, R., J. Hawoong and A. Barabasi (2000), 'Error and attack tolerance of complex networks.' *Nature*, **406**, 378–382.

Baker, R.G.V. (1994), 'An assessment of the space-time differential model for aggregate trip behaviour to planned suburban shopping centres.' *Geographical Analysis*, **26**, 341–362.

Baker, R.G.V. (1996), 'Multi-purpose shopping behaviour at planned suburban shopping centres: a space-time analysis.' *Environment and Planning A*, **28**, 611–630.

Baker, R.G.V. (2000), 'Towards a dynamic aggregate shopping model and its application to retail trading hour and market area analysis.' *Papers in Regional Science*, **79**, 413–434.

Barabasi, A. (2001), 'The physics of the Web.' *Physics World*, **14**(7).

Barabasi, A. and R. Albert (1999), 'Emerging of scaling in ransom networks.' *Science*, **286**, 509–512.

Blaut, J.M. (1961), 'Space and process.' *The Professional Geographer*, **13**, 1–7.

Boulding, K.E. (1985), 'Regions of time.' *Papers of the Regional Science Association*, **57**, 19–32.

Cheswick, B. (1999), Internet Mapping Project, http://www.cs.bell-labs.com.

Cohen, R., K. Erez, D. Be-Avraham and S. Havlin (2000), 'Resilience of the Internet to random breakdown.' *Physical Review Letters*, **85**, 4626–4628.

Erdos, P. and A. Renyi (1960), 'On the evolution of random graphs.' *Publications of Mathematical Institute of Hungarian Academy of Sciences*, **5**, 17–60.

Faloutsos, M., P. Faloutsos and C. Faloutsos (1999), 'On power law relationships of the Internet topology.' Proceedings ACM SIGCOMM, http//www.can.org/sigcomm99/papers.

Forer, P. (1978), 'Time-space and area in the city of the plains.' In T. Carlstein, D. Parkes and N. Thrift (eds.), *Timing Space and Spacing Time*, Vol. 1. Edward Arnold, London.

Gatrell, A. (1983), *Distance and Space*. Clarendon Press, Oxford.

Ghez, R. (1988), *A Primer of Diffusion Problems*. John Wiley, New York.

Internet Weather Report (2001), http://www.mids.org/weather.

Interent Traffic Report (2001), http://internettrafficreport.com/.

Janelle, D.G. (1968), 'Central place development in a time-space framework.' *The Professional Geographer*, **20**, 5–10.

Janelle, D.G. (1969), 'Spatial reorganisation: model and concept.' *Annals of the Association of American Geographers*, **59**, 348–364.

Jiang, B. and F. Ormeling (2000), 'Mapping Cyberspace: visualising, analysing and exploring virtual worlds.' *The Cartographic Journal*, **37**, 117–122.

Montroll, E.W and B.J. West (1979), 'Stochastic processes.' In E.W. Montroll and J.L. Lebowitz (eds.), *Studies in Statistical Mechanics, VII, Fluctuation Phenomena.* Elsevier, New York.

Padmanabhan, V.N. and L. Subramanian (2001), 'An investigation of geographic mapping techniques for Internet hosts.' Proceedings of ACM SIGCOMM, San Diego, USA.

Roy, J.R. and P.F. Less (1981), 'On appropriate microstate descriptions of entropy modelling.' *Transportation Research B*, **15**, 85–96.

Watts, D.J. and S.H. Strogatz (1998), 'Collective dynamics of "small world" networks.' *Nature*, **393**, 440–442.

Wilson, A.G. (1967), 'A statistical theory of spatial distribution models.' *Transportation Research*, **1**, 253–269.

Yujiro Ogawa, Hirotaka Suzuki, Bambang Rudyanto, Kiichi Hirono, Shigeya Yoshikawa and Masami Sugiura

Development of Disaster Information Network System in the Asian Region: Internet GIS for Disaster Information Management

1. Introduction

Due to recent advanced image processing technology, disaster information from satellite image data enables us to obtain various disaster information via remote sensing at anytime and place. Until this time, however, a system directly contributing to disaster management for disaster mitigation has not often been discussed. This is because much attention has been paid to technological breakthrough by satellite image data providers, and those who are working in disaster management have not been actively involved in this field. Also, it is hard to use disaster information from a satellite image for actual disaster management without overlaying it on general geographical data such as topography and natural conditions, and social data including population, buildings and infrastructure. Furthermore, the high cost and skills required for introducing Geographic Information System (GIS) technology has hindered its utilization for disaster information. Disaster management is required to deal with incidents in the real world. In this regard, GIS which can process two-dimensional as well as three-dimensional data, has a huge potential for utilization in emergency situations at the time of disasters, and can be used for evacuation planning. Nevertheless, GIS has not yet been widely used for the reasons mentioned above.

To solve these problems, we have developed VENTEN (Vehicle through Electric Network of disasTer gEographical informatioN), which is available to anyone. In this paper, the background of VENTEN development is explored, the VENTEN system and data to be processed by VENTEN are discussed, and the possibility of utilizing VENTEN for disaster management is analysed.

2. Background

2.1. *Disasters in the Asian Region*

The Asian region, with a number of earthquake epicenters, volcanoes and large-scale international rivers, and typhoon pathways, is prone to natural disasters. Furthermore, once a disaster occurs, human and property damage can

B. Boots et al. (eds.), Modelling Geographical Systems, 317–330.

expand in the many vulnerable megacities. The patterns of disasters in the Asian region, which will be processed by VENTEN, are analysed by using the Emergency Events Database (EM-DAT) administrated by the Centre for Research on the Epidemiology of Disasters, Universite Catholique de Louvain (UCL).

Figure 1 shows the transition of the amount of disaster occurrence by region. As damage caused by disasters varies year by year, the number of cases counted every ten years is recalculated into a yearly average value for the vertical axis. As shown in Figure 1, disaster occurrence in the Asian region always accounts for approximately 40–50% of the total in the world, which seems to be remarkably high considering the land area of this Asian region is approximately 20% of this total. It suggests that disaster countermeasures are given higher priority in the Asian region than in other regions. Since the death toll and the amount of damage are not analysed in this figure, the effect of disasters that caused extensive damage is not taken into consideration.

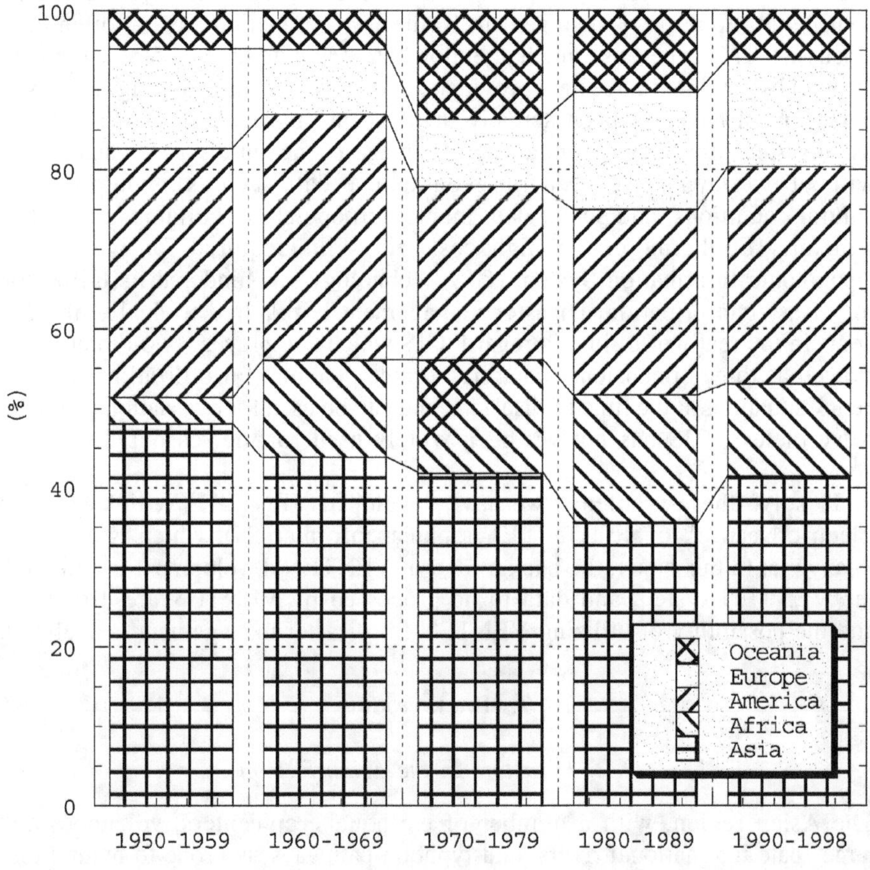

Figure 1. The transition of the amount of disasters occurred by region.

Figure 2 shows the death toll by region and by disaster type. The death toll in the Asian region is much higher than in other regions. The total number killed in the 48 years from 1950 to 1998 in Asia is 4,938,045 people, which is more than 70% of the total number killed in all regions combined. In comparison with the population of the Asian region of approximately 58% to the world population, the death toll in Asia is very high. Regarding the type of disasters, drought is the main cause of damage both in the Asian and African regions. In addition, floods, earthquakes, and typhoons result in the increase of the death toll in the Asian region. In particular, the number of dead caused by floods reached 45% of the total death toll in Asia, characterising the damage in this region. Many large rivers are located in this region, including the Yangtze, Ganges and Brahmaputra rivers. Most of them run through the area with a great deal of precipitation, resulting in more extensive flood damage than other disaster damages.

Wide areas are damaged by floods, frequently resulting in international problems. As floods also gradually damage the area for a long period of time,

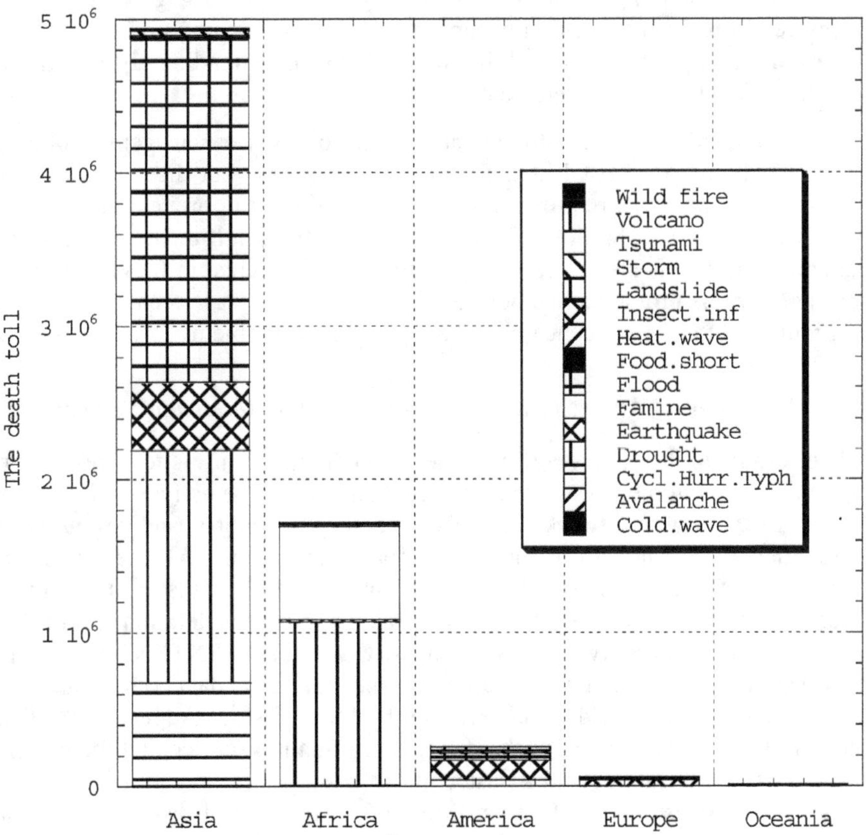

Figure 2. The death toll by region and by disaster type.

satellite image data will be useful for grasping the extent of damage and for disaster prevention planning to mitigate future damage. Taking all these factors into consideration, it is significant to introduce GIS to this field.

2.2. *Awareness of Administrators of Disaster Management in Asia*

Asian Disaster Reduction Center (ADRC) held the first ADRC Expert Meeting on February 17–18, 1999, with the participation of administrative officials in charge of disaster management in Asian countries discussing how to preserve the lessons we learned from past disasters for our children and grandchildren. The meeting had four workshops. The one titled "Utilization of technology" discussed utilization of remote sensing and GIS for disaster management and produced the following conclusions at the meeting after an active discussion.

- All the participating countries are aware of the value of GIS and remote sensing, and the advantages of their utilization for disaster information management.
- The future challenge is to give technological support for getting satellite images in real time and at a cheaper price, for introducing GIS and remote sensing, and to acquire the skills required for understanding abstract information for disaster management.

As mentioned above, the administrative office of disaster management of each country is highly interested in GIS and remote sensing, but the cost for introducing these and the required high skills needed still prevent them from introducing these technologies. The high price for satellite image data and geographical data is also a reason. Despite the high interest in GIS and remote sensing, various problems hamper them in introducing these technologies. All this indicates that those concerned truly want these problems to be solved.

2.3. *Internet Access to Disaster Information in the Asian Region*

It is desirable to develop a unified system which enables users to access information via the Internet, considering the cost and data to be delivered. However, the infrastructure for network varies from country to country, and despite good infrastructure it is not always available for other reasons. After establishing that the users in the Asian region use the Internet to get disaster information, Internet GIS is highly likely to be used, taking into consideration the further improvement of the network environment. Other than VENTEN, ADRC has provided disaster information of mainly text and numerical data both in Japanese and English, on the World Wide Web via the Internet since August 1998. This section discusses access log to the disaster information service and the current state of using the Internet for disaster information.

Between August 1st in 1999 and March 31st in 2000, disaster information service (both Japanese and English service together) was accessed 42,506 times. Among those, 28,294 were from Japan (from a host with a "jp" domain name),

Table 1. Analysis of access to the service.

ADRC member countries	15
ADRC advisory countries	4
Non-ADRC member countries in Asia	8
Domain with other countries	30
Others	6
Total	63

suggesting that the location of ADRC and service in Japanese have contributed to that number. Table 1 shows the analysis of access to the service by domain name, including access from Japan.

Although it suggests that most access to the service is from Japan, the service is accessed from wide area of the Asian region. Obviously, the service has been accessed most from the Asian countries because it mainly provides disaster information for the Asian region.

Considering the number of access hits from Western countries, where network infrastructure is well advanced and information provided in English, information service via the Internet is very much needed by Asian countries. In this regard, with an improved network infrastructure, the Internet has great potential for users to obtain geographical disaster information in the Asian region.

2.4. VENTEN Development Policy

Considering the background mentioned above, ADRC has developed a disaster information management system available to anyone from anywhere using the Internet, the progress of which has been remarkable.

The objective of VENTEN development is to provide the system and data. The system is available to any user via the Internet only if they have a PC connected to the Internet and a Web browser. Although various international institutions provide basic geographical information including topography and natural conditions, it has been difficult to obtain and hard to use because the data format may be changed according to the GIS application to be used. In developing VENTEN, ADRC has changed the format of the geographical data gathered so that it can be used on the VENTEN system, and provides the data as a part of the system.

Figure 3 shows the role of VENTEN. On the left of Figure 3, information providers, including various satellite data providers and aerial photograph agencies, make available the original primary data. Various image data processing and overlaying of data are necessary to obtain useful information for disaster management from this data. A pipeline is required to provide the processed information to those in charge of disaster management. Researchers in the disaster management can then study and analyse information on VENTEN

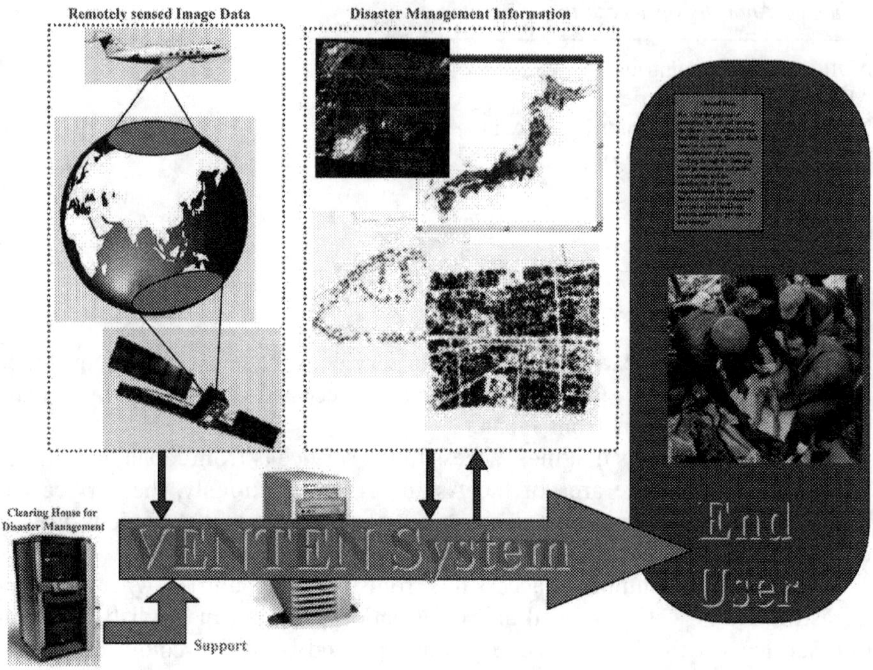

Figure 3. The role of VENTEN.

and add their results to it. VENTEN functions as a database, analyses geographical disaster information, and serves as an information distribution channel to those in charge of disaster management. In other words, VENTEN enables geographical disaster information to be fully utilised for disaster management, including planning and relief activities at the time of the disaster.

2.5. *Internet Services for Geographical Disaster Information*

Several national and international organisations have been established geographical disaster information providing service via the Internet. HAZPAC[1] has been established by U.S. Geological Survey to raise awareness of natural disaster risk. HAZPAC provides basic GIS functions via the Internet. Also it provides both basic geographical information and geographical disaster information in Pacific Rim region. ReliefWeb Map Centre[2] has been established by United Nations Office for the Coordination of Humanitarian Affairs. Though the service does not provide GIS functions, users can obtain a large number of map images which contain basic geographical information and geographical disaster information from all over the world. The map images are sorted by country and disaster event in the Web interface. The "Extremum" system, a Web-based decision support system for severe earthquake damage, has been established by Extreme Situations Research Center Ltd. and Russian governmental organisations. Pre-registered users can estimate loss and risk for actual or simulated earthquakes.

3. VENTEN SYSTEM

3.1. The Information-Processing Flow of VENTEN

The VENTEN system consists of a Web server, GIS server and database server. Figure 4 shows the information-processing flow in VENTEN. Upon requests from users, the Web server specifies the necessary information, including what geographical data and what area are needed (more than one geographical data selection is possible), for the GIS server. The GIS server refers as necessary to the database server and extracts the specified area from geographical data accumulated within itself, and then processes to send it to a Web server in a form of raster image data. The Web server adds a country selection menu, a geographical disaster information selection menu, some show/hide selection buttons for basic geographical data, and buttons for changing the scale and scope of the area to be displayed. Then it sends users information in hypertext format with the raster image data provided by the GIS server.

3.2. The Platform and Device for VENTEN

In the current VENTEN system, the Web server, GIS server and DB server shown in Figure 4, are located in the same machine, but these will be sepa-

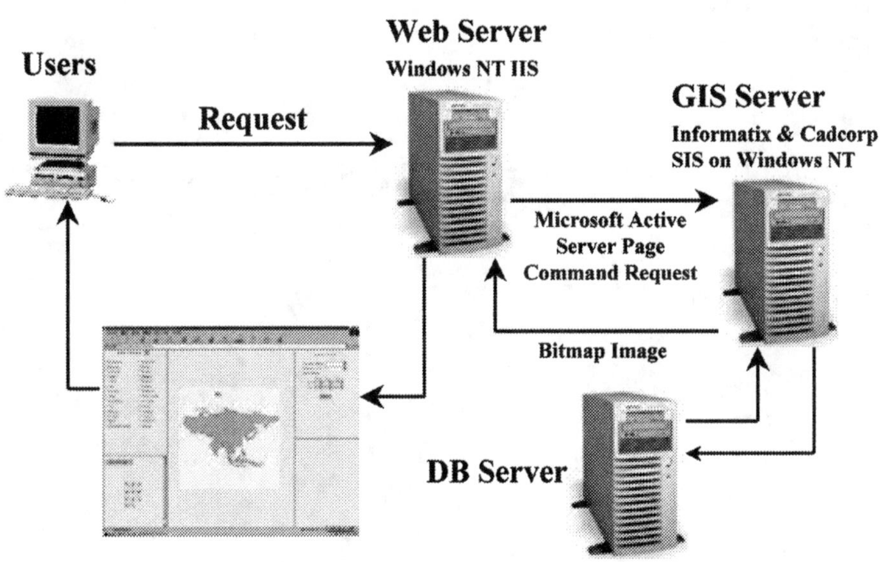

Figure 4. The information processing flow in VENTEN.

rated one by one in the future as both the data to be processed and the access load increase. Processing is controlled by a Microsoft Windows NT server. The Web server is controlled by Microsoft IIS (Internet Information Server). Databases, which are made by Oracle8, are accessible via Microsoft ODBC. For the GIS server engine, ASC (Active Server Component) of Informatix & Cadcorp SIS (Spatial Information System) is used. User requests are sent to the GIS server in a script embedded in an html document managed by the Web server. The GIS Server creates new image data upon the request and returns it to the Web server. Then the image data with peripheral menus and related information added by the Web server is sent to the users.

3.3. *How to Operate Venten VENTEN*

Figure 5 shows the initial screen of VENTEN. VENTEN consists of a map at the centre and four menus, and an additional window created when necessary. The map in the centre shows the ADRC member countries. The country to be displayed can be selected here. The country selection menu on the upper left of the screen also can be used to select a country. On the lower left is a control menu for the map scale, the displayed extent of the map, and the buffering. On the upper right is a geographical disaster information selection menu. The

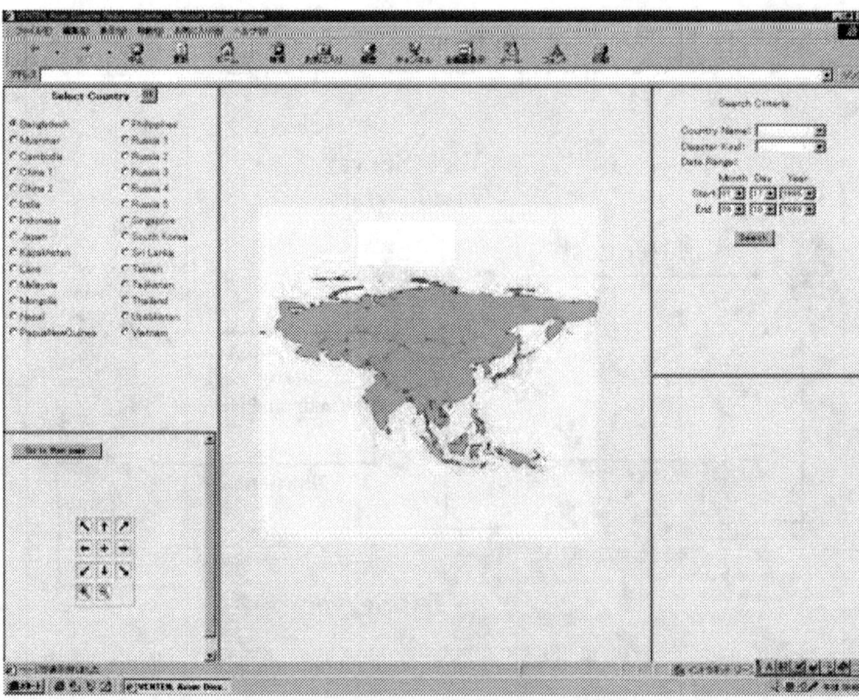

Figure 5. The initial screen of VENTEN.

information to be displayed can be specified by country name and disaster type.

When you specify a country name or geographical disaster information on the initial screen of VENTEN, the contents of the window change. As shown in Figures 6 and 7, the same lower left and upper right windows are displayed, but on the upper left is the property of the selected object, and in the lower right window is the show/hide toggle for the basic geographical information of the map shown in the centre. This window can be used to not only to select show/hide of the information, but also to set items to be selected/deselected as an object. This operation enables users to easily select an object by clicking on the map, which sometimes becomes a complicated process.

3.4. VENTEN Functions

VENTEN has GIS standard functions of "drawing map in any scale," "buffering," "overlaying" and "searching by location and attribute." Figure 6 shows a buffer enclosing an urban population. The buffer is set to 30 km with Osaka Itami Airport as its centre. The city names and their populations in that area are shown in another window. Likewise, VENTEN provides raster-based information to the end users, but the users can make various requests to process vector data on the server, enabling VENTEN to provide a service more advanced than that of Internet-based GIS with browsing as its main function.

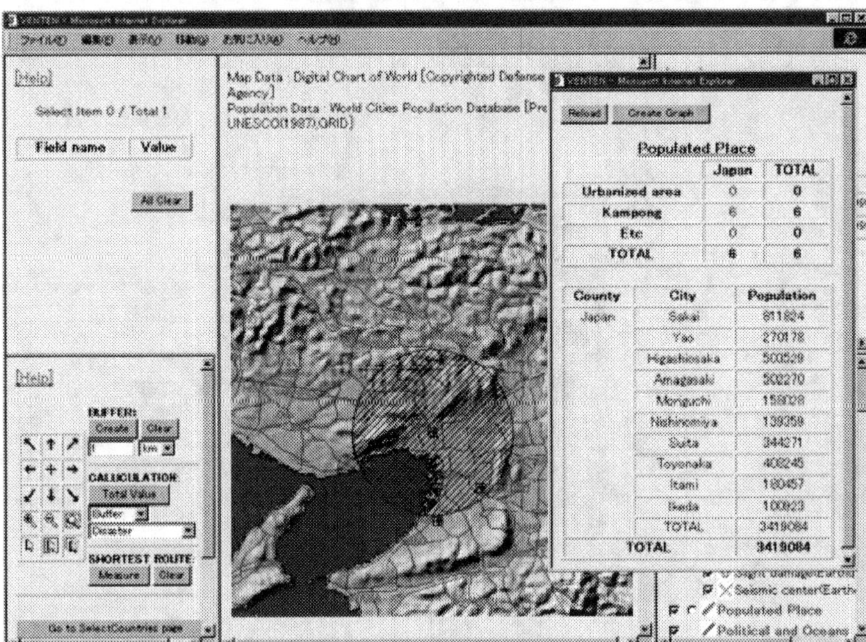

Figure 6. The buffer set to 30 km with Osaka Itami Airport as its centre.

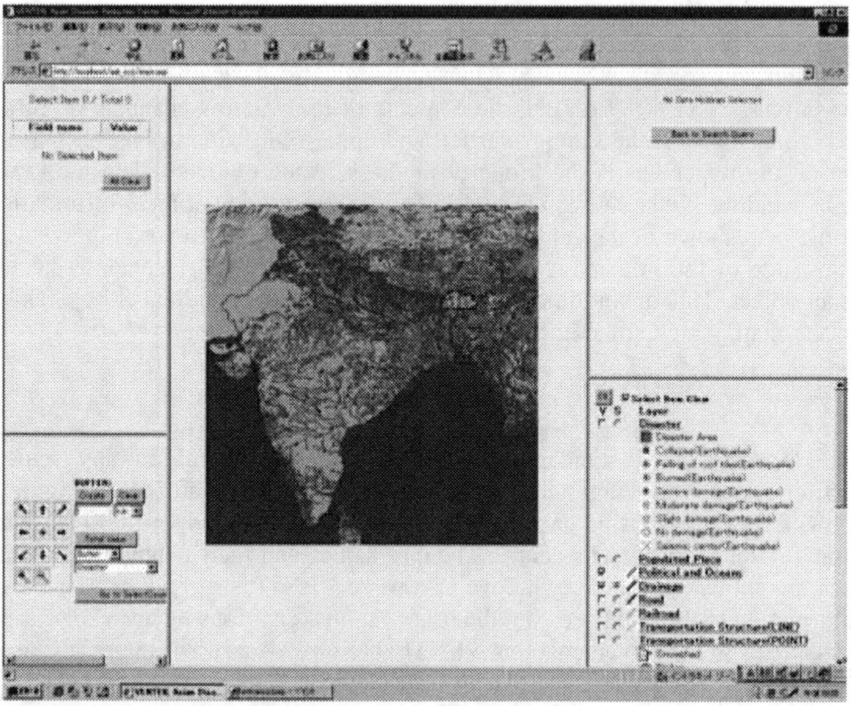

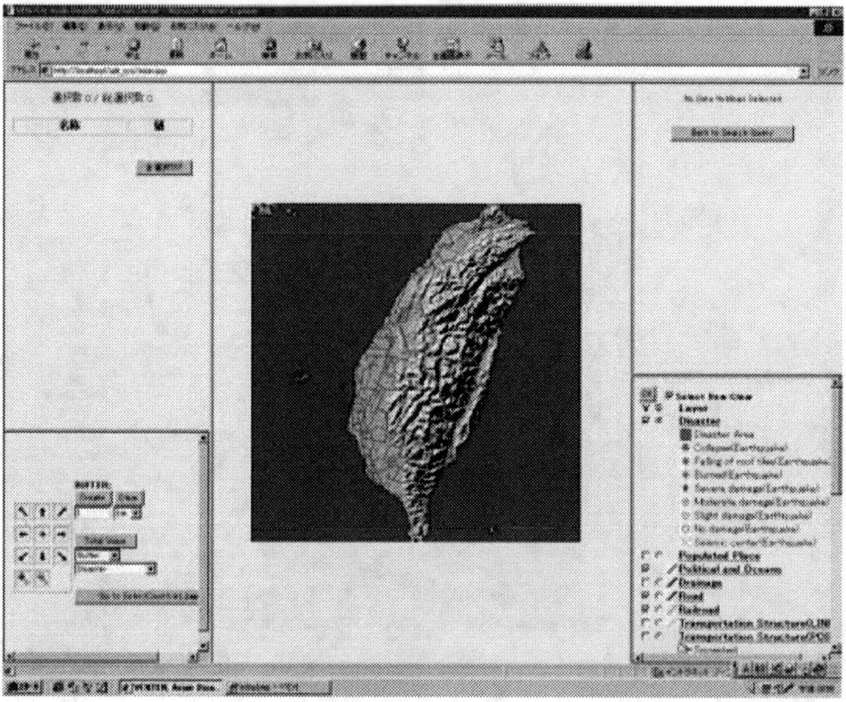

3.5. *Features of VENTEN System*

There are a couple of systems for Internet-based GIS. For example, one of them can be used by downloading an application program. Another system based on an image map only provides geographical information. VENTEN can be positioned in between those two systems. As for the systems mentioned in 2.5, both HAZPAC and the "Extremum" system can also be positioned same as VENTEN though ReliefWeb Map Centre is similar to later systems. Functions and type of data of VENTEN and HAZPAC are quite similar except their target region.

On VENTEN users can process vector data, but obtain only raster data based on the vector data. Although it restricts users to obtaining data, this system solves the problems of differences in response caused by different network environments and performance of the client machine, and of the property rights for data. For Internet GIS, the traffic load on the network when sending data is problematic. However, since VENTEN only sends a fixed scale of image of 470×470 pixels to be displayed at the centre of the VENTEN screen, it takes the server longer to process. The differences in network environments between VENTEN and end users do not have significant impact in sending data. Most data processing is done by the server machine and all the client machine should do is to display data it receives, so that differences in performance of client machines does not have a significant impact. It is easy to persuade many data providers to join in this system because vector data which is very close to the original information in its amount, is not given to users. The users will get only raster data, but users can process it the same as vector data.

Figure 7. The drainage area in India and Bangladesh (upper), and the road and the railway in Taiwan (lower).

Basic geographical information

- National borders (area)
- River systems (lines and area)
- Railways (lines)
- Roads (lines)
- Airports (point)
- Cities (point)
- City names (string)
- Populations (numerical value)
- Images with shading based on land height (raster image)
- Contour maps (raster image)

> Unesco (United Nations Educational, Scientific and Cultural Organization) and the World city population database of GRID (Global Resource Information Database) of UNEP (United Nations Environment Programme) are used for the population data. DCW (Digital Chart of World) by Defense Mapping Agency is used for other data.

Geographical disaster information

- Flood-stricken area of the Yantze River in 1998
- Damage on buildings in the area of Nishinomiya Station at the time of the Great Hanshin-Awaji Earthquake in 1995

4. DATA TO BE COVERED BY VENTEN

VENTEN covers data gathered from 22 ADRC member countries. ADRC has gathered two different data: general and basic geographical information to be used but not limited to disaster management, and geographical disaster information upon which the disaster management related information is reflected. So far the following data has been gathered:

As mentioned in 3.3, the basic geographical information can be shown or hidden by the menu on the lower right of the VENTEN screen. Also basic geographical information can be overlaid with geographical disaster information. Figure 7 is the data samples, showing image data with shading based on rivers and land height for India/Bangladesh to the top, and an image with shading based on the railways/road and land height for Taiwan to the bottom.

Figure 8 is a geographical disaster information sample, showing the flood-stricken area of the Yantze River in 1998, abstracted from a LANDSAT image overlaid with river, road and railway data at the usual time. As shown in Figure 8, VENTEN can be used to obtain geographical disaster information overlaid with other data, making the data extremely useful for disaster management.

5. COMMENTS AND SUGGESTIONS FOR VENTEN

Systems and data provision, which do not match user needs, have no meaning whatsoever. The ADRC is therefore calling the disaster management officers of member countries to use VENTEN and report on the results.

We also organize a training course called Basic GIS and Remote Sensing for Disaster Management[3] within the Seminar on Disaster Management training course for disaster management officers, held jointly with the Japan International Cooperation Agency, to have the trainees actually operate VENTEN and collect their opinions. As the trainees were given only a short time (about one hour) to use VENTEN, most of their comments were about data inadequacies (data is not the most recent, data is missing), required data (hospitals, military facilities, etc.), and the need to collect disaster management geographical information for not only ADRC member countries but for the whole world.

Data inadequacy is a problem requiring immediate solutions, and currently we are calling counterparts in member countries to provide the latest data. In the future, we hope to hold longer courses to seek opinions and comments on the functions and interface of the system, and to implement these improvements so that the system fully incorporates the opinions of users.

As for functions of VENTEN, some trainees pointed out the insufficient ability for buffering. At present users can set a buffer only for an existing object. Therefore users cannot estimate damage presuming given point is an epicenter. As the need for rough and early estimation of earthquake damages is very high, we have to realise the function of buffering for given object as fast as possible.

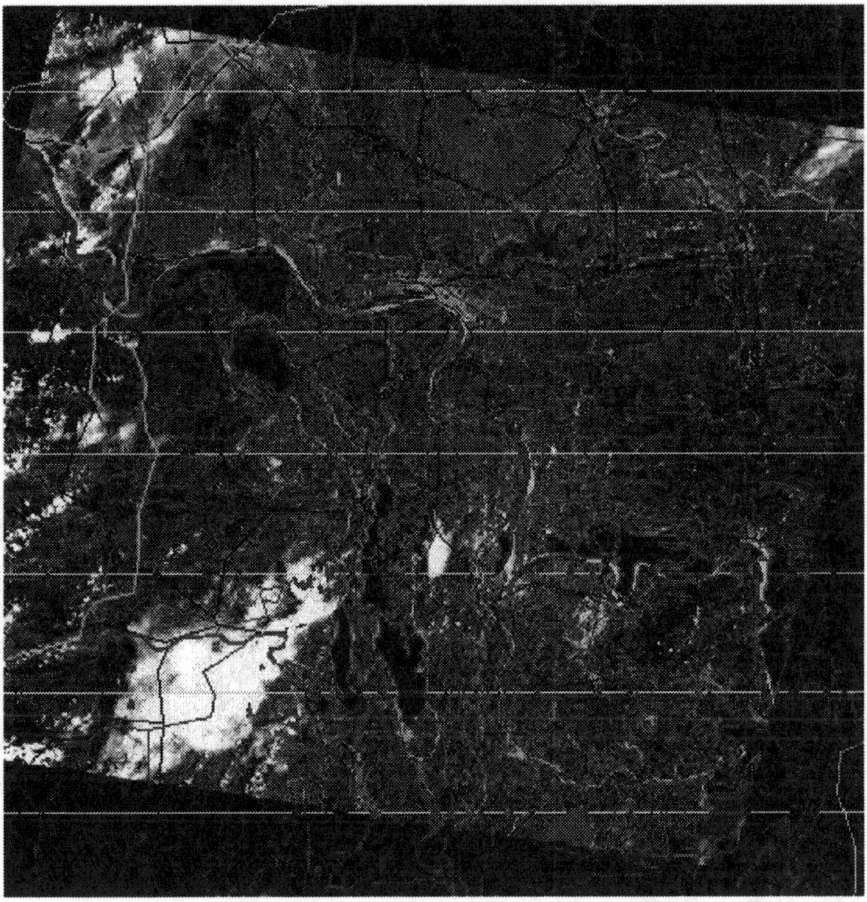

Figure 8. The flood-stricken area of the Yantze River.

6. CONCLUSIONS

This paper has presented the development of an Internet-based geographical disaster information system, its background and systematics, and data covered by the system. Current VENTEN is still rudimentary, with minimum functions available to the public. It is necessary to call on the related organisations to provide data in order to increase the geographical disaster data available for VENTEN. For the system, we will operate it on a trial basis for the administration of disaster management, and improve the interface and extend its functions based on the results. We will call on the satellite information providers, research institutes and universities studying remote sensing for disaster reduction to provide information. Some basic geographical data is old or doesn't fully cover the target area. ADRC tries to obtain the latest data and asks the ADRC member countries to cooperate in gathering the latest data.

Currently VENTEN can display and analyse the gathered basic geographical data and geographical disaster data, but in the future, we will connect it to disaster information services providing the text and numerical data mentioned in 2.3. The disaster information service system provides two entrances: by information category and by country, for a user to easily hit his/her target information. As disaster information increases, the current directory configuration may not be sufficient. Therefore, we plan to add the geographical search function for the huge amount of disaster information to the system, by using a part of VENTEN functions to substantially improve the user interface.

The data processing mentioned in 3.1 was applied in developing the system considering the network, the load on the client machine, and the copyright for data. However, other systems like JaMaPS[4] have already been proposed, under which client machines will process geographical data and, to some extent, distributed data will be referred or analysed. The main objective of developing VENTEN is to share the lessons we have learned from past disasters in a smooth information exchange. Through technological renovation and improvement of the network infrastructure, we will use the best possible method at each time to develop this system of sharing disaster information.

ACKNOWLEDGEMENTS

The authors would like to thank Dr. Richard Thomas, Dr. Atsuyuki Okabe and anonymous reviewer for their useful and detailed comments that greatly contributed to improve the quality of this paper. This project is supported by "Research and Development for Applying Advanced Computational Science and Technology" of Japan Science and Technology Corporation. We also wish to express our gratitude for its support.

Yujiro Ogawa, Hirotaka Suzuki and Bambang Rudyanto
Asian Disaster Reduction Center,
IHD Center Bldg., 1-5-1, Wakihama-kaigan-dori,Chuo-ku,
Kobe City, 651-0073 Japan

Kiichi Hirono, Shigeya Yoshikawa and Masami Sugiura
Asia Air Survey,
2-18, Shinjyuku-4, Shinjyuku-ku,
Tokyo, 160-0022 Japan

NOTES

1 See the Web page of HAZPAC(http://www.hazpac.org/).

2 See the Web page of ReliefWeb Map Centre(http://www.reliefweb.int/w/map.nsf/home).

3 Seminar on Disaster Management organized by Asian Disaster Reduction Center and Japan International Cooperation Agency, 1/11/2001–2/9/2001. Attended by 14 trainees from 13 countries. Basic GIS and Remote Sensing for Disaster Management was carried out on 1/31/2001.

4 See the Web page of JaMaPS(http://www.jamaps.org).

MASANAO TAKEYAMA

GEOGRAPHICAL CONCEPTUALIZATION OF CYBERPLACES

1. INTRODUCTION

Cities are becoming smart and intelligent as technologies for computing and communication are widely spread among people and also integrated into the urban infrastructure to support their activities. Some of these technologies are realized in the forms of desktop and handheld computers, mobile phones, or wearable information devices that empower each individual to behave and communicate efficiently. Some are beginning to be embedded invisibly in physical objects and environments such as home appliances, furniture, offices, buildings, automobiles, highways and stations. They are communicating with each other through collaborative wired or wireless communication networks to respond intelligently to various demands of people (Mitchell, 1999). Such technologies do not just make our life more convenient but also change patterns of our behaviour and activities as well as our conceptions of space and place.

Geographers have just begun to study cyberspace, i.e., space of electronically mediated communication, and its influence on societies of the real world as a new realm of geographical investigation. In particular, they analyze spatial aspects of cyberspace in the framework of abstract network diagrams by applying visualization techniques (Dodge, 1988; Dodge and Kitchin, 2001). Others try to understand cyberspace in terms of its representation and structure that simulate spatial properties of particular physical places (Adams, 1997, 1998). Some also discuss how cyberspace or the global information network is transforming human behaviour, socio-economic urban systems, and problems and approaches of city planning (Graham and Marvin, 1996; Kitchin, 1998; Wilson and Corey, 2000; Wheeler *et al.*, 2000). These studies show many ways in which both traditional and modern geographic perspectives and techniques can be applied to analyze spatiality of cyberspace and to understand the impact of virtual electronic space on real urban space.

Most of the researchers in geography of cyberspace usually assume physical and electronic space as two distinct spatial domains although they consider the interaction and influence between them. However, as technologies for computing and electronically mediated communication are being integrated into real world places and into human activities there, it is becoming more difficult to distinguish what is real from what is virtual. Indeed, electronically mediated communication is more and more 'simultaneously embedded in and sustained by systems of physical and material circulation, visual and acoustic communication, and remote interconnection' (Mitchell, 1999). One of the challenges

B. Boots et al. (eds.), Modelling Geographical Systems, 331–345.

for geographers in the age of electronically augmented space is to provide an appropriate spatial framework to understand and guide such geographical integrations of electronic and physical worlds.

By carefully considering various conceptions of electronically mediated space and place, Batty (1997) classifies these into four categories based on the distinctions between space and place and also between 'nodes' and 'nets', i.e., single and related places. In Figure 1, the upper left box represents the original domain of geography abstracting place into space. The upper right box labeled computer space or C-space is the domain of space that deals with abstractions of space into computers. Cyberspace located in the lower right box means new spaces that emerge from C-space through using computers to communicate. Finally, the box located in the lower left is named cyberplace, which is the embedding of the digital and electronic infrastructure of cyberspace into the physical infrastructure of traditional place.

Among these four spatial classes, Batty (1997) describes the properties of cyberplace as:

> Cyberplace consists of all the wires that comprise the networks that are being embedded into man-made structures such as roads and buildings. It extends to the material objects that are used to support this infrastructure such as machines for production, consumption and movement that are now quickly becoming a mix of the digital and the analog.

Although this explanation emphasizes the spatiality of cyberplace only in terms of its infrastructure, we could naturally extend this notion such that it would deal with other aspects of place of cyberplace such as sense, meaning, and even social settings.

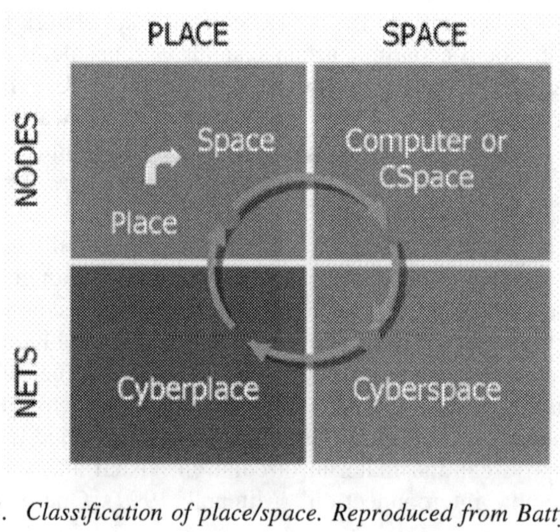

Figure 1. Classification of place/space. Reproduced from Batty (1997).

This paper focuses on the geography of cyberplace as introduced by Batty (1997) and tries to advance further the understanding of such a novel concept and domain of space. Rather than treating real and cyber space as two distinct domains, or representing one as some metaphor of the other, it aims at providing a conceptual framework for understanding geographical properties of cyberplace in terms of its own spatiality that is produced from the merger and the integration of physical place and electronically mediated communication and interaction. In particular, it studies how physical and social settings of traditional place are extended and augmented by the spatiality of cyberplace.

In the following, this paper first introduces two experiments to implement cyberplace, the Interactive Town and the HyperZoo, in order to get some sense of how a particular cyberplace looks and how people behave in such a place. Based on the findings from these experiments, it introduces the notions of smart place and hyper-place for understanding two levels of the electronic augmentation of human activity. It then shows an approach to deal with the locational duality of cyberplace, i.e. its partial and simultaneous presences in both physical and electronic spaces. Finally, the communicational structure realized in such an augmented place is represented in a topological expression to understand analytically the complex spatiality of cyberplace.

2. DESIGNING CYBERPLACES

2.1. The Interactive Town

The first experiment called the Interactive Town was carried out in 1998 to design cyberplace over a small town in a suburb of Yokohama city in Japan. The goal of this experiment is to develop an interactive gaming environment for school children to play in their living neighbourhood using mobile as well as fixed information technologies. The participants of the game explore several places in the town to find answers for questions they receive electronically from these places. In other words, it designs the town space as an interactive learning and playing environment. Figure 2 shows the information network infrastructure that realizes the Interactive Town.

Each player of this game carries a hand-held device, which allows the player to use E-mail communication, access web sites, and also take photo pictures with a digital camera attached to it. The player also has a type of mobile phone named PHS (Personal Handy-phone Systems) that is used for locational positioning in addition to its regular voice communication. The positioning capability of the PHS is one of the most critical functions of the information system of the Interactive Town. The positioning system tracks the current location of a player and sends the coordinate to a host server, which produces a digital map showing the location. Based on the location of a player, another server computer sends an E-mail message to the player asking a question regarding some place or object found around the location. For example, a player coming close to a shrine may receive a message with a photo image attached, asking the player

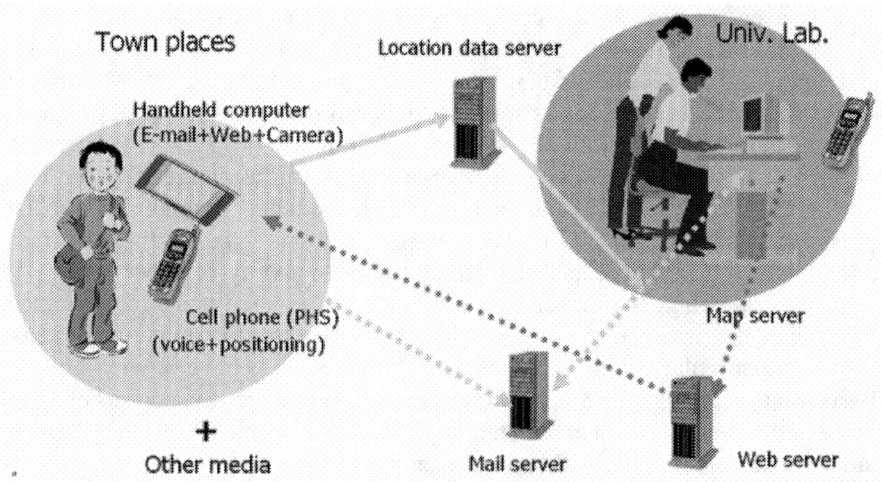

Figure 2. The information network of the Interactive Town.

to find some difference between the real architecture of the shrine and the one appearing in the image. Besides using E-mail communication, a player also receives and answers such a question using other communication technologies including a mobile phone and a video-conferencing system.

Most of the questions for the players are prepared to ask some knowledge about particular places in the town. Places for questions include a city square, rooftop of a building, shopping mall, park, shrine, office, and even space inside a train. Table 1 shows seven different places in the town to receive questions, media used to receive questions, and the types of questions asked in these places. Once a player finds the answer for a question, he or she sends it using E-mail

Table 1. Places, media, and questions.

	Places	Digital media	Question types
1	Subway train	Handheld + Cell phone	Question asking a view from a train
2	City square	Handheld + Camera + Cell phone	Look for a specific characters in town signs
3	Build. rooftop	Teleconference	Question regarding townscape
4	Mall	Handheld + Cell phone	Question asking a name of a product
5	Park	Image recognition system	Question asked by objects in a park
6	Shrine	Handheld + Digital camera +Cell phone	Question regarding the architecture of a shrine
7	Office	Local CATV	Question asked by a businessman

to the staff of the Interactive Town, who actually plays the role of the Interactive Town itself. If the answer is correct, the player scores a point that is represented by a circle symbol on the online game board (Figure 3). As the number of each circle corresponds to the number of the question, the system changes the colour of the circle with the number of the answered question into its taker's colour. The players compete with each other not simply in the number of obtained circles, but also in the number of gained triangles formed by three nearest circles.

After answering a question, each player needs to make a decision on which question to challenge next. For this, each player accesses, using the hand-held device, the online map (Figure 4) updated at every minute in order to check one's current location in the town (represented by square symbols), as well as the predefined locations of questions (represented by circle symbols), so that one can find an optimal target in terms of obtaining a triangle.

In this experiment of the Interactive Town, the player is not just exploring in real places in the town but is playing simultaneously in a symbolic (i.e., a game board) and geographic (i.e., a map) space. Using the computing and telecommunication technologies, these places and sites are mutually influenced and linked together in a way that an action of a player in a real place triggers another action or event in cyberspace and vice versa.

2.2. *The Hyper-Zoo*

The next experimental cyberplace, designed in the year 2000, is the HyperZoo, which also integrates real and virtual places through the activity of developing a virtual zoo using the photo images of real animals taken in real world zoological gardens. At a traditional zoo, people just watch and observe animals in cages. Although they may talk about animals with friends visiting there

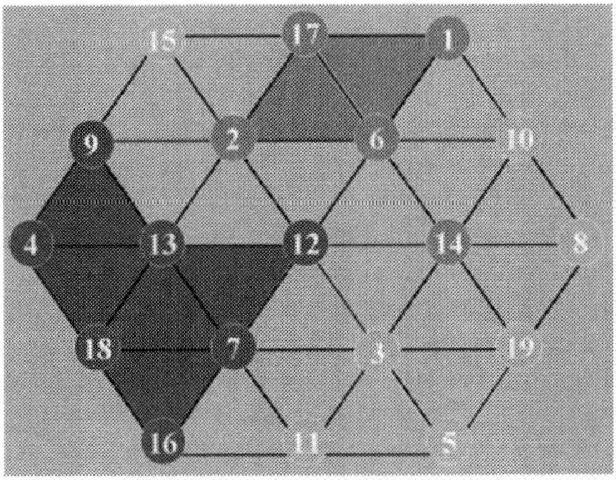

Figure 3. The online game board.

Figure 4. *The online map showing the locations of questions and players.*

together, or may take photographs of these animals for their memory, such interactions among people and animals are usually limited in a particular place and time. The major purpose of this experiment is to provide people visiting a zoo with the rich experience of expressing their own observation of animals and communicating it through the electronic network to other people visiting at another remote zoo.

The architecture of the HyperZoo is designed such that a virtual zoo constructed online networks two distant real zoological gardens, one of which is located in a suburb of Yokohama city and the other in Nago city of Okinawa prefecture in Japan. Each visitor to these real gardens carries a digital camera for taking photos of animals and also a laptop computer with wireless connection to access the website of the virtual zoo. The virtual zoo is a website to which a visitor of either zoo uploads photo images of observed animals and some texts on these observations such that other visitors accessing the virtual zoo can look at the images, read texts, and add some comments on these. When a visitor to the real zoological garden uploads an image of an animal to the virtual zoo, one can choose an icon of the animal from the animal icon menu and also its location in the virtual zoo such that the selected icon appears in a particular location in the files of virtual zoo (Figure 5).

The visitors of these two distant gardens collaborate over the Internet on developing a virtual zoo where the variety of animals increases as the visitors upload images of various kinds of animals found in each zoo. In this process,

Figure 5. The HyperZoo.

a visitor may ask another visitor to upload an image of some specific kind of animal by writing a request message in the bulletin board of the virtual zoo.

3. SPATIAL CONCEPTIONS OF CYBERPLACES

What are the common characteristics of the settings and experiences in these experimental cyberplaces? In both the experiments of the Interactive Town and the HyperZoo, activities are performed in and also distributed over the combinations of physical places and electronic space. From geographers' perspective, these activities extend the traditional notion of place. This section introduces several spatial concepts to characterize the properties of such an extended place.

3.1. Augmenting Place Experience

Cyberspace (represented by the shaded triangle in Figure 6) connects remote persons through the electronic telecommunication network. Various interactions over remote computers define forms of cyberspace. In cyberspace, however, communication and activities are basically separated from physical settings of real places. Although what a person types with one's keyboard in chatting with a distant friend may be influenced by some conditions of where the person is, say, the room temperature, such influence is usually neglected as being indirect for the interaction that defines cyberspace. What application software to use for the communication influences more directly on the interaction because the activity of communication is not happening in physical space but in electronic place. Thus, substituting physical settings for communication by settings

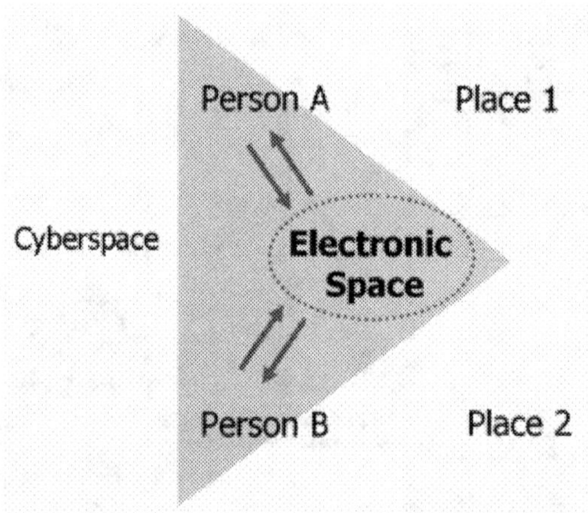

Figure 6. Cyberspace.

for electronic interaction, cyberspace always exists without any direct relationships with real places.

In addition to substituting physical by electronic, communication technologies are also used to support and extend a human activity at a physical place by sensing information of a person or one's environment to compute it for producing some response to them. For example, you might imagine a room equipped with some intelligent lighting facilities that change automatically the level and colour of lighting by detecting the kind of activities in the room. This electronic mediation between humans and environment distinguishes the spatiality of cyberplace from that of cyberspace. In other words, cyberplace is seen as electronically augmented place.

In the case of the Interactive Town, moving to a certain place in the town accompanies receiving a question in the form of text or image by an E-mail message. Thus, access to a particular place means simultaneously access to its correspondent electronic information. By applying technologies for mobile communication and positioning, a person interacts with one's environment not just through one's physical communication channels but also through the electronic communication channels. Such an experience of electronic communication and the various senses of place which one experiences physically are mutually integrated into a consistent activity of exploring the Interactive Town.

In this paper, the expression *smart place* is used for a type of cyberplace where such electronic augmentation of person-place interaction is limited in a particular physical place, say, a park, office, room and so on (represented by the dark shaded triangle in Figure 7). As a simple example, just overlaying some digital information onto a transparent head-mounted display, wall, or window to help one's navigation in a building produces a smart place.

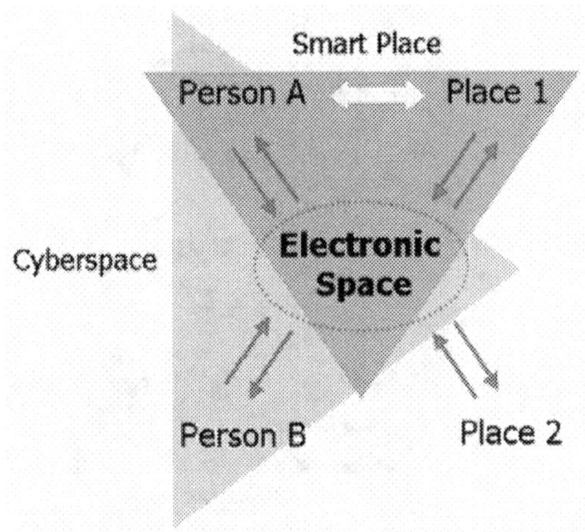

Figure 7. Place for Tele-Activity.

3.2. Place for Tele-Activity

Some places in the Interactive Town are mutually linked in the sense that one player's behaviour at one of these places influences in real-time the decision and strategy of another player at another place. Also in the experimental setting of the HyperZoo, the behaviour of people at two distant zoological gardens influence each other in the collaborative activity of developing a virtual zoo. In these examples, activities at geographically distant places are integrated coherently into a higher level of complex human activity by linking these places by an electronic communication network and some computing devices. In such remotely coupled activities, a person located in some physical place is not just able to interact with distant people, but possibly interact with objects, events, or environments of distant places.

A typical example of such a place augmentation is the situation of telemedicine where a telecommunication network connects two distant hospital rooms and surgery equipment such that a doctor in a hospital assists or controls an operation at another hospital. Here a doctor in a hospital influences people, activity and settings of another hospital at a distant place. Such electronic and physical settings for interaction provide people with the opportunity of tele-joining in some activities at some distant place. At the same time, how people are involved in such a tele-activity defines the form and nature of the place.

The notion of *hyper-place* is now introduced to represent such an extension of place and to distinguish it from the idea of smart place. Hyper-place (represented by the dark shaded square in Figure 8) is an electronic coupling of remote activities as well as environments that transcends the limit of functioning and settings of a physical place to support and accommodate a complex human activity of a higher order. The remote control of the Pathfinder on Mars

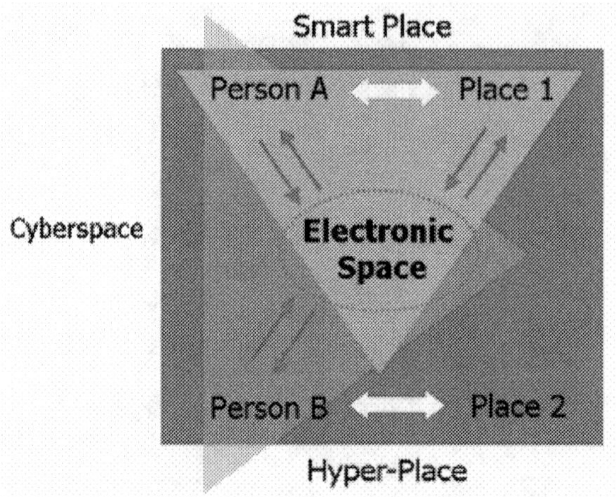

Figure 8. Hyper-Place.

from the control room at the NASA station is another activity realized in a kind of hyper-place that is connecting Mars and Earth. It is important to emphasize that the notion of hyper-place is not introduced especially for describing such space telecommunication or a futuristic science fiction world. The current rapid spread of mobile phones is actually producing various hyper-places in our everyday life. When a taxi driver talks using a mobile phone to another driver running at some remote place to exchange traffic information about their current places (Townsend, 2000), they are interacting in a type of hyper-place.

It is easy to realize that the notion of hyper-place is closely related with that of *telepresence*. According to Steuer (1993), telepresence is defined as the experience of presence in an environment by means of communication medium. Indeed, hyper-place is seen as a mixture of presence and telepresence. However, the notion of hyper-place emphasizes more the aspects of interaction as well as the integration of people, objects, and environments of geographically distributed places into a coherent activity while the concept of telepresence deals with the form of perception realized by telecommunication technology. Thus, tele-presence can be seen as a factor of hyper-place for realizing human tele-activities.

It is also worth noting that hyper-place is related to the sociological notion of *time-space distanciation* originally introduced by Anthony Giddens as the progressive stretching out of social relations across space (Giddens, 1984). Giddens (1990) explains that distanciation refers to 'the conditions under which time and space are organized so as to connect presence and absence'. Although both notions deal with extensions of human interaction over distant places, they can be distinguished from each other by the fact that the notion of hyper-place emphasizes the extension of placeness produced by embedding of electronic communication in physical settings of place while the idea of time-space distanciation focuses on the extension of functioning of social systems and their disembedding from physical places.

4. GEOGRAPHY OF CYBERPLACES

4.1. Real-Virtual Dual Presence

One of the fundamental questions about the spatial property of cyberplace is to ask where it is located. Such a question requires us to reconsider the geographical meaning of location as cyberplace exists partially and simultaneously in the real and electronic spaces. Moreover, since the interactivity between cyberspace and physical place defines cyberplace, we need some technique to relate a location in real space with a location in cyberspace. A possible approach to 'spatializing' cyberplace is to construct a mathematical space at a higher level of abstraction in which both physical and virtual locations of cyberplace can be identified. In short, the location of a particular cyberplace is defined by the pair of its real and "cyberspatial" locations. In a mathematical expression, a particular cyberplace is identified in the Cartesian product of real and virtual spaces.

This approach of dual spatiality is not just appropriate for providing us with a systematic way to locate every object and activity existing and present in cyberplace but also understanding this physical-electronic interweave. Figure 9 represents this idea of real/virtual duality of cyberplace location in a schematic mathematical expression. This space consists of two dimensions representing physical and electronic space. For the sake of simplicity, these spaces are assumed here to have just one dimension for each although it could be extended

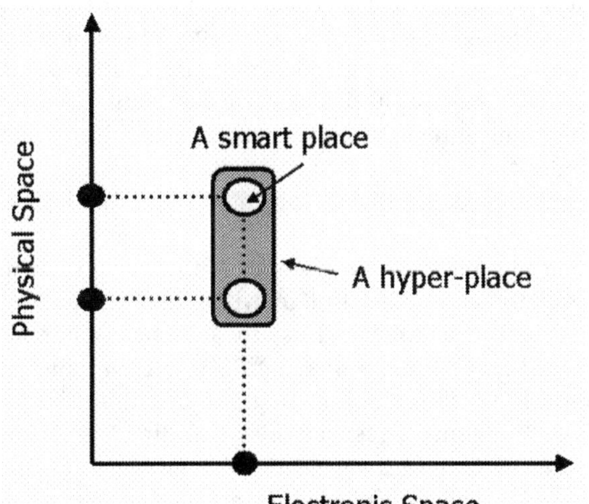

Figure 9. Dual location of cyberplace.

into higher dimensions. A point (shown as a white circle in Figure 9) in this space represents a location of a smart place that is actually defined by a coordinate pair of its physical and electronic addresses. For example, the location of a shrine in the Interactive Town is defined by its physical and electronic addresses. Whereas it is natural to use its geographical coordinate for the physical address of the shrine, there are several alternatives for the electronic address. It can be defined by the URL of the question message to be sent to a person approached there or by its address in the database or some hard disk. Or it might be more appropriate to use the URL of the website of the game board or the digital map. The choice in the representation of electronic location depends on the purpose of the representation.

The location of a simple hyper-place, which networks one electronic site with two physical places, is also represented in this mathematical space as a set of two points defined by a coordinate of a common electronic address and two different physical addresses. For example, the location of HyperZoo is defined in this way by the URL of virtual zoo website and two physical locations of zoological gardens in Japan.

It is certainly possible to go beyond such a schematic understanding of the location of cyberplace for a more realistic study. For example, a distance between different cyberplaces can be defined. Moreover, the accessibility to a particular cyberplace can also be measured. However, it is beyond the scope of this research and is reserved for future study.

4.2. Spatial Representation of Cyberplace

Another approach to represent the real-virtual integration of cyberplace is to look at the patterning of communication or the flow of information in cyberplace. In relation to this concern, Paul Adams (1998) shows a topological technique to represent and analyze the patterning of communication that characterizes the geography of physical and virtual places. In particular, he succeeds in understanding the concept of virtual place by relating the structure of links and nodes in communication systems with the structuration of individual-society interactions. Indeed he 'mapped' virtual places of various kinds onto the topology and social structuration of physical places and processes (Adams, 1998).

Following this approach, it now becomes possible for us to represent the structural patterning of communication in cyberplace. The idea is explained with the case of the HyperZoo as follows. First, the topology of communication structure at a real world zoo is drawn in Figure 10 (left) in which the center node and its surrounding nodes represent a particular animal like an elephant and persons watching the animal, respectively. The directed arrows among these nodes depict the flow of information from the animal to persons representing the observations of the animal by the persons.

Then, we can apply the same approach to representing the communication structure of the virtual zoo constructed in cyberspace. Contrary to the real world

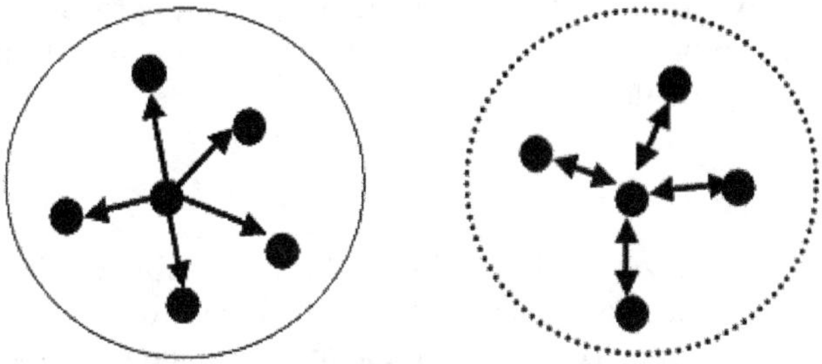

Figure 10. Topologies for a real zoo (left) and the virtual zoo (right).

zoo, the visitors to the virtual zoo accessing it through the Internet do not just observe animals passively. They might add some information about some particular animal in the virtual zoo using its electronic bulletin board. They also could exchange their ideas about animals with other people visiting the virtual zoo. This interactive nature of the virtual zoo is represented by the bi-directional arrows in Figure 10 (right).

Finally, by integrating the topological representations of communication patterns at the real world zoo and the virtual zoo, we have the topology of the HyperZoo (Figure 11), which represents the interaction among people visiting different real world zoological gardens as well as persons electronically accessing the virtual zoo. Such an expression helps us in recognizing how the patterning of communication of the real world zoo is extended and developed into the more complex and interactive structure of the HyperZoo. Thus the topological approach works effectively as a technique to grasp an aspect

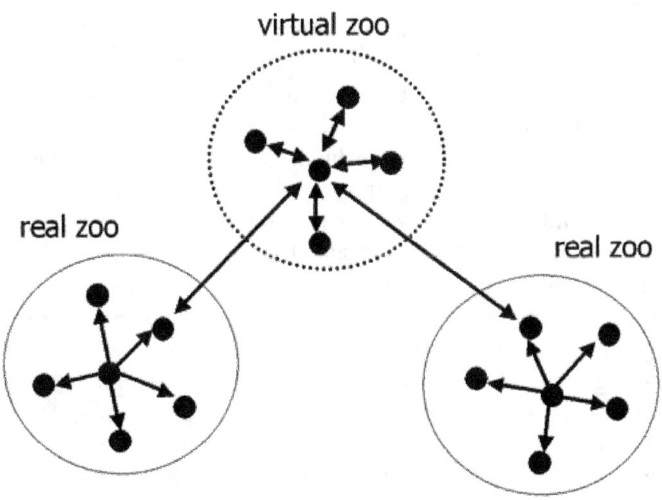

Figure 11. The topology of HyperZoo.

of spatiality of cyberplace. We can also expect this approach to help architects in designing a particular cyberplace through the combination of physical and computational architectures.

5. CONCLUSION

Cyberplace as introduced in Batty (1997) does not substitute the interaction between person and place as cyberspace or virtual place often does. Cyberplace rather extends and augments the interaction and communication among people and places in the real world. This property of place augmentation requires us to distinguish geography of cyberplace from geography of cyberspace as well as real world place. Based on the findings from some experiments to implement cyberplaces, this research introduced a conceptual framework for characterizing geographically the integration between physical places and electronically mediated spaces. In particular, the notions of smart place and hyper-place were defined as two different levels of place augmentation. It also showed an approach to deal with the locational duality of cyberplace that is its partial and simultaneous presence in both physical and electronic spaces. In addition, the communicational structure realized in such an integrated place was represented in a topological expression to understand analytically the complex spatiality of cyberplace.

This research is a first attempt to claim the importance and possibility to begin a geography of cyberplace. For future research several significant questions regarding cyberplace geography are posed as follows.

First, it is important to emphasize behavioural characteristics in geography of cyberplaces because human behaviour and activities actually integrate physical functions of place and electronically mediated communication into placeness of cyberplace. Therefore it is necessary to study how it is different from human geographic behaviour in the real world space and from that in cyberspace. For example, it is interesting to ask how experiences of playing, living, and working in a particular cyberplace influence human senses of place as well as their future behaviour.

A second question concerns the influences of cyberplace on the future transformation of real world geography and geography of cyberspace. The implementation and construction of various types of cyberplace might bring about changes in spatial structures and functions of real world cities. This problem can be approached in economic and social aspects of cities or from architectural perspectives. We can also ask a similar question about societies and architectures of cyberspace.

Finally, geography of cyberplace necessitates the development of an approach and methodologies of its spatial representations and analyses. The topological approach as proposed in this research is one possibility for such an endeavour. It is also worth asking how geographical information systems can be applied and extended to study cyberplace.

ACKNOWLEDGEMENTS

I thank Atsuyuki Okabe and Barry Boots for their valuable comments and guidance. I also appreciate the comments from the anonymous reviewer.

Faculty of Economics
Keio University
2-15-45 Mita, Minato-Ku, Tokyo 108-8345, Japan

REFERENCES

Adams, P. (1997), 'Cyberspace and virtual places.' *Geographical Review*, **87**(2), 155–171.

Adams, P. (1998), 'Network topologies and virtual place.' *Annals of the Association of American Geographers*, **88**(1), 88–106.

Batty, M. (1997), 'Virtual geography.' *Futures*, **29**(4/5), 337–352.

Dodge, M. (1998), The geographies of cyberspace. Paper presented at the *94th Annual Meeting of the Association of American Geographers*, 25–29th March 1998, Boston, USA.

Dodge, M. and R. Kitchin (2001), *Mapping Cyberspace*. Routledge, London.

Giddens, A. (1984), *The Construction of Society: Outline of the Theory of Structuration*. University of California Press, Berkeley.

Giddens, A. (1990), *The Consequence of Modernity*. University of California Press, Stanford.

Graham, S. and S. Marvin (1996), *Telecommunications and the City: Electronic Spaces, Urban Places*. Routledge, London.

Kitchin, R. (1998), *Cyberspace: The World in the Wires*. John Wiley & Sons Ltd, Chichester.

Steuer, J. (1992), 'Defining virtual reality: dimensions determining telepresence.' *Journal of Communication*, **42**(4), 73–93.

Townsend, M. (2000), 'Life in the real-time city: mobile telephones and urban metabolism.' *Journal of Urban Technology*, **7**(2), 85–104.

Wheeler, J.O., Y. Aoyama and B. Warf, eds. (2000), *Cities in the Telecommunications Age: The Fracturing of Geographies*. Routledge, New York.

Wilson, M.I. and K.E. Corey (2000), *Information Tectonics*. John Wiley & Sons Ltd., Chichester.

CONTRIBUTORS

Yoshio Arai is Professor of Human Geography at the University of Tokyo at Komaba. He received his Ph.D. degree from the University of Tokyo in 1993. His research interests stem from a broad area of human geography with particular emphasis on time-geography of human activities, geography of information society and historical analyses of regional change using GIS.

Theo Arentze received his Ph.D. in urban planning from the Eindhoven University of Technology. He is now an Assistant Professor in the Urban Planning group at the same university. His main fields of expertise and current research interests are decision support systems, activity-based modelling, discrete choice modelling, knowledge discovery and learning-based systems with applications in urban and transport planning.

Robert Baker is Senior Lecturer in Geography and Planning at the University of New England (UNE), Australia. He received his Ph.D. at the University of New South Wales in 1991. He has a wide interest in applied mathematical modelling in human and physical geography. He received the UNE Vice-Chancellor's Award for Excellence in Research in 2000 for Retail Modelling and Quaternary Studies.

Dimitris Ballas is Research Fellow in the School of Geography at the University of Leeds. He received his Ph.D. from the University of Leeds in 2001. His main research interests include the development and use of spatial microsimulation methodologies for human systems modelling and for urban, regional and social policy analysis and evaluation. He is also interested in studying the spatial dimensions of income and wealth inequalities, socio-economic polarisation and local labour market segmentation.

Barry Boots is Professor of Geography and Environmental Studies at Wilfrid Laurier University. He received a Ph.D. degree from Rutgers University in 1972 and a D.Sc. degree from the University of Bristol in 1995. His main interests are measuring spatial structure and spatial relationships, spatial statistics, and Voronoi diagrams.

Graham Clarke is Senior Lecturer in the School of Geography at the University of Leeds. He received his Ph.D. from the University of Leeds in 1986. His research interests include the geography of public services, retail and business geography, and GIS and modelling. He has also more recently become interested in the geography of crime and policing.

Robert Feick is Assistant Professor appointed jointly to the School of Planning and the Department of Geography at the University of Waterloo. He received his Ph.D. in Geography from the University of Waterloo in 2000. His research focuses on GIS and decision support systems with particular emphasis on collaborative spatial decision making, multi-criteria analysis methods, and public facility systems planning.

Brent Hall is Professor of Planning and Associate Dean, Computing in the Faculty of Environmental Studies at the University of Waterloo. He received his Ph.D. degree from McMaster University in 1980. His main interests are geographic information systems with a focus on decision support problems. He is also interested in the representation of geographic boundaries and data visualisation.

Kiichi Hirono is Subsection Chief of Geographic Information Technology Development at Asia Air Survey Co., Ltd. His main interests are GIS with a focus on urban disaster prevention. He is also interested in remote sensing of marine physical phenomena by using satellite or airborne images.

Chang-Hyeon Joh is a Ph.D. candidate at the Urban Planning Group of the Eindhoven University of Technology. He received his M.A. in Geography from the Seoul National University. His research interests are activity pattern analysis and activity-travel behaviour modelling. He is also interested in studying the relationships between urban structure and individual spatial behaviour.

Takeki Izumi is Assistant Professor of Geography at Tokyo Metropolitan University. He received his Ph.D. degree in urban engineering from the University of Tokyo in 2001. His research focuses on urban climate system with numerical simulation and remote sensing.

Shiro Koike is a Graduate Student of Human Geography at the University of Tokyo. He received his Ph.D. degree from the University of Tokyo in 2002. His main interests are spatio-temporal and historical analyses of geographic phenomena using GIS.

Kazuyuki Konagaya is Associate Professor of Regional Economics and Urban Geography at the Institute for Economic Research of Osaka City University. He received his M.D. in Physics from University of Tokyo in 1985 and M.D. in Geography from Kyoto University in 1989. His interests are modelling spacio-temporal aspects of economic phenomena (such as land price and land use) and GIS.

Jay Lee is Associate Professor of Geography at Kent State University. He received his Ph.D. from the University of Western Ontario in 1989. His research interests stem from a broad area of quantitative modelling of geographic phenomena over space and over time.

Isao Mizuno is Associate Professor of Geography at Ochanomizu University. He received his Ph.D. from Tokyo Metropolitan University in 1998. His research focuses on modelling uneven regional development from a self-organisation perspective, analysing topological structure of transportation networks, and reinterpreting classical location theory.

Tomoki Nakaya is Associate Professor of Geography at Ritsumeikan University. He received his Ph.D. degree from the Tokyo Metropolitan University in 1997. His main research fields are mathematical modelling, spatial statistics and geographic information systems in human geography with special interests in spatial interaction and medical geography.

Yujiro Ogawa is an Executive Director of Asian Disaster Reduction Center and a Visiting Professor of Kyushu University. He received his Doctor of Engineering degree from the University of Tokyo in 1993. His main interest field is disaster management in the context of urban planning.

Atsuyuki Okabe received his Ph.D. from the University of Pennsylvania in 1975 and the degree of Doctor of Engineering from the University of Tokyo in 1977. Previously he has held the position of Associate Professor at the Institute of Socio-Economic Planning, University of Tsukuba. He is currently Director of the Centre for Spatial Information Science and Professor of the Department of Urban Engineering, University of Tokyo.

Bambang Rudyanto is a Visiting Researcher to the Asian Disaster Reduction Center. He received a Ph.D. degree from the University of Tokyo in 1996. His main interests are regional/urban planning, infrastructure planning, environmental sciences and macro economics.

Yukio Sadahiro is an Associate Professor of Urban Planning in the Faculty of Engineering at the University of Tokyo. He received his Ph.D. degree from the University of Tokyo in 1994. His main interests are spatial analysis, spatio-temporal analysis, quality of spatial data, spatial data mining, and visualization of spatial information.

Narushige Shiode is a Research Fellow at University College London. He is completing a Ph.D. on comparative analysis of urban and cyberspaces and its implications for urban and spatial planning.

Phillip Smith is a Postgraduate Student at the School of Geography, University of Manchester undertaking a Ph.D. dissertation on modelling the HIV/AIDS epidemic in Europe.

Masami Sugiura is Section Chief of the Disaster Prevention Consultancy at Asia Air Survey Co., Ltd. He is a registered consulting engineer of civil engineering (rivers, erosion control and seashores). Especially he is an expert on urban disasters and geomorphological hazards.

Hirotaka Suzuki is a Visiting Researcher of Asian Disaster Reduction Centre and Assistant Professor of Graphic Science at Osaka City University. He received his Doctor of Engineering degree from the University of Tokyo in 1996. His main interests are spatial analysis in visual environment and graphic science education.

Masanao Takeyama is Associate Professor of Economics at Keio University. He received his Ph.D. in Geography from the University of California at Santa Barbara in 1996. His main interests are spatial analyses of information societies with emphasis on the influence of digital telecommunication technologies on urban space. He is also interested in computational modelling of complex spatial systems.

Toshikazu Tamura is Professor of Geomorphology and Environmental Studies at Tohoku University. He received a D.Sc. degree from Tohoku University in 1979. His main interests are studying hillslope classification and chronology, environmental changes in the humid tropical and temperate zones, and watershed environmental management.

Richard Thomas is Professor of Geography at the University of Manchester. He received his Ph.D. from the University of Nottingham in 1972. His current research interests are in the construction and application of multiregion epidemic models and the influence of urban regeneration initiatives on levels of mental distress.

Harry Timmermans is Professor of Urban Planning at the Eindhoven University of Technology. He received his Ph.D. from Nijmegen University in 1980. His main interests are modelling spatial choice behaviour and developing decision support systems.

Ian Turton is Principal Research Fellow (Computational Geography) of the School of Geography at the University of Leeds. He received a Ph.D. degree from the University of Edinburgh in 1992. His research interests include intelligent agent-based methodologies, visualisation of geographic data especially over the Web, crime mapping and spatial data mining, and interoperability standards for GIS.

Teruhisa Watanabe is an Engineer for the Hitachi Software Engineering Co., Ltd. He received his master's degree in environmental studies from the University of Tokyo in 2001. His main interests are sustainable city planning using geographic information systems.

Shigeya Yoshikawa belongs to Geographic Information Technology Development Section at Asia Air Survey Co., Ltd. His main responsibilities are GIS and WEB-GIS systems (disaster prevention, sightseeing information, environment resources and so on) planning and development.

Shipeng Zhao is GIS Technical Manager at SuperMap Japan Co., Ltd. He received a D.Sc. degree from the Institute of Geography, Chinese Academy of Sciences in 1995. His main interests are studying mountainous disasters using GIS and providing GIS-based solutions for various resource and environmental problems.

INDEX

The GeoJournal Library

1. B. Currey and G. Hugo (eds.): *Famine as Geographical Phenomenon.* 1984 ISBN 90-277-1762-1
2. S.H.U. Bowie, F.R.S. and I. Thornton (eds.): *Environmental Geochemistry and Health.* Report of the Royal Society's British National Committee for Problems of the Environment. 1985 ISBN 90-277-1879-2
3. L.A. Kosiński and K.M. Elahi (eds.): *Population Redistribution and Development in South Asia.* 1985 ISBN 90-277-1938-1
4. Y. Gradus (ed.): *Desert Development.* Man and Technology in Sparselands. 1985 ISBN 90-277-2043-6
5. F.J. Calzonetti and B.D. Solomon (eds.): *Geographical Dimensions of Energy.* 1985 ISBN 90-277-2061-4
6. J. Lundqvist, U. Lohm and M. Falkenmark (eds.): *Strategies for River Basin Management.* Environmental Integration of Land and Water in River Basin. 1985 ISBN 90-277-2111-4
7. A. Rogers and F.J. Willekens (eds.): *Migration and Settlement.* A Multiregional Comparative Study. 1986 ISBN 90-277-2119-X
8. R. Laulajainen: *Spatial Strategies in Retailing.* 1987 ISBN 90-277-2595-0
9. T.H. Lee, H.R. Linden, D.A. Dreyfus and T. Vasko (eds.): *The Methane Age.* 1988 ISBN 90-277-2745-7
10. H.J. Walker (ed.): *Artificial Structures and Shorelines.* 1988 ISBN 90-277-2746-5
11. A. Kellerman: *Time, Space, and Society.* Geographical Societal Perspectives. 1989 ISBN 0-7923-0123-4
12. P. Fabbri (ed.): *Recreational Uses of Coastal Areas.* A Research Project of the Commission on the Coastal Environment, International Geographical Union. 1990 ISBN 0-7923-0279-6
13. L.M. Brush, M.G. Wolman and Huang Bing-Wei (eds.): *Taming the Yellow River: Silt and Floods.* Proceedings of a Bilateral Seminar on Problems in the Lower Reaches of the Yellow River, China. 1989 ISBN 0-7923-0416-0
14. J. Stillwell and H.J. Scholten (eds.): *Contemporary Research in Population Geography.* A Comparison of the United Kingdom and the Netherlands. 1990 ISBN 0-7923-0431-4
15. M.S. Kenzer (ed.): *Applied Geography.* Issues, Questions, and Concerns. 1989 ISBN 0-7923-0438-1
16. D. Nir: *Region as a Socio-environmental System.* An Introduction to a Systemic Regional Geography. 1990 ISBN 0-7923-0516-7
17. H.J. Scholten and J.C.H. Stillwell (eds.): *Geographical Information Systems for Urban and Regional Planning.* 1990 ISBN 0-7923-0793-3
18. F.M. Brouwer, A.J. Thomas and M.J. Chadwick (eds.): *Land Use Changes in Europe.* Processes of Change, Environmental Transformations and Future Patterns. 1991 ISBN 0-7923-1099-3

The GeoJournal Library

---

19. C.J. Campbell: *The Golden Century of Oil 1950–2050.* The Depletion of a Resource. 1991 ISBN 0-7923-1442-5
20. F.M. Dieleman and S. Musterd (eds.): *The Randstad: A Research and Policy Laboratory.* 1992 ISBN 0-7923-1649-5
21. V.I. Ilyichev and V.V. Anikiev (eds.): *Oceanic and Anthropogenic Controls of Life in the Pacific Ocean.* 1992 ISBN 0-7923-1854-4
22. A.K. Dutt and F.J. Costa (eds.): *Perspectives on Planning and Urban Development in Belgium.* 1992 ISBN 0-7923-1885-4
23. J. Portugali: *Implicate Relations. Society and Space in the Israeli-Palestinian Conflict. 1993* ISBN 0-7923-1886-2
24. M.J.C. de Lepper, H.J. Scholten and R.M. Stern (eds.): *The Added Value of Geographical Information Systems in Public and Environmental Health.* 1995 ISBN 0-7923-1887-0
25. J.P. Dorian, P.A. Minakir and V.T. Borisovich (eds.): *CIS Energy and Minerals Development.* Prospects, Problems and Opportunities for International Cooperation. 1993 ISBN 0-7923-2323-8
26. P.P. Wong (ed.): *Tourism vs Environment: The Case for Coastal Areas.* 1993 ISBN 0-7923-2404-8
27. G.B. Benko and U. Strohmayer (eds.): *Geography, History and Social Sciences.* 1995 ISBN 0-7923-2543-5
28. A. Faludi and A. der Valk: *Rule and Order. Dutch Planning Doctrine in the Twentieth Century.* 1994 ISBN 0-7923-2619-9
29. B.C. Hewitson and R.G. Crane (eds.): *Neural Nets: Applications in Geography.* 1994 ISBN 0-7923-2746-2
30. A.K. Dutt, F.J. Costa, S. Aggarwal and A.G. Noble (eds.): *The Asian City: Processes of Development, Characteristics and Planning.* 1994 ISBN 0-7923-3135-4
31. R. Laulajainen and H.A. Stafford: *Corporate Geography.* Business Location Principles and Cases. 1995 ISBN 0-7923-3326-8
32. J. Portugali (ed.): *The Construction of Cognitive Maps.* 1996 ISBN 0-7923-3949-5
33. E. Biagini: *Northern Ireland and Beyond.* Social and Geographical Issues. 1996 ISBN 0-7923-4046-9
34. A.K. Dutt (ed.): *Southeast Asia: A Ten Nation Region.* 1996 ISBN 0-7923-4171-6
35. J. Settele, C. Margules, P. Poschlod and K. Henle (eds.): *Species Survival in Fragmented Landscapes.* 1996 ISBN 0-7923-4239-9
36. M. Yoshino, M. Domrös, A. Douguédroit, J. Paszynski and L.D. Nkemdirim (eds.): *Climates and Societies – A Climatological Perspective.* A Contribution on Global Change and Related Problems Prepared by the Commission on Climatology of the International Geographical Union. 1997 ISBN 0-7923-4324-7
37. D. Borri, A. Khakee and C. Lacirignola (eds.): *Evaluating Theory-Practice and Urban-Rural Interplay in Planning.* 1997 ISBN 0-7923-4326-3

The GeoJournal Library

38. J.A.A. Jones, C. Liu, M-K. Woo and H-T. Kung (eds.): *Regional Hydrological Response to Climate Change.* 1996 ISBN 0-7923-4329-8
39. R. Lloyd: *Spatial Cognition.* Geographic Environments. 1997 ISBN 0-7923-4375-1
40. I. Lyons Murphy: *The Danube: A River Basin in Transition.* 1997 ISBN 0-7923-4558-4
41. H.J. Bruins and H. Lithwick (eds.): *The Arid Frontier. Interactive Management of Environment and Development.* 1998 ISBN 0-7923-4227-5
42. G. Lipshitz: *Country on the Move: Migration to and within Israel, 1948–1995.* 1998 ISBN 0-7923-4850-8
43. S. Musterd, W. Ostendorf and M. Breebaart: *Multi-Ethnic Metropolis: Patterns and Policies.* 1998 ISBN 0-7923-4854-0
44. B.K. Maloney (ed.): *Human Activities and the Tropical Rainforest.* Past, Present and Possible Future. 1998 ISBN 0-7923-4858-3
45. H. van der Wusten (ed.): *The Urban University and its Identity.* Roots, Location, Roles. 1998 ISBN 0-7923-4870-2
46. J. Kalvoda and C.L. Rosenfeld (eds.): *Geomorphological Hazards in High Mountain Areas.* 1998 ISBN 0-7923-4961-X
47. N. Lichfield, A. Barbanente, D. Borri, A. Khakee and A. Prat (eds.): *Evaluation in Planning.* Facing the Challenge of Complexity. 1998 ISBN 0-7923-4870-2
48. A. Buttimer and L. Wallin (eds.): *Nature and Identity in Cross-Cultural Perspective.* 1999 ISBN 0-7923-5651-9
49. A. Vallega: *Fundamentals of Integrated Coastal Management.* 1999 ISBN 0-7923-5875-9
50. D. Rumley: *The Geopolitics of Australia's Regional Relations.* 1999 ISBN 0-7923-5916-X
51. H. Stevens: *The Institutional Position of Seaports.* An International Comparison. 1999 ISBN 0-7923-5979-8
52. H. Lithwick and Y. Gradus (eds.): *Developing Frontier Cities.* Global Perspectives – Regional Contexts. 2000 ISBN 0-7923-6061-3
53. H. Knippenberg and J. Markusse (eds.): *Nationalising and Denationalising European Border Regions, 1800–2000.* Views from Geography and History. 2000 ISBN 0-7923-6066-4
54. R. Gerber and G.K. Chuan (eds.): *Fieldwork in Geography: Reflections, Perspectives and Actions.* 2000 ISBN 0-7923-6329-9
55. M. Dobry (ed.): *Democratic and Capitalist Transitions in Eastern Europe.* Lessons for the Social Sciences. 2000 ISBN 0-7923-6331-0
56. Y. Murayama: *Japanese Urban System.* 2000 ISBN 0-7923-6600-X
57. D. Zheng, Q. Zhang and S. Wu (eds.): *Mountain Geoecology and Sustainable Development of the Tibetan Plateau.* 2000 ISBN 0-7923-6688-3
58. A.J. Conacher (ed.): *Land Degradation.* Papers selected from Contributions to the Sixth Meeting of the International Geographical Union's Commission on Land Degradation and Desertification, Perth, Western Australia, 20–28 September 1999. 2001 ISBN 0-7923-6770-7

The GeoJournal Library

59. S. Conti and P. Giaccaria: *Local Development and Competitiveness.* 2001 ISBN 0-7923-6829-0
60. P. Miao (ed.): *Public Places in Asia Pacific Cities.* Current Issues and Strategies. 2001 ISBN 0-7923-7083-X
61. N. Maiellaro (ed.): *Towards Sustainable Buiding.* 2001 ISBN 1-4020-0012-X
62. G.S. Dunbar (ed.): *Geography: Discipline, Profession and Subject since 1870.* An International Survey. 2001 ISBN 1-4020-0019-7
63. J. Stillwell and H.J. Scholten (eds.): *Land Use Simulation for Europe.* 2001 ISBN 1-4020-0213-0
64. P. Doyle and M.R. Bennett (eds.): *Fields of Battle.* Terrain in Military History. 2002 ISBN 1-4020-0433-8
65. C.M. Hall and A.M. Williams (eds.): *Tourism and Migration.* New Relationships between Production and Consumption. 2002 ISBN 1-4020-0454-0
66. I.R. Bowler, C.R. Bryant and C. Cocklin (eds.): *The Sustainability of Rural Systems.* Geographical Interpretations. 2002 ISBN 1-4020-0513-X
67. O. Yiftachel, J. Little, D. Hedgcock and I. Alexander (eds.): *The Power of Planning.* Spaces of Control and Transformation. 2001 ISBN Hb; 1-4020-0533-4 ISBN Pb; 1-4020-0534-2
68. K. Hewitt, M.-L. Byrne, M. English and G. Young (eds.): *Landscapes of Transition.* Landform Assemblages and Transformations in Cold Regions. 2002 ISBN 1-4020-0663-2
69. M. Romanos and C. Auffrey (eds.): *Managing Intermediate Size Cities.* Sustainable Development in a Growth Region of Thailand. 2002 ISBN 1-4020-0818-X
70. B. Boots, A. Okabe and R. Thomas (eds.): *Modelling Geographical Systems.* Statistical and Computational Applications. 2003 ISBN 1-4020-0821-X
71. R. Gerber and M. Williams (eds.): *Geography, Culture and Education.* 2002 ISBN 1-4020-0878-3
72. D. Felsenstein, E.W. Schamp and A. Shachar (eds.): *Emerging Nodes in the Global Economy: Frankfurt and Tel Aviv Compared.* 2002 ISBN 1-4020-0924-0
73. R. Gerber (ed.): *International Handbook on Geographical Education.* 2003 ISBN 1-4020-1019-2
74. M. de Jong, K. Lalenis and V. Mamadouh (eds.): *The Theory and Practice of Institutional Transplantation.* Experiences with the Transfer of Policy Institutions. 2002 ISBN 1-4020-1049-4

KLUWER ACADEMIC PUBLISHERS – DORDRECHT / BOSTON / LONDON

GPSR Compliance
The European Union's (EU) General Product Safety Regulation (GPSR) is a set of rules that requires consumer products to be safe and our obligations to ensure this.

If you have any concerns about our products, you can contact us on

ProductSafety@springernature.com

In case Publisher is established outside the EU, the EU authorized representative is:

Springer Nature Customer Service Center GmbH
Europaplatz 3
69115 Heidelberg, Germany

www.ingramcontent.com/pod-product-compliance
Lightning Source LLC
LaVergne TN
LVHW020056220926
848484LV00002BB/560
* 9 7 8 1 4 0 2 0 0 8 2 1 4 *